Lozi, Hénon and Belykh Chaotic Attractors

Over the past fifty years, the development of chaotic dynamical systems theory and its subsequent wide applicability in science and technology has been an extremely important achievement of modern mathematics. Chaotic attractors are not a fleeting curiosity, and their continued study is important for the progress of mathematics.

This book collects several of the new relevant results on the most important of them: the Lozi, Hénon, and Belykh attractors. Existence proofs for strange attractors in piecewise-smooth nonlinear Lozi-Hénon and Belykh maps are given. Generalization of Lozi map in higher dimensions, Markov partition, or embedding into the 2D border collision normal form of this map are considered. K-symbol fractional order discrete-time and relationship between this map and maxtype difference equations are explored. Statistical self-similarity, control of chaotic transients, and target-oriented control of Hénon and Lozi attractors are presented. Controlling chimera and solitary states by additive noise in networks of chaotic maps, detecting invariant expanding cones for generating word sets to identify chaos in piecewise-linear maps, and studying border collision bifurcations in a piecewise linear duopoly model complete this book.

This book is an essential companion for students and researchers in mathematics, physics, engineering, and related disciplines seeking to deepen their understanding of chaotic dynamical systems and their applications.

The chapters in this book were originally published in *Journal of Difference Equations and Applications.*

René Lozi is Emeritus Professor at Université Côte d'Azur, France, and Vice-President of the International Society of Difference Equations. His research areas include complexity and emergence theory, dynamical systems, bifurcations, control of chaos, cryptography based on chaos, and memristors.

Lyudmila Efremova is Professor at Lobachevsky State University of Nizhny Novgorod and Moscow Institute of Physics and Technology, Russia. Her scientific interests include regular and chaotic properties of low-dimensional discrete dynamical systems.

Mohammed-Salah Abdelouahab is Professor at Abdelhafid Boussouf University Center of Mila, Algeria. He is the head of the research team in fractional calculus and its applications at the laboratory of mathematics and their interactions, and the Editor-in-Chief of the *Journal of Innovative Applied Mathematics and Computational Sciences.*

Safwan El Assad is Professor at Polytech Nantes (Nantes Université), Nantes, France. Between 1988 and 2005, his research activities concerned radar imagery and digital communications. Nowadays his research largely focuses on chaos-based cryptography, encryption, crypto-compression, steganography, hash functions, and authenticated encryption.

Lozi, Hénon and Belykh Chaotic Attractors

New Results Fifty Years On

Edited by
René Lozi, Lyudmila Efremova,
Mohammed-Salah Abdelouahab
and Safwan El Assad

CRC Press
Taylor & Francis Group
Boca Raton London New York

First edition published 2025
by CRC Press
4 Park Square, Milton Park, Abingdon, Oxon OX14 4RN

and by CRC Press
2385 NW Executive Center Drive, Suite 320, Boca Raton FL 33431

British Library Cataloguing-in-Publication Data
A catalogue record for this book is available from the British Library

ISBN: 978-1-032-94042-7 (hbk)
ISBN: 978-1-032-94045-8 (pbk)
ISBN: 978-1-003-56864-3 (ebk)

DOI: 10.1201/9781003568643

Typeset in Minion Pro
by Newgen Publishing UK

Publisher's Note
The publisher accepts responsibility for any inconsistencies that may have arisen during the conversion of this book from journal articles to book chapters, namely the inclusion of journal terminology.

Disclaimer
Every effort has been made to contact copyright holders for their permission to reprint material in this book. The publishers would be grateful to hear from any copyright holder who is not here acknowledged and will undertake to rectify any errors or omissions in future editions of this book.

Contents

Citation Information

The following chapters in this book were originally published in the *Journal of Difference Equations and Applications*, volume 29, issue 9–12 (2023). When citing this material, please use the original page numbering for each article, as follows:

Chapter 1

Controlling chaotic transients in the Hénon and the Lozi map with the safety function
Rubén Capeáns and Miguel A. F. Sanjuán
Journal of Difference Equations and Applications, volume 29, issue 9–12 (2023), pp. 876–884

Chapter 2

On target-oriented control of Hénon and Lozi maps
E. Braverman and A. Rodkina
Journal of Difference Equations and Applications, volume 29, issue 9–12 (2023), pp. 885–908

Chapter 3

Controlling chimera and solitary states by additive noise in networks of chaotic maps
Elena Rybalova, Eckehard Schöll and Galina Strelkova
Journal of Difference Equations and Applications, volume 29, issue 9–12 (2023), pp. 909–930

Chapter 4

Statistical self-similarity in Lozi and Hénon's strange attractors
Alexandre M. de Paula Viveiros
Journal of Difference Equations and Applications, volume 29, issue 9–12 (2023), pp. 931–951

Chapter 5

Markov partition in the attractor of Lozi maps
Diogo Baptista
Journal of Difference Equations and Applications, volume 29, issue 9–12 (2023), pp. 952–964

Chapter 6

Lozi map embedded into the 2D border collision normal form
I. Sushko, V. Avrutin and L. Gardini
Journal of Difference Equations and Applications, volume 29, issue 9–12 (2023), pp. 965–981

Chapter 7

A higher-dimensional generalization of the Lozi map: bifurcations and dynamics
Shakir Bilal and Ramakrishna Ramaswamy
Journal of Difference Equations and Applications, volume 29, issue 9–12 (2023), pp. 982–993

Chapter 8

Existence proofs for strange attractors in piecewise-smooth nonlinear Lozi-Hénon and Belykh maps
Vladimir N. Belykh, Nikita V. Barabash and Dina A. Grechko
Journal of Difference Equations and Applications, volume 29, issue 9–12 (2023), pp. 994–1014

Chapter 9

On the relationship between Lozi maps and max-type difference equations
A. Linero Bas and D. Nieves Roldán
Journal of Difference Equations and Applications, volume 29, issue 9–12 (2023), pp. 1015–1044

Chapter 10

K-symbol fractional order discrete-time models of Lozi system
Rabha W. Ibrahim
Journal of Difference Equations and Applications, volume 29, issue 9–12 (2023), pp. 1045–1064

Chapter 11

Border collision bifurcations in a piecewise linear duopoly model
Laura Gardini and Davide Radi
Journal of Difference Equations and Applications, volume 29, issue 9–12 (2023), pp. 1065–1093

Chapter 12

Detecting invariant expanding cones for generating word sets to identify chaos in piecewise-linear maps
D. J. W. Simpson
Journal of Difference Equations and Applications, volume 29, issue 9–12 (2023), pp. 1094–1126

Notes on Contributors

Mohammed-Salah Abdelouahab, Laboratory of Mathematics and Their Interactions, Abdelhafid Boussouf University Center of Mila, Mila, Algeria.

Viktor Avrutin, IST, University of Stuttgart, Stuttgart, Germany.

Diogo Baptista, ESTG, Polytechnic of Leiria, Leiria, Portugal; INESC Coimbra, Coimbra, Portugal; CIMA-UE, University of Évora, Évora, Portugal.

Nikita V. Barabash, Volga State University of Water Transport, Nizhny Novgorod, Russia; Lobachevsky State University of Nizhny Novgorod, Nizhny Novgorod, Russia.

Vladimir N. Belykh, Volga State University of Water Transport, Nizhny Novgorod, Russia; Lobachevsky State University of Nizhny Novgorod, Nizhny Novgorod, Russia.

Shakir Bilal, Center for Global Health Infectious Disease Research Interdisciplinary Research Building (IDRB), University of South Florida, Tampa, FL, USA; Amity Institute of Integrative Sciences and Health, Amity University Haryana, Gurugram (Manesar), India.

Elena Braverman, Department of Mathematics and Statistics, University of Calgary, Canada.

Rubén Capeáns, Nonlinear Dynamics, Chaos and Complex Systems Group, Departamento de Física, Universidad Rey Juan, Carlos, Móstoles, Madrid, Spain.

Alexandre M. de Paula Viveiros, Departamento de Física, CCEN, Universidade Federal da Paraíba, Cidade Universitária, João Pessoa, Paraíba, Brazil; Departamento de Formação Geral, Centro Federal de Educação Tecnológica de Minas Gerais – Unidade III, Leopoldina, Minas Gerais, Brazil.

Lyudmila Efremova, Department of Differential Equations, Mathematical and Numerical Analysis, Lobachevsky State University of Nizhny Novgorod, Nizhny Novgorod, Russia; Department of Fundamental Mathematics, Moscow Institute of Physics and Technology, Dolgoprudny, Russia.

Safwan El Assad, Polytech Nantes (Nantes Université), Nantes, France.

Laura Gardini, DESP, University of Urbino Carlo Bo, Urbino, Italy.

Dina A. Grechko, Volga State University of Water Transport, Nizhny Novgorod, Russia; Lobachevsky State University of Nizhny Novgorod, Nizhny Novgorod, Russia.

Rabha W. Ibrahim, Department of Computer Science and Mathematics, Lebanese American University, Beirut, Lebanon.

Antonio Linero Bas, Mathematics, Facultad de Matemáticas, University of Murcia, Murcia, Spain.

René Lozi, Université Côte d'Azur, Laboratoire J. A. Dieudonné, Nice Cedex, France.

Daniel Nieves Roldán, Mathematics, Facultad de Matemáticas, University of Murcia, Murcia, Spain.

Davide Radi, DiMSEFA, Catholic University of the Sacred Heart, Milan, Italy; Department of Finance, VŠB – Technical University of Ostrava, Ostrava, Czech Republic.

Ramakrishna Ramaswamy, Department of Chemistry, Indian Institute of Technology, New Delhi, India.

Alexandra Rodkina, Department of Mathematics, The University of the West Indies, Kingston, Jamaica.

Elena Rybalova, Institute of Physics, Saratov State University, Saratov, Russia.

Miguel A. F. Sanjuán, Nonlinear Dynamics, Chaos and Complex Systems Group, Departamento de Física, Universidad Rey Juan, Carlos, Móstoles, Madrid, Spain.

Eckehard Schöll, Institut für Theoretische Physik, Technische Universität Berlin, Berlin, Germany; Bernstein Center for Computational Neuroscience Berlin, Berlin, Germany; Potsdam Institute for Climate Impact Research, Potsdam, Germany.

David J. W. Simpson, School of Mathematical and Computational Sciences, Massey University, Palmerston North, New Zealand.

Galina Strelkova, Institute of Physics, Saratov State University, Saratov, Russia.

Irina Sushko, Institute of Mathematics, NAS of Ukraine, Kyiv, Ukraine.

Introduction – Lozi, Hénon, and Belykh chaotic attractors: new results fifty years on

René Lozi, Lyudmila Efremova, Mohammed-Salah Abdelouahab, and Safwan El Assad

Since the first publication of a continuous chaotic attractor, the Lorenz attractor, in 1963 [23], and a discrete chaotic attractor, the Hénon attractor in 1976 [14], the study of chaotic dynamical systems has been developed along several directions in the second half of twentieth century, followed since the early 90s by unexpected applications of these pure mathematical objects to many fields, including biology, cryptography, optimization, economics, and finance, etc.

This book collects original research papers addressing such themes, in particular chaotic attractors. Nowadays, Belykh, Hénon and Lozi attractors are the most studied and used for application purposes among discrete chaotic attractors. It offers new insight into the analysis of systems that appear in different contexts. For two years (from Spring 2021 to Spring 2023), 23 authors from 13 countries (Brazil, Canada, Germany, India, Italy, Jamaica, Lebanon, New Zealand, Portugal, Russia, Spain, Ukraine, USA) worked jointly to publish new results on these chaotic attractors and new ones.

In this book theoretical research on the corresponding maps are considered. Not all the papers publishing new research on these attractors have been collected during the two years of conception of this special issue due to various reasons (some are PhD dissertations, some authors having published more than one article on the topic, only one being in this book, etc.). A partial list of these articles is briefly commented at the end of this introduction.

Different aspects of control of the Hénon and Lozi maps are first presented together.

In Chapter 1 Capeáns and Sanjuán[10] considered both maps for a choice of parameters where they show transient chaos. They applied a new approach of the partial control method that they have previously developed to prevent the escape of orbits close to the chaotic saddle that behave chaotically for a while to eventually escape to an external attractor. This new approach is based on finding a special function called the *safety function* that allows to automatically find the minimum control necessary to avoid this escape. They highlighted the strong connection between the safety function and the classical *safe set*. To illustrate that, they computed the safety function for both maps and also showed the strong dependence of this function on the magnitude of disturbances affecting the map, and how this change drastically impacts the controlled orbits.

Braverman and Rodkina, in Chapter 2 [9], explored the stabilization for nonlinear systems of difference equations with modified Target-Oriented Control and a chosen equilibrium as a target, both in deterministic and stochastic settings. The influence of stochastic components in the control parameters was explored. The results were tested on the Hénon and the Lozi maps.

In Chapter 3 Rybalova, Schöll and Strelkova[27] studied numerically the spatio-temporal dynamics of ring networks of coupled discrete-time systems in the presence of additive noise. Ther obustness of chimera states with respect to noise perturbations was explored for two ensembles in which the individual elements are described by either logistic maps or Hénon maps in the chaotic regime. The influence of noise on the behavior of solitary states has been investigated for a ring of non-locally coupled Lozi maps. They found that there is a finite and sufficiently wide region with respect to the coupling strength and the noise intensity where the probability of observing the chimeras and solitary states is high.

Statistical methods including Markov partition are also involved in understanding the chaotic dynamics of these two maps.

In Chapter 4 Paula Viveiros (de) [26] considered non-classical characterizations of strange attractors that assign occupancy times and numbers to each of the hypercubes that cover them, provide local information, and generate statistical distributions over such quantities, such as global information, at the expense of losing all the local ones. He presented statistical distributions in the Lozi map of the mentioned times and numbers, and quantities derived from them. Such quantities are linked to each hypercube of a set that corresponds to a small fraction of their total. Such distributions confirm the correspondence between local and global statistical properties.

Baptista [4], in Chapter 5, studied iterations of 2D maps, in particular iterations of Lozi maps in the region of the parameter space where it has a strange attractor. Using symbolic dynamics techniques for 2D maps, based on the kneading theory of Milnor and Thurston and also in the symbolic dynamic formalism developed by Sousa Ramos, through the kneading sequence for the Lozi maps, he characterized the region in the parameter space that contains the kneading curves and present a method to define a Markov partition for the Lozi attractors. Consequently, the topological entropy for the Lozi map was computed.

Lozi, Hénon, and Belykh maps have been generalized in many ways since their initial publication. In this special issue, four articles are linked to these generalizations.

In Chapter 6 Sushko, Avrutin, and Gardini [29] investigated how the bifurcation structure of the Lozi map is incorporated into the bifurcation structure of the 2D border-collision normal form using an analytical representation of the boundaries of the largest periodicity regions related to the cycles with rotation number $1/n$, $n \geq 3$. At the center bifurcation boundary of the stability domain of the fixed point both maps are conservative that leads to a quite intricate bifurcation structure near this boundary.

In Chapter 7 Bilal and Ramaswamy [6] generalized the 2D Lozi map in order to systematically obtain piecewise continuous maps in three and higher dimensions. Similar to higher dimensional generalizations of the related Hénon map, these higher dimensional Lozi maps support hyperchaotic dynamics. They carried out a bifurcation analysis and investigated the dynamics through both numerical and analytical means. The analysis was extended to a sequence of approximations that smooth the discontinuity of the derivatives in the Lozi map.

Belykh, Barabash, and Grechko [5] in Chapter 8 introduced a 2D map with a single piecewise-smooth nonlinear function that is suitable for analytic studies of its strange attractor. This 2D piecewise-smooth nonlinear map represents a broad range of chaotic maps, including a hybrid Lozi-Hénon map and the Belykh map. To prove the existence of a strange attractor in the 2D map, they (i) constructed its trapping region that contains all limit sets and (ii) demonstrated that the invariant set's trajectories have negative and positive Lyapunov exponents. Moreover, they developed an invariant cone approach to establish the latter property, which involves constructing expanding and contracting cones as bounds for the eigenvectors of the variational equations along the chaotic trajectories. They applied their approach to analyze the chaotic hybrid Lozi-Hénon map, the original Lozi map, and the Belykh map.

In Chapter 9 Linero Bas and Nieves Roldàn [22] revised a transformation that links generalized Lozi maps with max-type difference equations. In this view, according to the technique of topological conjugation, they related the dynamics of a concrete Lozi map with a complete uniparametric family of max-equations, and they applied this fact to investigate the dynamics of two particular families.

In Chapter 10 Ibrahim [16] suggested three different systems: the fractional Lozi system (FLS), the k-symbol FLS, and the generalized k-symbol FLS. He examined several of these systems' crucial dynamics and explored the sufficient and necessary conditions for the k-symbol fractional dynamic systems to be stable and asymptotically stable, providing the sufficient conditions on the k-symbol FLS to attain steady and asymptotically steady zero resolutions. Additionally, he suggested control laws designed to stabilize a combination of these systems.

Chapters 11 and 12 are devoted to topological properties of 2D-maps.

In Chapter 11 Gardini and Radi [11] considered a duopoly model characterized by a 2D non-invertible continuous map T given by piecewise linear functions, with several partitions, defining a duopoly game. The structure of the game is such that it has separate second iterate so that its dynamics can be studied via a one-dimensional composite function; that is, piecewise linear with multiple partitions in which the definition of the map changes. The number of partitions may change from 2 to 5, depending on the parameters. The dynamics are characterized by degenerate bifurcations and border-collision bifurcations, which are typical in maps having kink points. Here, the peculiarity is the multiplicity of the partitions, which leads to bifurcations different from those occurring in maps with only one kink point. They showed several bifurcations, coexistence of cycles, attracting and superstable, as well chaotic attractors and chaotic repellors, related to the outcome of particular border collision bifurcations.

In Chapter 12 [28] Simpson detected invariant expanding cones for generating word sets to identify chaos in piecewise-linear maps. He showed how to formally identify chaotic attractors in continuous, piecewise-linear maps on $\mathbb{R}^N$. For such a map f, this is achieved by constructing three objects. First, $\Omega_{trap} \subset \mathbb{R}^N$ is trapping region for f. Second, $\mathbf{W}$ is a finite set of words that encodes the forward orbits of all points in Ω_{trap}. Finally, $C \subset T\mathbb{R}^N$ is an invariant expanding cone for derivatives of compositions of f formed by the words in $\mathbf{W}$. The existence of Ω_{trap}, $\mathbf{W}$, and C implies f has a topological attractor with a positive Lyapunov exponent. He developed an algorithm that identifies these objects for 2D homeomorphisms comprised of two affine pieces. The main effort is in the explicit construction of Ω_{trap} and C. Their existence is equated to a set of computable conditions in a general way. This results in a computer-assisted proof of chaos throughout a relatively large region of parameter space. He also observed how the failure of C to be expanding can coincide with a bifurcation of f.

Research on chaotic attractors is a hot topic today. Between April 2021 and June 2024, significant literature has been published. We give a brief overview of the contents of those articles that were not published in this book, although related to the same topic.

Kucharski [19] studied the case $b < 0$ of the Lozi map and presented a rigorous proof that for a set of parameters quasi-symmetric to the set $\mathcal{M}$ defined by Misurewicz in 1978 in the case $b > 0$, this map has a strange attractor. With Boroński and Ou [8] he rigorously determined an open region $\mathcal{B}$ in the parameter space for which this map exhibits periodic points of least period n, for all $n > 13$.

Ou [25] proved that all periodic orbits of this map can be classified by using symbolic dynamics and that the bifurcation parameters of some types of periodic orbits form analytic curves in the parameter space. Kilassa Kvaternik [18] in his doctoral dissertation studied the tangential homoclinic points locus of the Lozi map and applications.

Wormell [31] defined a new propery of chaotic attractor: the conditional mixing. She has conjectured that for Lozi map, this property is verified for generic parameters (a,b) and Lebesgue-almost all real variable x, with respect to level curves $x = x_0$ (that is, the measures $\rho(\cdot|x_0)$). She has shown strong and direct numerical evidence in favor of this conjecture.

When high-dimensional non-uniformly hyperbolic chaotic systems undergo dynamical perturbations, their long-time statistics are generally observed to respond differentiably with respect to the perturbation. In [30], considering this map as a family of the mathematically tractable class of piecewise-smooth hyperbolic maps, she proved that the existence of a formal derivative of the response reduces to an exponential mixing property of the SRB measure when conditioned on the map's singularity set.

Li and Gu [21], in the conservative case, studied the hierarchical structure of a chaotic sea and the existence condition and delimitation method of island chain for this map.

Many new generalizations of the Hénon and the Lozi maps have been proposed.

Boroński and Stimac [7] showed that Hénon-like and Lozi-like maps on their strange attractors are conjugate to natural extensions of maps on metric trees with dense set of branch points. In consequence, these trees approximate very well the topology of the attractors, and the maps on them give good models of the dynamics. Kucharski [20] generalized the Lozi-like family introduced by Misiurewicz and Stimac that encompasses, in particular, certain Lozi-like maps, orientation preserving or reversing Lozi maps or large parameter regions of 2D border collision normal forms.

Ibrahim and Baleanu [15] using a linear interpolation of quadratic and absolute value functions defined a combined Hénon-Lozi fractional map and studied it. Al-Saidi et al. [3] extended the definition of fractional map to the case where the order of derivation is variable using Lozi map as an example. Joshi et al. [17] considered the definition of fractional map to the case where the order of derivation v is a complex number with $0 < Re(v) < 1$.

A canonical example of 2D border-collision normal form family of difference equations was presented by Glendinning and Simpson [12], which reduces to the Lozi map for certain values of parameters. Ghosh, Mc Lachlan, and Simpson [13] extended their results to the non-orientation-preserving cases. In [1] Aliwi and Ajeena proposed an unconventional generalization of this map based on the Sine Chaotic Map concept, and on the Cosine Chaotic map in [2].

Menasri [24] replaced the piecewise linear term in the first equation of the Lozi map by the max of two arbitrary functions.

Conclusion

Since the first example of a chaotic attractor, sixty years ago, research on such objects is still relevant. This book is only one step in this process which will continue for a long time.

References

[1] B. H. Aliwi, R. K. K. Ajeena: On fuzzy sine chaotic based model in security communications, *JPSP* 6(4) (2022) pp. 8127–8133. https://journalppw.com/index.php/jpsp/article/view/5169

[2] B. H. Aliwi, R. K. K. Ajeena: A performed knapsack problem on the fuzzy chaos cryptosystem with cosine Lozi chaotic map, *AIP Conference Proceedings* 2414 (2023), pp. 040047-1–040047-10. https://doi.org/10.1063/5.0114840

[3] N. M. G. Al-Saidi, H. Natiq, D. Baleanu, R. W. Ibrahim: The dynamic and discrete systems of variable fractional order in the sense of the Lozi structure map, *AIMS Math.* 8 (2023), pp. 733–751. https://doi.org/10.3934/math.2023035

[4] D. Baptista: Markov partition in the attractor of Lozi maps, *J. Differ. Equ. Appl.* 29(9-12) (2023), pp. 952–964. doi:10.1080/10236198.2023.2186155

[5] V. Belykh, N. Barabash, D. Grechko: Existence proofs for strange attractors in piecewise-smooth nonlinear Lozi-Hénon and Belykh maps, *J. Differ. Equ. Appl.* 29(9-12) (2023), pp. 994–1014. doi:10.1080/10236198.2023.2193653

[6] S. Bilal, R. Ramaswamy: A higher-dimensional generalization of the Lozi map: bifurcations and dynamics, *J. Differ. Equ. Appl.* 29(9-12) (2023), pp. 982–993. doi: 10.1080/10236198.2022.2041625

[7] J. P. Boroński, M.-S. Stimac: Densely branching trees as models for Hénon-like and Lozi-like attractors, *Adv. Math.*, 429 (2023) p. 109191. doi:10.1016/j.aim. 2023.109191

[8] J. P. Boroński, P. Kucharski, D.-S. Ou: Lozi maps with periodic points of all periods n > 13. *preprint*, (2022). Available online: www.researchgate.net/publication/366740872_Lozi_maps_with_periodic_points_of_all_periods_n_13 (accessed on 27 April 2023).

[9] E. Braverman, A. Rodkina: On target-oriented control of Hénon and Lozi maps, *J. Differ. Equ. Appl.* 29(9-12) (2023), pp. 885–908. doi:10.1080/10236198.2022.2088286

[10] R. Capeáns, M. Sanjuán: Controlling chaotic transients in the Hénon and the Lozi map with the safety function, *J. Differ. Equ. Appl.* 29(9-12) (2023), pp. 876–884. doi:10.1080/10236198.2022.2063050

[11] L. Gardini, D. Radi: Border collision bifurcations in a piecewise linear duopoly model, *J. Differ. Equ. Appl.* 29(9-12) (2023), pp. 1065–1093. doi:10.1080/10236198.2023.2203276

[12] P. A. Glendinning, D. J. W Simpson: Chaos in the border-collision normal form: A computer-assisted proof using induced maps and invariant expanding cones, *Appl. Math. Comput.* 434 (2022), p. 127357. https://doi.org/10.1016/j.amc.2022.127357.

[13] I. Ghosh, R. McLachlan, D. Simpson: Robust chaos in orientation-reversing and non-invertible two-dimensional piecewise-linear maps, *arXiv*, (2022), https://arxiv.org/pdf/2307.05144v1.pdf.

[14] M. Hénon: A two-dimensional mapping with a strange attractor. *Commun. Math. Phys.* 50, (1976), pp. 69–77. https://doi.org/10.1007/BF01608556.

[15] R. W. Ibrahim, D. Baleanu: Global stability of local fractional Hénon-Lozi map using fixed point theory, *AIMS Math.* 7(6) (2022), pp. 11399–11416. doi:10.3934/math.2022636

[16] R. W. Ibrahim: K-symbol fractional order discrete time models of Lozi system, *J. Differ. Equ. Appl.* 29(9-12) (2023), pp. 1045–1064. doi:10.1080/10236198.2022.2158736

[17] D. D. Joshi, P. M. Gade, S. Bhalekar. Study of low-dimensional nonlinear fractional difference equations of complex order, *Chaos* 32(11) (2022), p. 113101. https://doi.org/10.1063/5.0095939

[18] K. Kilassa Kvaternik: Tangential homoclinic points locus of the Lozi maps and applications, Doctoral thesis (2022), University of Zagreb, Faculty of Science. https://repozitorij.pmf.unizg.hr/

[19] P. Kucharski: Strange attractors for the family of orientation preserving Lozi Maps, *Chaos* 33 (2023), p. 113121. https://doi.org/10.1063/5.0139893

[20] P. Kucharski: Strange attractors trees for the generalized Lozi-like family, *arXiv*, (2023). arXiv:2302.04641v2

[21] B. Li, E.-G. Gu: The stratification and invariant region in the sea of an area-preserving map, Preprint, *Research Square*, (2023). https://doi.org/10.21203/rs.3.rs-2744467/v1

[22] A. Linero Bas, D. Nieves Roldán: On the relationship between Lozi maps and max-type difference equations, *J. Differ. Equ. Appl.* 29(9-12) (2023), pp. 1015–1044. doi: 10.1080/10236198.2022.2120393

[23] E. N. Lorenz: Deterministic nonperiodic flow, *J. Atmos. Sci.* 20, (1963), pp. 130–141. https://doi.org/10.1175/1520-0469(1963)020⟨0130:DNF⟩2.0.CO;2.

[24] A. Menasri: Lozi maps with max function and its application, *Aust. J. Math. Anal. Appl.* 20(1), (2023) Art. 6, p.13. https://ajmaa.org/searchroot/files/pdf/v20n1/v20i1p6.pdf

[25] D.-S. Ou: Critical points in higher dimensions, I: Reverse order of periodic orbit creations in the Lozi family, arXiv:2203.02326v2. https://doi.org/10.48550/arXiv.2203.02326

[26] A. de Paula Viveiros: Statistical self-similarity in Lozi and Hénon's strange attractors, *J. Differ. Equ. Appl.* 29(9-12) (2023), pp. 931–951. doi:10.1080/10236198.2023.2213439

[27] E. Rybalova, E. Schöll, G. Strelkova: Controlling chimera and solitary states by additive noise in networks of chaotic maps, *J. Differ. Equ. Appl.* 29(9-12) (2023), pp. 909–930. doi:10.1080/10236198.2022.2118580

[28] D. J. W. Simpson: Detecting invariant expanding cones for generating word sets to identify chaos in piecewise-linear maps, *J. Differ. Equ. Appl.* 29(9-12) (2023), pp. 1094–1126. doi:10.1080/10236198.2022.2070009

[29] I. Sushko, V. Avrutin, L. Gardini: Lozi map embedded into the 2D border collision normal form, *J. Differ. Equ. Appl.* 29(9-12) (2023), pp. 965–981. doi:10.1080/10236198.2023.2205960

[30] C. L. Wormell: On convergence of linear response formulae in some piecewise hyperbolic maps, *Nonlinearity* 37, (2024), 125001, (36pp.). https://doi.org/10.1088/1361-6544/ad85bd

[31] C. L. Wormell: Conditional mixing in deterministic chaos, *Ergod. Th. & Dynam. Sys.*, (2023), online, pp. 1–31. doi:10.1017/etds.2023.55

Controlling chaotic transients in the Hénon and the Lozi map with the safety function

Rubén Capeáns and Miguel A. F. Sanjuán

ABSTRACT

In this work, we deal with the Hénon and the Lozi map for a choice of parameters where they show transient chaos. Orbits close to the chaotic saddle behave chaotically for a while to eventually escape to an external attractor. Traditionally, to prevent such an escape, the partial control technique has been applied. This method stands out for considering disturbances (noise) affecting the map and for finding a special region of the phase space, called *the safe set*, where the control required to sustain the orbits is small. However, in this work, we will apply a new approach of the partial control method that has been recently developed. This new approach is based on finding a special function called *the safety function* which allows to automatically find the minimum control necessary to avoid the escape of the orbits. Furthermore, we will show the strong connection between the safety function and the classical safe set. To illustrate that, we will compute for the first time, safety functions for the two-dimensional Hénon and Lozi maps, where we also show the strong dependence of this function with the magnitude of disturbances affecting the map, and how this change drastically impacts the controlled orbits.

1. Introduction

Chaotic transient behaviour [8,9] usually arises when due to the change of a parameter of the system, a chaotic attractor collides with his own boundary basin causing a boundary crisis. The chaotic set becomes a non-attracting chaotic set (i.e. chaotic repellor or a chaotic saddle), and almost all orbits in the neighbourhood of the nonattracting chaotic set are free to escape to an external attractor. If in addition we consider that the dynamics are affected for some disturbances (noise), all trajectories sooner or later eventually escape.

Traditionally, to avoid the escape of the orbits and sustain them around the nonattracting chaotic set, we have used the classical partial control method. This technique is applied on maps and is based on finding certain special set in the phase space, called the safe set, where the control needed is small. Unlike other control methods [6,7,12,13], partial control is designed from the ground up to deal with disturbances. Furthermore, it takes into account the magnitude of the disturbances, to find the best safe set. One of the more remarkable result is that the control used to sustain the orbits in the safe set, is always smaller than the magnitude of disturbances affecting the orbits.

However, in this work, we will not use the classical approach of the partial control method based on finding a safe set. Instead, we will use the new approach based on the computation of a special functions called the safety function from which we can obtain the minimum control necessary to sustain the orbits and also the minimum safe set. In this sense, this new approach is a generalization of the classical partial control method.

In the next sections, we will illustrate the application of this new approach based on the computation of the safety function. First with the one-dimensional tent map, and then, for the first time, with the two-dimensional Hénon and Lozi map.

2. The partial control: from the classical approach to the new approach

The first step to apply the partial control technique is to define a region Q in the phase space containing the chaotic saddle. In this region Q, we assume that the dynamics can be described by the following map:

$$q_{n+1} = f(q_n) + \xi_n + u_n, \quad \text{with } |\xi_n| \leq \xi_0, \quad |u_n| \leq u_0 < \xi_0,$$

where q describes the state vector of the system, ξ_n is the disturbance affecting the map and we assume it is limited by an upper bound value ξ_0. The control u_n is applied every iteration of the map with the knowledge of $f(q_n) + \xi_n$. This control is also limited by the upper bound value u_0, which can be previously fixed by the controller.

The points $q \in Q$ that can be sustained in Q without exceed the control u_0, defines the safe set (which is a subset of Q). To do that, there is an algorithm called the Sculpting Algorithm in [10,11]. This algorithm takes as input, the map f defined in Q and the values ξ_0 and u_0, and compute the corresponding safe set. However, if the value u_0 introduced is too small, no safe set exists. Therefore to find the minimum u_0 value, we have to gradually increasing u_0 until the safe set appears. This task can be rather unpractical since u_0 is manually set for every computation. In a recent work [3], it was presented a new approach of partial control, that automatically computes the minimum u_0 value. This approach takes as input the map f defined in Q and the value ξ_0, to then compute a special function in the region Q called the safety function U. The minimum of this function is called u_0. The set of points $q \in Q$ that satisfy $U(q) = u_0$ define what we call the minimum safe set. Transient chaotic orbits can be sustained in the minimum safe set forever by applying every iteration a control $|u_n| \leq u_0 < \xi_0$. The algorithm to compute the safety function is presented and explained in [3]. To make this work self-contained and make it easier for the reader, we also include the algorithm at the end of this manuscript, in the Appendix.

Before illustrating, for the first time, the application of this control method in two-dimensional maps, we believe that it would be helpful for the reader an initial example using a one-dimensional map to show how this method works. To do that we will take a previous work presented in [5] where we use the well known tent map slope-three given by

$$x_{n+1} = \begin{cases} \mu x_n + \xi_n + u_n & \text{for } x_n \leq \dfrac{1}{2}, \\ \mu(1 - x_n) + \xi_n + u_n & \text{for } x_n > \dfrac{1}{2}. \end{cases} \tag{1}$$

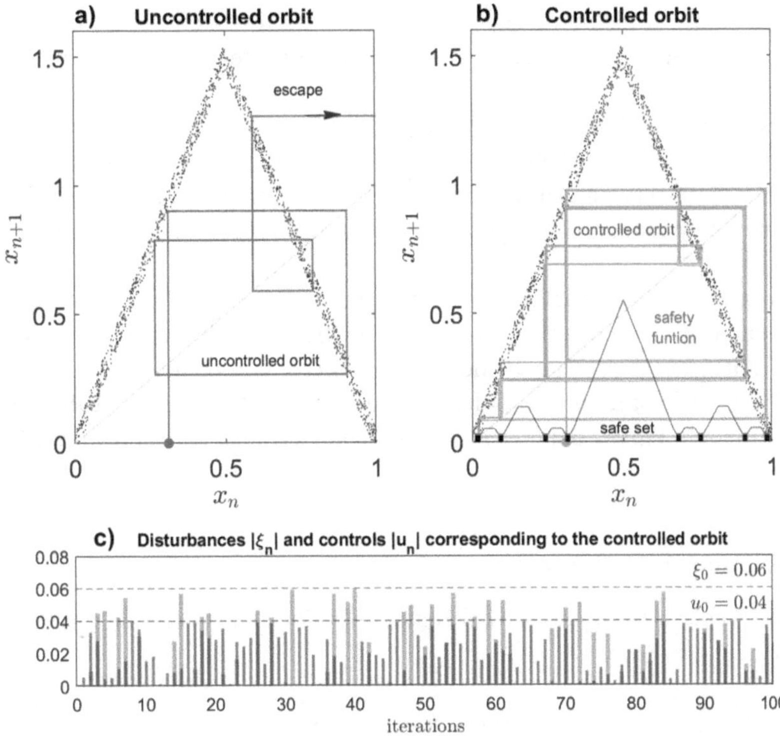

Figure 1. Partial control method. The slope-three tent map is represented in this figure. The map is affected by a uniform disturbance bounded by $\xi_0 = 0.06$. The small dots help to visualize the intensity and distribution of the disturbance. (a) An uncontrolled orbit that escapes from the interval $Q = [0, 1]$ after a few iterations. (b) In red, the safety function that has 8 minima with value $u_0 = 0.04$. These 8 minima defines the safe set, which is represented with the black pieces at the bottom. In blue, a controlled orbit is shown. This controlled orbit starts in the point $x_0 = 0.3$ that belongs to the safe set. At every iteration of the map, the orbit is forced to pass through the safe set to remain forever in $Q = [0, 1]$. (c) Disturbances (gray) $|\xi_n| \leq \xi_0 = 0.06$ affecting the controlled orbit and controls (blue) $|u_n| \leq u_0 = 0.04$ applied during the first 100 iterations of the map.

This map with $\mu = 3$ exhibits transient chaos in the interval $Q = [0, 1]$ (see Figure 1(a)). We consider in this case that that orbits of this map are affected by disturbances $\xi_n \leq \xi_0 = 0.06$, then, we compute the corresponding safety function shown in Figure 1(b) in red line. As shown, the safety function has eight minima with the value $u_0 = 0.04$. The minimum safe set corresponds with the location of this minima indicated in Figure 1(b) by the small black pieces. Finally, orbits starting in the the safe set can be sustained inside it by applying every iteration a control $|u_n| \leq u_0 = 0.04$. In Figure 1(b), 100 iterations of a controlled orbit (blue line) is represented and the corresponding 100 $|\xi_n|$ disturbances and $|u_n|$ controls in Figure 1(c).

3. Partial control applied to the Hénon map and the Lozi map

In previous works [1,2,4,11], safe sets have been computed from 1D, 2D and 3D maps, by applying the classical algorithm described in [10,11]. However, none of these safe sets have been computed by means of the safety function. So far, the safety function approach presented in [3] has been mainly applied to one-dimensional maps as the tent map presented before.

In this work, we want to show for the first time the computation of the safety functions in two-dimensional maps. The extra dimension just adds more computation since now the safety function will be a two-dimensional surface $U(q_x, q_y)$. However the algorithm to compute the safety function (see the Appendix) remains the same since it applies to any dimension.

To illustrate the control method, we choose the well known Hénon map and the Lozi map for a choice of parameters where they show transient chaos. In both cases, we will follow the same procedure: First define the region Q containing the chaotic saddle and set the upper bound of disturbance xi_0 affecting the map in Q. Next, compute the corresponding safety function. Finally obtain the minimum safe set and control the orbits to remain in it.

3.1. Application to the Hénon map

In 1976, the French astronomer Michel Hénon introduced the map later named after him, defined as

$$\begin{aligned} x_{n+1} &= a - by_n - x_n^2, \\ y_{n+1} &= x_n. \end{aligned} \tag{2}$$

Hénon proved that this is the most general form of quadratic maps, which shows transient chaos for a wide range of parameters a and b. Here, we have have chosen the parameter values $a=6$ and $b=0.4$. For these values, the trajectories with initial conditions in the square $Q = [-4, 4] \times [-4, 4]$ have a short chaotic transient, before finally escaping this region towards infinity.

The corresponding Hénon map including the disturbance and the control is

$$\begin{aligned} x_{n+1} &= a - by_n - x_n^2 + \xi_n^x + u_n^x, \\ y_{n+1} &= x_n + \xi_n^y + u_n^y, \end{aligned} \tag{3}$$

where we choose the bound of disturbance $\xi_0 = 0.20$ so that $\| \xi_n^x, \xi_n^y \| \leq \xi_0$. The control is also limited so that $\| u_n^x, u_n^y \| \leq u_0$, where the minimum u_0 possible, will be defined by the minimum value of the safety function. Below this value, no safe set exists.

The corresponding safety function is shown in Figure 2(a), where it has been plotted in logarithmic scale for a better visualization. For this case, the minimum value is found to be $\min(U) = 0.15 = u_0$. The points $q \in Q$ with $U(q) = 0.15$ defines the minimum safe set, which is represented in the Figure 2(b).

Finally, we use the safe set to control the orbits inside it. Every iteration of the map, the control (u_n^x, u_n^y) with $\| u_n^x, u_n^y \| \leq 0.15$ is applied to put the orbit back in the nearest safe point. A controlled orbit of 10,000 iterations is represented in Figure 2(b) by the red dots.

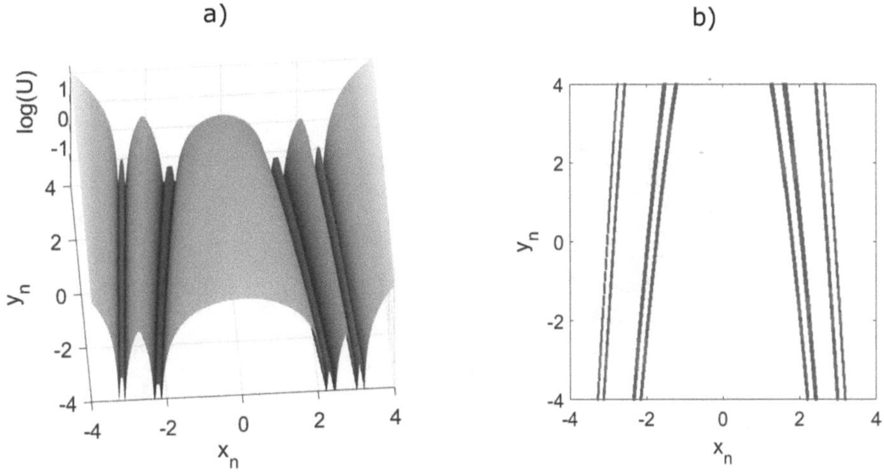

Figure 2. The 2D safety function for the Hénon map. The bound of a disturbance is $\xi_0 = 0.20$. (a) The safety function was computed in the square $Q = [-4, 4] \times [-4, 4]$ using a grid of 1000×1000 points. The safety function takes 18 iterations to converge and it has the minimum value $\min(U) = 0.15 = u_0$. The function is represented logarithmic to enhance the visualization. (b) The points $q \in Q$ that satisfy $U(q) = 0.15$ define the safe set (in blue). We also represent 10,000 iterations of a controlled orbit marked with the red dots.

3.2. Application to the Lozi map

A piecewise linear version of the Hénon map was given by the French mathematician René Lozi in the form,

$$\begin{aligned} x_{n+1} &= 1 - a|x_n| + by_n, \\ y_{n+1} &= x_n. \end{aligned} \tag{4}$$

For the choice of parameters $a = 2$, $b = 0.5$, the orbits in the square $Q = [-4, 4] \times [-4, 4]$ behave chaotic for a while to eventually escape from it. In order to avoid this escape, we will applied the partial control technique following the same steps as with the Hénon map. In this case the controlled maps is the following:

$$\begin{aligned} x_{n+1} &= 1 - a|x_n| + by_n + \xi_n^x + u_n^x, \\ y_{n+1} &= x_n + \xi_n^y + u_n^y, \end{aligned} \tag{5}$$

with $\| \xi_n^x, \xi_n^y \| \leq \xi_0$ and $\| u_n^x, u_n^y \| \leq u_0$.

To compute an example, we consider the upper bound of disturbance $\xi_0 = 0.050$. Then, we compute the safety function U shown in Figure 3(a), where the function is plotted logarithmic for a better visualization. The minimum of U is found to be $\min(U) = 0.035 = u_0$. Therefore, the minimum safe set is the set of point $q \in Q$ for which $U(q) = 0.035$. This safe set is shown in the in Figure 3(b), where we also draw 10,000 iterations of a controlled orbit (red points).

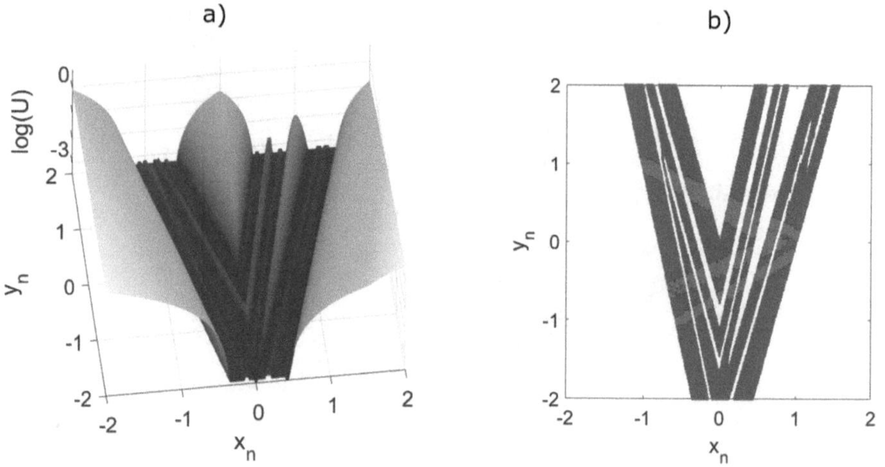

Figure 3. The 2D safety function for the Lozi map. The bound of a disturbance selected is $\xi_0 = 0.050$. (a) The safety function was computed in the square $Q = [-4, 4] \times [-4, 4]$ using a grid of 1000×1000 points. The safety function takes 23 iterations to converge and it has the minimum value $\min(U) = 0.035 = u_0$. The function is represented logarithmic to improve the visualization. (b) By taking the points $q \in Q$ that satisfy $U(q) = 0.035$, we obtain the safe set (blue). We also represent 10,000 iterations of a controlled orbit marked with the red dots.

4. Safe set variation with ξ_0

The partial control method takes advantage of the fractal structure of the non-attracting chaotic set responsible for the chaotic transients. As shown in the two-dimensional maps used here, the safe sets resembles the coarse grained structure of the stable manifold of the chaotic saddle. The grain size is mainly determined by the magnitude of disturbance affecting the map. Smaller values of ξ_0 leads to finer safe sets like a Cantor set. In Figure 4, different safe sets for the Hénon and the Lozi map has been computed. Top safe sets correspond to the Hénon map and bottom figures to the Lozi map. In both cases, the bound of disturbance ξ_0 affecting the map, decreases from left to right figures.

It each safe set of Figure 4, it has been also plotted a controlled orbits consisting of 10,000 iterations. We can observe that any starting orbit in the safe set, quickly converges to a smaller region called the asymptotic safe set [10], that resembles the coarse grained structure of the chaotic saddle. Once the controlled orbit enters in the asymptotic safe set, it remains inside forever.

Finally, we want to point out that, although the safe sets computed here were the smaller possible, there is no restriction to compute safe sets with bigger values of u_0. These safe sets will be a fattened version of the minimum safe set. This feature can be of interest if we want that the controlled orbit visit more points of the region Q, at the expense of applying bigger controls $|u_n| \leq u_0$.

5. Conclusions

In this work, we have shown the application of the new approach of partial control technique based on the safety functions. This technique is applied to maps showing transient

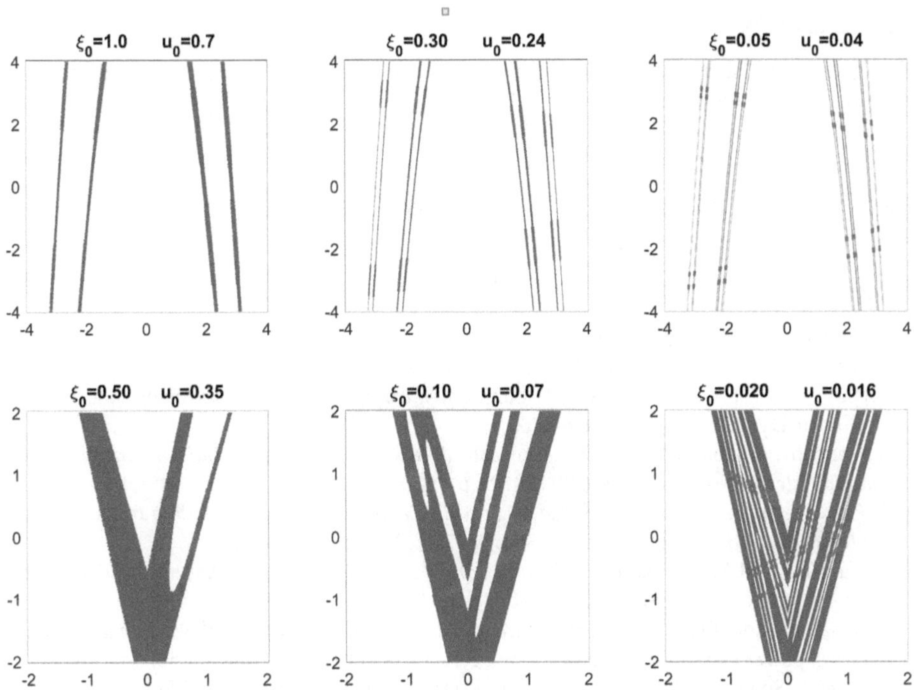

Figure 4. Safe sets for different ξ_0. The top figure correspond to safe sets computed for the Hénon map affected by different upper disturbance values ξ_0 that decreases from left to right. On the bottom, the safe sets corresponding to the Lozi map. All the safe sets are minimum with the minimum bound of control $u_0 < \xi_0$ indicated on the top. Below this control value, no safe set exists. Note that the safe sets computed with biguer ξ_0, consist of a few and fat strips, while the safe sets computed with smaller ξ_0, are made of many and thin strips. This is very clear in the case of the Hénon safe sets where we have 4,8 and 16 strips respectively. The red dots plotted over each safe set correspond to 10,000 iterations of a controlled orbit.

chaos with the goal to avoid the escape of the orbits from the non-attractive chaotic set. For the first time, we applied this technique to the two-dimensional Hénon map and the Lozi map, both affected by disturbance that is considered bounded. In each case, we define a region Q containing the chaotic saddle, where we compute the safety function. Then, we have extracted the minimum safe set, where the orbits can be sustaining using a minimum control bound. Finally, we have shown how the minimum safe sets change depending on the bound of disturbance affecting the map, which has a drastic impact in the controlled orbits.

Disclosure statement

The authors declare that they have no conflict of interest concerning the publication of this manuscript.

Funding

This work was supported by the Spanish State Research Agency Agencia Estatal de Investigación (AEI) and the European Regional Development Fund (ERDF) under Project No. PID2019-105554GB-100.

Data availability

All data generated or analysed during this study have been obtained through numerical simulations that are included in this article

ORCID

Miguel A. F. Sanjuán http://orcid.org/0000-0003-3515-0837

References

[1] R. Capeáns, J. Sabuco, and M.A.F. Sanjuán, *When less is more: partial control to avoid extinction of predators in an ecological model*, Ecol. Complex. 19 (2014), pp. 1–8.
[2] R. Capeáns, J. Sabuco, and M.A.F. Sanjuán, *Parametric partial control of chaotic systems*, Nonlinear Dyn. 2 (2016), pp. 869–876.
[3] R. Capeáns, J. Sabuco, and M.A.F. Sanjuán, *A new approach of the partial control method in chaotic systems*, Nonlinear Dyn. 98 (2019), pp. 873–887.
[4] R. Capeáns, J. Sabuco, M.A.F. Sanjuán, and J.A. Yorke, *Partially controlling transient chaos in the Lorenz equations*, Phil. Trans. R. Soc. A 375 (2016), p. 20160211.
[5] R. Capeáns and M.A.F. Sanjuán, *Beyond partial control: controlling chaotic transients with the safety function*, Nonlinear Dyn. 107 (2022), pp. 2903–2910.
[6] M. Dhamala and Y.C. Lai, *Controlling transient chaos in deterministic flows with applications to electrical power systems and ecology*, Phys. Rev. E. 59 (1999), pp. 1646–1655.
[7] V. In, M.L. Spano, J.D. Neff, W.L. Ditto, C.S. Daw, K.D. Edwards, and K. Nguyen, *Maintenance of chaos in a computational model of thermal pulse combustor*, Chaos. 7 (1997), pp. 605–613.
[8] Y.C. Lai and T. Tél, Transient Chaos: Complex Dynamics on Finite Time Scales, Cambridge University Press, Cambridge, 2011.
[9] E. Ott, Chaos in Dynamical Systems, 2nd ed., Cambridge University Press, Cambridge, 2012.
[10] J. Sabuco, M.A.F. Sanjuán, and J.A. Yorke, *Dynamics of partial control*, Chaos. 22 (2012), pp. 047507.
[11] J. Sabuco, S. Zambrano, M.A.F. Sanjuán, and J.A. Yorke, *Finding safety in partially controllable chaotic systems*, Commun. Nonlinear Sci. Numer. Simul. 17 (2012), pp. 4274–4280.
[12] I.B. Schwartz and I. Triandaf, *Sustainning chaos by using basin boundary saddles*, Phys. Rev. Lett. 77 (1996), pp. 4740–4743.
[13] W. Yang, M. Ding, A.J. Mandell, and E. Ott, *Preserving chaos: control strategies to preserve complex dynamics with potential relevance to biological disorders*, Phys. Rev. E. 51 (1995), pp. 102–110.

Appendix

Appendix: The safety function algorithm

Notation:
$i \equiv$ index of the starting point $q[i]$, $i = 1 : N$ where $N =$ total number of grid points in Q.
$s \equiv$ index of the disturbance $\xi[s]$, $s = 1 : M$ where $M =$ number of disturbed images.
$j \equiv$ index of the arrival point $q[j]$, $j = 1 : N$.
Map with the index notation: $q_{n+1} = f(q_n) + \xi_n + u_n \quad \rightarrow \quad q[j] = f(q[i]) + \xi[s] + u[i, s, j]$.
Computation of the safety function U:

- Initially set $U_0[j] = 0$, $\forall j = 1 : N$, $\quad k = 0$.

while $U_{k+1} \neq U_k$ **do**

for $i = 1$ to N **do**
  for $s = 1$ to M_i **do**
    for $j = 1$ to N **do**
      $u[i, s, j] = \left| f(q[i]) + \xi[s] - q[j] \right|$ ▷ Distance between the disturbed image $f(q[i]) + \xi[s]$ and the arrival point $q[j]$. Note that the $u[i, s, j]$ values remain unchanged every iteration of the while loop so compute them once and save them.

      $u^*[i, s, j] = \max_j \left(u[i, s, j], U_k[j] \right)$
    end for
    $u^{**}[i, s] = \min_j \left(u^*[i, s, j] \right)$
  end for
  $U_{k+1}[i] = \max_s \left(u^{**}[i, s] \right)$
end for
$k = k + 1$

end while

Compact formula: $U_{k+1}[i] = \max_{1 \le s \le M_i} \left(\min_{1 \le j \le N} \left(\max_j \left(u[i, s, j], U_k[j] \right) \right) \right)$

On target-oriented control of Hénon and Lozi maps

Elena Braverman and Alexandra Rodkina

ABSTRACT
We explore stabilization for nonlinear systems of difference equations with modified Target-Oriented Control and a chosen equilibrium as a target, both in deterministic and stochastic settings. The influence of stochastic components in the control parameters is explored. The results are tested on the Hénon and the Lozi maps.

1. Introduction

Unstable and chaotic behaviour of dynamical systems gave rise to the problem of their stabilization and synchronization actual in many areas of science and engineering. The idea to apply control designed by the difference of the current and some other, for instance, approximated state, appeared in the seminal paper introducing Ott–Grebogi–Yorke (OGY) method [35]. The review of chaos control methods both in continuous and discrete cases can be found in [38]. The application of delayed feedback control is going back to the pioneering paper of Pyragas [37] which produced a simple and efficient method to stabilize unstable equilibrium points for a system of differential equations. The method of delayed control was further developed and improved, see the recent papers [19–21] and references therein. For discrete dynamical systems, the application of controls using the difference of the state variable with a computed value at the next step, or after several steps, appeared to be more efficient than incorporating delays. This idea, coherent with the OGY method, is implemented in the Prediction Based Control (PBC) [30,41] where the controlled value of the map is a weighted average between the state variable and a computed value which, in the simplest case, evaluates the map at the next step. Though there are common ideas in the areas of differential and difference equations, there are some substantial differences in approaches and methods. Here we focus on discrete dynamical systems, requiring for stability of autonomous maps that all the eigenvalues are inside the unit circle (compared to negative real parts for systems of differential equations).

The purpose of the present paper is to investigate deterministic and stochastic chaos control for nonlinear systems, as well as test and compare them using the examples of the Hénon and the Lozi maps. Here we focus on Target-Oriented Control (TOC) with a fixed point chosen to stabilize as a target. TOC with a chosen and fixed target becomes a one-parameter method in the deterministic scalar case. In two-dimensional Hénon and Lozi maps, with noise involved in the control, the number of parameters increases, where some are more important than the others, and their interplay is illustrated in examples and simulations.

The Hénon [23] and Lozi [32] maps, their generalizations, limit orbits, dynamical properties and various applications continue to attract the attention of researchers, see, for example, recent publications [2,26–28,33,34,42] and references therein. Unlike these papers, we do not deal with complex dynamics and structure of the attracting sets (though these sets are presented in simulations) but dedicate our efforts to stabilization, illustrating general statements with these two types of maps. The issue of stabilization is of special interest in population dynamics [18,29–31]. Another focus is on the influence of stochastic perturbations on stabilization, which also is of significant importance in Mathematical Biology [15,39] due to the fact that noise is ubiquitous in ecological systems, and is sometimes considered to be an important factor in sustaining biodiversity. The idea of stabilization by noise originates in the work of Khasminskii [25] and was employed in many publications during the following years. Among other works, stabilization of difference equations was studied in [1,5,12], stabilization for certain higher-dimensional models in [24], and finite difference schemes for two-dimensional linear systems in [4,13]. The approach of our paper is closest to [9–11], where prediction-based, proportional feedback and target-oriented controls with stochastically perturbed parameters were studied for scalar difference equations.

One of our purposes in this paper is to establish the range of control parameters in TOC and noise amplitudes for which a system without noise is unstable, but the addition of noise to the control component leads to stabilization.

The Hénon system

$$\begin{aligned} x_{n+1} &= y_n + 1 - a x_n^2, \\ y_{n+1} &= b x_n, \end{aligned} \tag{1}$$

for $n \in \mathbb{N}_0 := \{0\} \cup \mathbb{N}$, has two equilibrium points

$$(x^*, y^*) = \left(\frac{1}{2a} \left[b - 1 \pm \sqrt{4a + (b-1)^2} \right], bx^* \right), \quad a, b > 0.$$

The Lozi map leads to a system of difference equations

$$\begin{aligned} x_{n+1} &= y_n + 1 - a|x_n|, \\ y_{n+1} &= b x_n, \end{aligned} \tag{2}$$

for $n \in \mathbb{N}_0$, with two equilibrium points

$$(x^*, y^*) = \left(\frac{1}{1 \pm a - b}, bx^* \right), \quad -a < 1 - b < a, \ a, b > 0.$$

Following the tradition, we consider $a = 1.4, b = 0.3$ in most of the examples, but also discuss the influence of other values for a and b. It is well known that for the values $a = 1.4$,

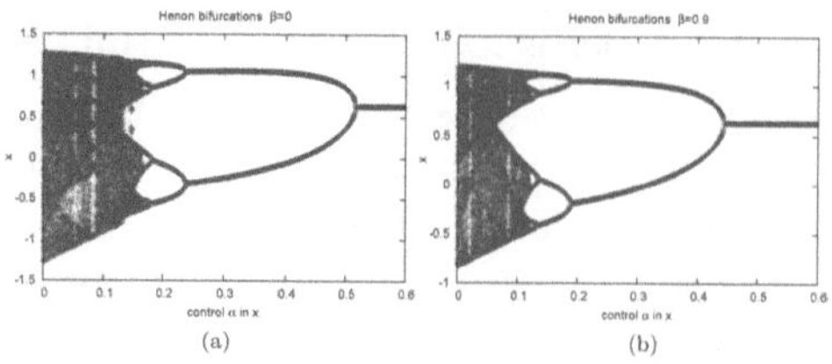

Figure 1. Bifurcation diagram of the Hénon map for x and (a) no control in y, (b) $\beta = 0.9$ control in y. The range of initial values is $x_0 \in [0.1, 0.8]$ and $y_0 \in [0.1, 0.2]$.

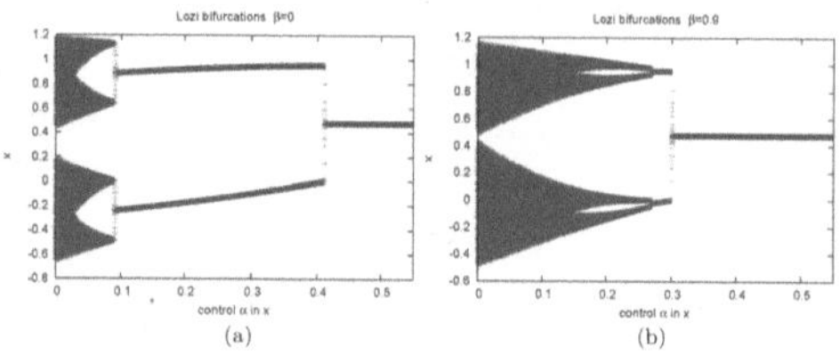

Figure 2. Bifurcation diagram of the Lozi map for x and (a) no control in y, (b) $\beta = 0.9$ control in y. The range of initial values is $x_0 \in [0.1, 0.8]$ and $y_0 \in [0.1, 0.2]$.

$b = 0.3$, both (1) and (2) are chaotic. Looking at the bifurcation diagram in Figure 1(a), with $\alpha = 0$, we see that the limit set of the uncontrolled Hénon map includes the segment $[-1.2, 1.2]$, no stability is observed. Similar chaotic properties of the Lozi map can be observed on Figure 2(a), for $\alpha = 0$.

Both systems can be described as a nonlinear autonomous system

$$X_{n+1} = F(X_n), \quad X_n \in \mathbb{R}^m, \quad n \in \mathbb{N}_0.$$

Scalar Target-Oriented Control (TOC) introduced first in [16] establishes a target state T and implements an increase or a decrease of the state variable each time step, depending on whether its value exceeds or is below the target state. For a one-dimensional equation $x_{n+1} = f(x_n)$ with an arbitrary continuous function $f : \mathbb{R} \to \mathbb{R}$, TOC has the form

$$x_{n+1} = f(cT + (1-c)x_n), \quad T \geq 0, \quad c \in [0, 1). \tag{3}$$

Here c is the control intensity, for $c = 0$ we get the original map, and as $c \to 1$, the right-hand side in (3) approaches $f(T)$. Once the target T is chosen as a fixed point of f, obviously it is also a fixed point of (3), and TOC becomes an efficient method of its stabilization. If $T = 0$ we get an earlier known Proportional Feedback (PF) control method [22] with a proportional reduction of the state variable at each step. PF control of scalar maps was studied in detail in [14,29]. Under certain assumptions, it stabilizes the zero equilibrium which is important in pest control problems and is aligned with TOC stabilization using a stabilized fixed point as a target. In other cases, a non-zero equilibrium can be stabilized. For an arbitrary T, a fixed point of (3) is a weighted average of a positive fixed point of f [18], assuming it exists and is unique, and T, with the weight depending on the control intensity c.

In [8], we considered vector extensions of TOC for systems (VTOC)

$$X_{n+1} = F(cT + (1 - c)X_n), \quad c \in [0, 1),$$

as well as Vector Modified Target Oriented Control (VMTOC)

$$X_{n+1} = cT + (1 - c)F(X_n), \quad c \in [0, 1). \tag{4}$$

In this paper, we apply VMTOC to (1) and (2) with $F = F_1$ and $F = F_2$, respectively, where

$$F_1((x, y)') = (y + 1 - ax^2, bx)', \quad F_2((x, y)') = (y + 1 - a|x|, bx)',$$

we use $\cdot'$ for transposition.

In addition to VMTOC [8] with a constant control intensity (4), we describe some important modifications.

- We consider variable control intensity. For TOC, a point being stabilized is a weighted average of the target and a fixed point of the original map. Thus, to stabilize a chosen equilibrium, we have to choose this fixed point as a target.
- Using the results of the previous item, a stochastic component of the control can be introduced and studied, with the noise amplitude ℓ. Moreover, with $\ell > 0$ stabilization can be achieved in the cases, when for $\ell = 0$, a two-cycle is stable rather than the target equilibrium.
- In (4), the control is implemented using left multiplication by a scalar matrix cI. We consider more general matrices. The choice of diagonal matrices gives a good trade-off of simplicity and a differential approach to the variables. For both Hénon and Lozi maps, x is more 'sensitive' to control than y, as expected.

Chaos control of nonlinear applied systems is a more challenging problem than for scalar maps. The present paper considers VMTOC [6–8,16,18] with a chosen equilibrium point for the Hénon and the Lozi maps. The theory applies to any of the two points, however, in examples and simulations, we reduce ourselves to stabilization of $(x^*, y^*) = (\frac{1}{2a}[b - 1 + \sqrt{4a + (b-1)^2}], bx^*)$ for the Hénon and $(x^*, y^*) = (\frac{1}{1+a-b}, bx^*)$ for the Lozi map, respectively, using it as a target.

Constant, variable and stochastic control types are investigated. For constant control intensity, estimation of convergence with an initial point in a neighbourhood of the target is considered. First, the influence of the control intensity for x is much more significant than the contribution of control in y. Second, while theoretically control level should increase with the increase of the domain for initial values, in simulations we see that the control value providing local stabilization will work for a larger domain in a certain neighbourhood of the chosen equilibrium point. Next, variable control is important in itself, describing a possibility of changing intensity, as well as for further application to a stochastically perturbed control. The introduction of stochastic control is two-fold:

(1) to demonstrate the range of noise which keeps stability for the same interval of parameters and the initial domain, where a deterministic system with variable deterministic control stabilizes a chosen equilibrium;

(2) to improve deterministic results in a sense that there is stability in a stochastic case, while no stability in the deterministic case with the mean control values is observed (compare with [12,24]).

For the first part, results for variable coefficients provide a required background, the same method was applied in [9]. For the second part, sometimes a neighbourhood of the equilibrium point in which stabilization can be achieved is quite small [12]. However, here we get robust results (illustrated with simulations and bifurcation diagrams) for the Hénon map and a global type estimate for the Lozi map with more restrictions on the control parameters than in the right half-plane.

The rest of the paper has the following structure. In Section 2, we consider stabilization of (1) and (2) with a modified deterministic version of TOC. In Section 3, a control is assumed to be perturbed by some noise. In both sections, theoretical results are illustrated with examples. A brief discussion of the results and comparison to other control methods in Section 4 conclude the paper.

2. Stabilization of equilibrium points by VMTOC

2.1. General deterministic step-dependent control

We generalize VMTOC (4) in $\mathbb{R}^m$ to a system with a matrix control

$$X_{n+1} = UT + (I - U)F(X_n), \quad X_0 \in B_R(X^*), \quad n \in \mathbb{N}_0, \tag{5}$$

where $X_0 \in \mathbb{R}^m$, U is a constant $m \times m$ matrix, I is the identity matrix, a certain norm $\|\cdot\|$ is assumed in $\mathbb{R}^m$, the same notation will be used for an induced matrix norm,

$$B_R(X^*) := \left\{X \in \mathbb{R}^m : \|X - X^*\| < R\right\}$$

is an open ball. For stabilization of a chosen equilibrium X^*, we use it as a target $T = X^*$ and take U as a diagonal matrix such that $U = \mathrm{diag}(d_1, \ldots, d_m)$.

In addition to constant diagonal U in (5), we consider step-dependent VMTOC with an infinite family of diagonal matrices $\{U_n\}$ with the entries $d_{1,n}, \ldots, d_{m,n} \in [0, 1)$ on the diagonal

$$X_{n+1} = U_{n+1}X^* + (I - U_{n+1})F(X_n), \quad X_0 \in B_R(X^*), \quad n \in \mathbb{N}_0. \tag{6}$$

Then, VMTOC with different matrices at various steps becomes

$$X_{n+1} = F(X_n) - U_{n+1}(F(X_n) - X^*), \quad X_0 \in B_R(X^*), \quad n \in \mathbb{N}_0,$$

with $F(X^*) = X^*$, leading to

$$X_{n+1} - X^* = (I - U_{n+1})(F(X_n) - X^*).$$

Theorem 2.1: *Assume that there exist $R > 0$, the diagonal matrices U_n and $\nu^* = \nu^*(R) \in (0, 1)$ such that*

$$\|(I - U_n)[F(X) - X^*]\| \leq \nu^* \|X - X^*\|, \quad n \in \mathbb{N}, \; X \in B_R(X^*). \tag{7}$$

Then any solution X_n of (6) with $X_0 \in B_R(X^*)$ converges to X^*, and X^* is asymptotically stable in $B_R(X^*)$.

Proof: If $X_0 \in B_R(X^*)$, we have $\|X_0 - X^*\| < R$ and

$$\|X_1 - X^*\| = \|U_1 X^* + (I - U_1)F(X_0) - X^*\|$$
$$= \|(I - U_1)[F(X_0) - X^*]\| \le \nu^* \|X_0 - X^*\| < R.$$

Using induction, assume that $X_n \in B_R(X^*)$ and $\|X_n - X^*\| \le (\nu^*)^n \|X_0 - X^*\|$. Then, by (7),

$$\|X_{n+1} - X^*\| = \|U_{n+1} X^* + (I - U_{n+1})F(X_n) - X^*\|$$
$$= \|(I - U_{n+1})[F(X_n) - X^*]\| \le \nu^* \|X_n - X^*\| \le (\nu^*)^{n+1} \|X_0 - X^*\|,$$

which implies that all $X_n \in B_R(X^*)$ and $\|X_n - X^*\| \le (\nu^*)^n \|X_0 - X^*\| \to 0$, as $n \to \infty$, for any $X_0 \in B_R(X^*)$. So we get convergence to X^*, with a guaranteed convergence rate, thus X^* is asymptotically stable in $B_R(X^*)$. ■

Further, Theorem 2.1 allows stabilizing any locally Lipschitz map in a chosen domain.

Corollary 2.2: *Suppose that there are constants $R > 0$ and $L > 0$ such that*

$$\|F(X) - X^*\| \le L \|X - X^*\|, \quad X \in B_R(X^*). \tag{8}$$

Then, for each $\nu^* \in (0,1)$, there exists a diagonal matrix $U = \mathrm{diag}(d_1, \ldots, d_m)$ with the entries $d_1, \ldots, d_m \in [0,1)$ such that (7) holds, and X^* is an asymptotically stable point of (6) in $B_R(X^*)$.

Proof: For $U = (1 - \frac{\nu^*}{\max\{1, L\|I\|\}})I$, we get

$$\|(I - U)[F(X) - X^*]\| \le \|I - U\| \|F(X) - X^*\|$$
$$\le \frac{\nu^*}{\max\{1, L\|I\|\}} L \|X - X^*\| \le \nu^* \|X - X^*\|.$$

For $U_n = \mathrm{diag}(d_{1,n}, \ldots, d_{m,n})$ with $1 - \frac{\nu^*}{\max\{1, L\|I\|\}} \le d_{j,n} < 1$, we also get

$$\|(I - U_n)[F(X) - X^*]\| \le \|I - U_n\| \|F(X) - X^*\| \le \nu^* \|X - X^*\|,$$

so (7) holds, and reference to Theorem 2.1 concludes the proof. ■

Sharper control estimates with U different from a scalar matrix determined in the proof of Corollary 2.2 are possible for specific F and chosen norms. A particular case of F and $\|\cdot\|_\infty$ norm, $\|(x_1, \ldots, x_m)\|_\infty = \max_{k=1,\ldots,m} |x_k|$, is outlined below.

Corollary 2.3: *Let $F = (f_1, \ldots, f_m)'$, $X = (x_1, \ldots, x_m)'$, $T = X^* = (x_1^*, \ldots, x_m^*)'$ be an equilibrium point, $R > 0$ and positive constants A_{ij}, $i, j = 1, \ldots, m$, $\mathcal{A} = (A_{ij})_{i,j=1}^m$ be such that*

$$|f_i(X) - x_i^*| \le \sum_{j=1}^m A_{ij} \left| x_j - x_j^* \right|, \quad X \in B_R(X^*). \tag{9}$$

Then, for each $\nu^* \in (0,1)$, there exists $U = \mathrm{diag}(d_1, \ldots, d_m)$ with $d_1, \ldots, d_m \in [0,1)$ such that in the induced norm $\|\cdot\|_\infty$, $\|(I - U)\mathcal{A}\| \le \nu^*$.

Also, for any diagonal matrices $U_n = \text{diag}(d_{1,n}, \dots, d_{m,n})$ *with* $d_j \le d_{j,n} < 1$, $j = 1, \dots, m$, X^* *is an asymptotically stable point of* (6) *in* $B_R(X^*)$.

Proof: Let (9) hold. Denote

$$L_i := \sum_{j=1}^{m} A_{ij}, \quad L := \max_{i=1,\dots,m} L_i. \tag{10}$$

Let us fix any $\nu^* \in (0,1)$ and define $d_i = \max\{1 - \frac{\nu^*}{L_i}, 0\}$. Then, $(1 - d_i)L_i \le \nu^*$, for each $i = 1, \dots, m$, which implies $\|(I - U)\mathcal{A}\| \le \nu^*$.

Consider now any diagonal matrix $U_n = \text{diag}(d_{1,n}, \dots, d_{m,n})$ with $d_j \le d_{j,n} < 1$, $j = 1, \dots, m$. Then the ith entry of $(I - U_n)[F(X) - X^*]$ satisfies

$$\begin{aligned} |(1 - d_{i,n})(f_i(X) - X^*)| &\le (1 - d_{i,n}) \sum_{j=1}^{m} A_{ij} \left| x_j - x_j^* \right| \\ &\le (1 - d_i)L_i \max_{j=1,\dots,m} \left| x_j - x_j^* \right| \le \nu^* \max_{j=1,\dots,m} \left| x_j - x_j^* \right|. \end{aligned}$$

Thus (7) holds, and the reference to Theorem 2.1 concludes the proof. ■

Remark 2.1: Assume that all f_i are continuously differentiable. Then in (9), by the Mean Value Theorem, for some $\bar{X} \in B_R(X^*)$,

$$|f_i(X) - x_i^*| = \left| \nabla f(\bar{X})(X - X^*) \right| \le \sum_{j=1}^{m} \max_{(x_1,\dots,x_m) \in B_R(X^*)} \left| \frac{\partial f_i}{\partial x_j} \right| \left| x_j - x_j^* \right|.$$

Thus, as R in (9) becomes small, for a smooth map F, the values of A_{ij} in (9) approach $|\frac{\partial f_i}{\partial x_j}(X^*)|$.

In Sections 2.2–2.3, we consider examples of the Hénon map, $a = 1.4$, $b = 0.3$, with the equilibrium $(x^*, y^*) \approx (0.6314, 0.1894)$ and of the Lozi map, $a = 1.4$, $b = 0.3$, with $(x^*, y^*) \approx (0.4762, 0.1429)$.

2.2. Step-independent Hénon and Lozi maps

First, we illustrate that stabilization with constant VMTOC and the diagonal matrix $U = \text{diag}(\alpha, \beta)$ is possible. The computation is supported by numerical simulations. Everywhere in simulations, we choose stabilization of the equilibrium in the right half-plane with $|x| = x$, and the initial values satisfying $x_0 > 0$.

2.2.1. *Hénon*

Consider constant VMTOC of the Hénon map

$$\begin{aligned} x_{n+1} &= \alpha x^* + (1-\alpha)(y_n + 1 - a x_n^2), \\ y_{n+1} &= \beta y^* + (1-\beta) b x_n, \end{aligned} \tag{11}$$

where the matrix of the controlled autonomous system is

$$(I-U)\mathcal{A} = \begin{bmatrix} -2(1-\alpha)ax^* & 1-\alpha \\ (1-\beta)b & 0 \end{bmatrix}$$

with two constant values $\beta = 0$ and $\beta = 0.9$, to outline the role of β. For $\beta = 0$, all the eigenvalues of $(I-U)\mathcal{A}$ are in the unit circle if and only if [17, p. 188] the trace and the determinant satisfy

$$|\mathrm{tr}((I-U)\mathcal{A})| < 1 + \det((I-U)\mathcal{A}) < 2. \tag{12}$$

Note that, if both eigenvalues of $(1-U)\mathcal{A}$ are in the unit circle, there exists a norm $\|\cdot\|$ such that $\|(1-U)\mathcal{A}\| < 1$.

Since $\det((I-U)\mathcal{A}) < 0$, the right inequality in (12) is valid, and the left one is equivalent to

$$\begin{aligned} 2(1-\alpha)ax^* < 1-(1-\alpha)b \iff 1-\alpha < \frac{1}{2ax^*+b} \\ \approx \frac{1}{2.8\cdot 0.631354 + 0.3} = 1-\alpha^* \approx 0.48361. \end{aligned}$$

Thus, for $\alpha > \alpha^* \approx 0.51639$, the equilibrium $(x^*, y^*) \approx (0.6314, 0.1894)$ of (11) is locally asymptotically stable. In Figure 1(a), we can observe in the bifurcation diagram that the last period-halving bifurcation is quite close to the theoretically predicted value for local stability and is $\alpha^* \approx 0.5164$.

Further, assume that some $\alpha > \alpha^* \approx 0.51639$ is set and notice that $\frac{1}{2a(1-\alpha)} - x^* > 0$ due to the choice of $\alpha > \alpha^*$. Then, for any $R \in (0, \frac{1}{2a(1-\alpha)} - x^*)$ and $(x_0, y_0) \in B_R((x^*, y^*))$, the solution tends to the equilibrium in the first quadrant.

For the bifurcation diagram in Figure 1(a), we took quite a large $R > 0.5$.

For $\beta = 0.9$, all the eigenvalues of $(I-U)\mathcal{A} = \begin{bmatrix} -2(1-\alpha)ax^* & 1-\alpha \\ 0.1b & 0 \end{bmatrix}$ satisfy $|\lambda| < 1$ for $1-\alpha < 1/(2ax^* + 0.1b) \approx 0.55624$. Therefore, for $\beta = 0.9$ and $\alpha > \alpha^* \approx 0.44376$, the positive equilibrium (x^*, y^*) of (11) is locally asymptotically stable. In Figure 1(b), we see that for $\beta = 0.9$, the last period-halving bifurcation with stability for larger values is at $\alpha \approx 0.444$, which is aligned with predicted local stability.

2.2.2. *Lozi*

Consider constant VMTOC of the Lozi map

$$\begin{aligned} x_{n+1} &= \alpha x^* + (1-\alpha)(y_n + 1 - a|x_n|), \\ y_{n+1} &= \beta y^* + (1-\beta) b x_n, \end{aligned} \tag{13}$$

with $x_0, x_n > 0$ and $(I - U)\mathcal{A}$ being the Jacobian matrix of controlled system (13)

$$(I - U)\mathcal{A} = \begin{bmatrix} -(1-\alpha)a & 1-\alpha \\ (1-\beta)b & 0 \end{bmatrix}.$$

For $\beta = 0$, conditions (12) are equivalent to

$$(1-\alpha)a < 1 - (1-\alpha)b \iff 1 - \alpha < \frac{1}{a+b} \approx 0.588235,$$

and for $\alpha > \alpha^* \approx 0.411765$, the positive equilibrium of (13) is locally asymptotically stable. In Figure 2(a), we see that the last period-halving bifurcation is at $\alpha \approx 0.412$, as predicted.

Similarly, once $\beta = 0.9$, the inequality $1 - \alpha < \frac{1}{a+0.1b} \approx 0.6993$ holds for $\alpha > \alpha^* \approx 0.3007$ which is illustrated with Figure 2(b), where stability starts at $\alpha \approx 0.301$.

Remark 2.2: Assume now that $x^* > 0$ and we do not restrict ourselves to the right half-plane. For $x_n < 0$ in $\mathbf{l}_\infty$-norm, for example, we have

$$\begin{aligned} \|X_{n+1} - X^*\| &= \left\| \begin{bmatrix} (1-\alpha)a(|x_n| - x^*) + (1-\alpha)(y_n - y^*) \\ (1-\beta)b(x_n - x^*) \end{bmatrix} \right\| \\ &\le \max\{(1-\alpha)a|x_n - x^*| + (1-\alpha)|y_n - y^*|, (1-\beta)b|x_n - x^*|\} \\ &\le \max\{(1-\alpha)(1+a), (1-\beta)b\}\|X_n - X^*\|. \end{aligned}$$

Therefore, for $b = 0.3 \in (0,1)$, $a = 1.4$, we have $(1-\beta)b < 0.3$, so $\nu^* = (1-\alpha)(1+a) < 1$ if $\alpha > 1 - \frac{1}{1+a} = 0.5833$, which is a larger lower estimate than above. So stability in all the plane is guaranteed if $\alpha > 0.5833$. We later illustrate this fact with simulations for controls with noise.

2.3. Variable Hénon and Lozi maps

Further, we proceed to the variable VMTOC and specify constants in the conditions of Corollary 2.3 for the Hénon and the Lozi maps. Let

$$U_n = \begin{bmatrix} d_{1,n} & 0 \\ 0 & d_{2,n} \end{bmatrix}, \quad d_{1,n}, d_{2,n} \in [0,1), \quad n \in \mathbb{N}.$$

This leads to VMTOC of the Hénon map

$$\begin{aligned} x_{n+1} &= d_{1,n+1}x^* + \left(1 - d_{1,n+1}\right)(y_n + 1 - ax_n^2), \\ y_{n+1} &= d_{2,n+1}y^* + \left(1 - d_{2,n+1}\right)bx_n, \end{aligned} \tag{14}$$

and the Lozi map

$$\begin{aligned} x_{n+1} &= d_{1,n+1}x^* + \left(1 - d_{1,n+1}\right)(y_n + 1 - a|x_n|), \\ y_{n+1} &= d_{2,n+1}y^* + \left(1 - d_{2,n+1}\right)bx_n. \end{aligned} \tag{15}$$

For the sake of brevity in the following two sections, we denote $\alpha_n = d_{1,n+1}$, $\beta_n = d_{2,n+1}$.

2.3.1. Hénon

We have

$$\begin{bmatrix} 1-ax^2+y-x^* \\ bx-y^* \end{bmatrix} = \begin{bmatrix} (1-ax^2+y)-(1-a(x^*)^2+y^*) \\ b(x-x^*) \end{bmatrix}$$
$$= \begin{bmatrix} -a(x+x^*) & 1 \\ b & 0 \end{bmatrix} \begin{bmatrix} x-x^* \\ y-y^* \end{bmatrix} =: \bar{A}(X-X^*),$$

where $x \in (x^*-R, x^*+R)$ and $R > 0$ is a fixed constant. Thus (9)–(10) hold with

$$A_{11} = \max_{x\in[x^*-R,x^*+R]} a|x+x^*|, \quad A_{12} = 1, \quad A_{21} = b, \quad A_{22} = 0, \quad \mathcal{A} = (A_{ij})_{i,j=1}^2, \tag{16}$$

$L_1 = \max_{x\in[x^*-R,x^*+R]} a|x+x^*| + 1$ and $L_2 = b$. Note that $L_2 < 1$, for $b = 0.3$. Let us follow the computations in the proof of Corollary 2.3 to find possible diagonal entries of U_n guaranteeing that the norms of the matrices

$$(I-U_n)\mathcal{A} = \begin{bmatrix} (1-\alpha_n) \max\limits_{x\in[x^*-R,x^*+R]} a|x+x^*| & 1-\alpha_n \\ (1-\beta_n)b & 0 \end{bmatrix}$$
$$= \begin{bmatrix} (1-\alpha_n)a(2x^*+R) & 1-\alpha_n \\ (1-\beta_n)b & 0 \end{bmatrix}$$

do not exceed $\nu^* < 1$. Let us illustrate the dependency of the lower bound on α_n on the choice of the matrix norm. In $\mathbf{l}_\infty$-norm, the numbers $\beta_n \in [0,1)$ can be arbitrary as $b = 0.3 \in (0,1)$. Consider, for example, a particular case of no control in y (all $\beta_n = 0$) and distinguish between the two cases:

(1) small $R \approx 0.01$. Then $0 < x + x^* < 1.273$. Let us fix $\alpha_* > 1 - \frac{1}{a(2x^*+R)+1} \approx 0.641$. For $\alpha_n \in (\alpha_*, 1)$ and $x_0 \in (x^* - 0.01, x^* + 0.01)$, local stabilization is achieved.
(2) larger $R \approx 0.36$. Then $0 < x + x^* < 1.63$. We fix $\alpha_* \in (0.694, 1)$ and get stabilization for all $\alpha_n \in [\alpha_*, 1)$.

Recall that in Figure 1(a), we observe stabilization for no control in y and any $\alpha \approx 0.516$ or higher. To illustrate the role of y-control and sensitivity of the estimates to the choice of a norm, we note that, while in $\mathbf{l}_\infty$-norm, the sum of the moduli of the entries in each row of $(I-U_n)\mathcal{A}$ should be less than $\nu^* < 1$, $\nu^* \approx 1$, in $\mathbf{l}_1$-norm $\|(x_1,\ldots,x_m)'\|_1$, the required condition is

$$(1-\alpha_n)a(2|x^*|+R) + (1-\beta_n)b < \nu^*, \quad 1-\alpha_n < \nu^*.$$

The second inequality holds for any $\alpha_n \in [\alpha_*, 1)$, while the first one for a small $R \approx 0.01$ and $\beta_n = 0$ becomes $\alpha_n \geq \alpha^* > 1 - 0.7/(2ax^*) \approx 0.6041$, which is a better estimate than in $\mathbf{l}_\infty$-norm, while for $\beta_n = 0.9$, we get stabilization for $\alpha_n \in (0.4513, 1)$. Finally, for $\mathbf{l}_2$ vector norm, the induced norm is a spectral norm, with $\|B\|$ for a matrix B being the largest value $\sqrt{\lambda}$, where λ is an eigenvalue of $B'B$. For a fixed $\nu^* < 1$, $\nu^* \to 1^-$, both eigenvalues

λ_i of

$$M_n = ((I - U_n)\mathcal{A})'((I - U_n)\mathcal{A})$$
$$= \begin{bmatrix} (1-\alpha_n)^2 a^2(2x^* + R)^2 + (1-\beta_n)^2 b^2 & (1-\alpha_n)^2 a(2x^* + R) \\ (1-\alpha_n)^2 a(2x^* + R) & (1-\alpha_n)^2 \end{bmatrix}$$

should satisfy (the eigenvalues are real and non-negative) $\lambda_i \le (\nu^*)^2 < 1$.

It is sufficient to check for the larger eigenvalue $\operatorname{tr}(M_n) + \sqrt{\operatorname{tr}^2(M_n) - 4\det(M_n)} < 2(\nu^*)^2$, which, under $\operatorname{tr}(M_n) \le 2(\nu^*)^2 < 2$, is equivalent to $\operatorname{tr}(M_n) \le (\nu^*)^2 + \det M_n/(\nu^*)^2$, or

$$(1-\alpha_n)^2 a^2(2x^* + R)^2 + (1-\beta_n)^2 b^2 + (1-\alpha_n)^2$$
$$\le (\nu^*)^2 + (1-\beta_n)^2 b^2 (1-\alpha_n)^2 (\nu^*)^{-2} < 2. \tag{17}$$

Since $\nu^* < 1$, for $\nu^* \to 1^-$, we have $(1-\beta_n)^2 b^2 (1-\alpha_n)^2 (\nu^*)^{-2} < 1$, then (17) is valid if

$$(1-\alpha_n)^2 \left[a^2(2x^* + R)^2 + 1 - b^2(1-\beta_n)^2\right] + b^2(1-\beta_n)^2 \le \mu^* < 1$$

for some $\mu^* \in (0, 1)$, which gives the estimate, in the absence of y-control ($\beta_n \equiv 0$), that the control intensity required is $\alpha_n \ge \alpha_* > 0.53$ for small R and $\alpha_* > 0.613$ for $R = 0.36$. Let $\beta_n \equiv 0.9$, this slightly improves to smaller $\alpha_* \approx 0.51$ and $\alpha_* \approx 0.6$, respectively.

The constants defined in (16) imply (9) for (14) with $(x^*, y^*) \approx (0.6314, 0.1894)$ and also apply for $a \in (0, 1.4)$. Similar estimates can be developed for the other equilibrium, as well as for arbitrary $a, b > 0$.

Proposition 2.4: *Let (x^*, y^*) be an equilibrium of (1), and R satisfy $R < |x^*|$. Then there exist constants $\alpha_* = \alpha_*(R)$ and $\beta_* = \beta_*(R)$ such that for any $\alpha_n \in (\alpha_*, 1)$, $\beta_n \in (\beta_*, 1)$, a solution of (14) converges to (x^*, y^*) for any initial value $(x_0, y_0) \in B_R(X^*)$.*

Proof: For an arbitrary norm and $R < |x^*|$, map (1) satisfies (9) for $x \in B_R(X^*)$ with the Lipschitz constant $L(R)$. The constant L is also norm-dependent but the norms are equivalent in $\mathbb{R}^2$, we fix the norm, the radius R and the prescribed convergence rate $\nu^* \in (0, 1)$. Denote $\alpha_* = 1 - \frac{\nu^*}{\max\{1, L\|I\|\}} \in (0, 1)$, $\beta_* = \alpha_*$. Then for any $\alpha_n \in (\alpha_*, 1)$, $\beta_n \in (\beta_*, 1)$, a solution of (14) with $(x_0, y_0) \in B_R(X^*)$ satisfies

$$\|(I - U_n)[F(X) - X^*]\| \le \|I - U_n\| \|F(X) - X^*\| \le \nu^* \|X - X^*\|,$$

so (7) holds, and application of Theorem 2.1 concludes the proof. ■

2.3.2. Lozi

Computation similar to Section 2.3.1 leads to

$$\begin{bmatrix} 1 - a|x| + y - x^* \\ bx - y^* \end{bmatrix} = \begin{bmatrix} (1 - ax + y) - (1 - ax^* + y^*) \\ b(x - x^*) \end{bmatrix}$$
$$= \begin{bmatrix} -a & 1 \\ b & 0 \end{bmatrix} \begin{bmatrix} x - x^* \\ y - y^* \end{bmatrix} =: \bar{A}(X - X^*),$$

where $x \in (x^* - R, x^* + R)$, $R < x^* \approx 0.4762$. Thus (9) holds for the Lozi map with

$$A_{11} = a, \quad A_{12} = 1, \quad A_{21} = b, \quad A_{22} = 0, \quad L_1 = a + 1, \quad L_2 = b. \tag{18}$$

The norm of $(I - U_n)\mathcal{A} = \begin{bmatrix} a(1-\alpha_n) & 1-\alpha_n \\ (1-\beta_n)b & 0 \end{bmatrix}$ should not exceed $\nu^* < 1$. Further, let us study the dependency of the lower bound on α_n on the choice of the matrix norm. In $\mathbf{l}_\infty$-norm, $\beta_n \in [0, 1)$ can be arbitrary as $b = 0.3 \in (0, 1)$. For α_n, to achieve convergence, $(1 - \alpha_n)(a + 1) \le \nu^* < 1$ should be satisfied, or for $\alpha_n \ge \alpha_* > 1 - 1/2.4 \approx 0.584$ stabilization is achieved.

For $\mathbf{l}_1$-norm, we should have $(1 - \alpha_n)a + (1 - \beta_n)b < 1$, the condition $1 - \alpha_n < 1$ is true for any non-zero control level. Thus, for $\beta_n = 0$, we get $\alpha_n > 1 - 0.7/1.4 = 0.5$, so the equilibrium is stable for any $\alpha_n > 0.5$. If $\beta_n = 0.9$, $\alpha_n > 1 - 0.97/1.4 \approx 0.31$.

As for the spectral norm, the matrix

$$M_n = ((I - U_n)\mathcal{A})'((I - U_n)\mathcal{A}) = \begin{bmatrix} (1-\alpha_n)^2 a^2 + (1-\beta_n)^2 b^2 & (1-\alpha_n)^2 a \\ (1-\alpha_n)^2 a & (1-\alpha_n)^2 \end{bmatrix}$$

has the trace $(1 - \alpha_n)^2 a^2 + (1 - \beta_n)^2 b^2 + (1 - \alpha_n)^2$ and the determinant $(1 - \beta_n)^2 b^2 (1-\alpha_n)^2$. The spectral norm does not exceed $\nu^* < 1$ if the larger of two real nonnegative eigenvalues of the quadratic equation $\lambda^2 - (\mathrm{tr}\ \mathrm{M_n})\lambda + \det \mathrm{M_n} = 0$ is less than $(\nu^*)^2 < 1$. Thus $\mathrm{tr}\ \mathrm{M_n} + \sqrt{\mathrm{tr}^2\ \mathrm{M_n} - 4\det \mathrm{M_n}} \le 2(\nu^*)^2$, which after some simplifications is equivalent to $\mathrm{tr}\ \mathrm{M_n} < \det \mathrm{M_n}/(\nu^*)^2 + (\nu^*)^2$, or

$$(1 - \alpha_n)^2 a^2 + (1 - \beta_n)^2 b^2 + (1 - \alpha_n)^2 < \frac{(1 - \beta_n)^2 b^2 (1 - \alpha_n)^2}{(\nu^*)^2} + (\nu^*)^2. \tag{19}$$

If $(1 - \alpha_n)^2 < \frac{1-(1-\beta_n)^2 b^2}{a^2+1-(1-\beta_n)^2 b^2}$, or (19) holds with $\nu^* = 1$, from the continuity of the right-hand side of (19) in ν^* and strictness of the inequality, there is $\nu^* < 1$ such that (19) holds, and the spectral norm is *notgreaterthan*.

For $\beta_n \equiv 0$, we get $\alpha_n \ge \alpha_* > 0.44$, while $\beta_n \equiv 0.9$ leads to $\alpha_n \ge \alpha_* > 0.42$.

The constants defined in (18) imply (9) for (15), and they are exactly the same in the case $x^* < 0$ and $x < 0$. The proof of the following result is similar to the proof of Proposition 2.4.

Proposition 2.5: *Let (x^*, y^*) be a fixed equilibrium of (2), $R < |x^*|$. Then there exist constants $\alpha_* = \alpha_*(R)$ and $\beta_* = \beta_*(R)$ such that for any $\alpha_n \in (\alpha_*, 1)$, $\beta_n \in (\beta_*, 1)$, a solution of (15) converges to (x^*, y^*) for any initial value $(x_0, y_0) \in B_R(X^*)$.*

Remark 2.3: Note that the matrix $\mathcal{A}$ defined in (16) depends on the radius $R < |x^*|$ in the domain $B_R((x^*, y^*))$, due to x being involved; therefore, the bounds for the control parameters guaranteeing stability can become close to one with the growth of the radius R. In contrast to the Hénon case, the matrix $\mathcal{A}$ in (18) for any radius $R < |x^*|$ does not include x explicitly and thus is uniform, see Remark 2.2.

3. Stabilization with stochastically perturbed control

3.1. General stochastically perturbed control

Stochastic control is a well-developed area, especially for parametric optimization, optimal stochastic control, dynamic programming etc. (see [3] and references therein).

Our approach is based on the application of Kolmogorov's Law of Large Numbers, see Lemma 3.2, and is closest to the methods developed and applied in [10,11,24] (see also [5,12]). Note that we prove only local stability with any a priori fixed probability from $(0, 1)$. However, in some cases, computer simulations demonstrate that for chosen parameters, stability actually happens in a relatively wide area of initial parameters, which gives us hope to extend our local results. However, this extension is left for future research.

We consider a complete filtered probability space $(\Omega, \mathcal{F}, \{\mathcal{F}_n\}_{n\in\mathbb{N}}, \mathbb{P})$, where the filtration $(\mathcal{F}_n)_{n\in\mathbb{N}}$ is naturally generated by m sequences of mutually independent identically distributed random variables $\chi_i := (\chi_{i,n})_{n\in\mathbb{N}}$, $i = 1, \ldots, m$, i.e. $\mathcal{F}_n = \sigma\{\chi_{i,j}, j = 1, 2, \ldots, n, i = 1, 2, \ldots, m\}$. The standard abbreviation 'a.s.' is used for either 'almost sure' or 'almost surely' with respect to a fixed probability measure $\mathbb{P}$. For a detailed introduction of stochastic concepts and notations, we refer the reader to [40].

In this paper, we reduce our investigation only to the case when control parameters in VMTOC model (6) are perturbed by a bounded noise, since in many real-world models, in particular, in population dynamics, noise amplitudes are bounded.

Assumption 3.1: *$\chi_i := (\chi_{i,n})_{n\in\mathbb{N}}$, $i = 1, \ldots, m$ are m sequences of mutually independent identically distributed random variables such that $|\chi_{i,n}| \le 1$, $i = 1, \ldots, m$, $n \in \mathbb{N}$.*

Note that, in general, random variables χ_i can have different distributions.

We study VMTOC (6), with an infinite family of diagonal $m \times m$ matrices $\{U_n\}$, $U_n = \mathrm{diag}(d_{1,n}, \ldots, d_{m,n})$, where controls $d_{i,n}$ are stochastically perturbed

$$d_{i,n} = \alpha_i + \ell_i \chi_{i,n}, \quad \alpha_i \in (0, 1), \quad \ell_i \in [0, \min\{\alpha_i, 1 - \alpha_i\}), \quad i = 1, \ldots, m. \tag{20}$$

Here $\chi_{i,n}$ are random variables satisfying Assumption 3.1, ℓ_i are the intensities of noises. Note that (20) guarantees that $d_{i,n} = \alpha_i + \ell_i \chi_{i,n} \in (0, 1)$, $i = 1, \ldots, m$, $n \in \mathbb{N}$.

Instead of condition (7), we assume now that

$$\|(I - U_n)[F(X) - X^*]\| \le \nu(n)\|X - X^*\|, \quad n \in \mathbb{N}, \ X \in B_R(X^*), \tag{21}$$

where

$$\nu(n) = \phi(d_{1,n}, \ldots, d_{m,n}), \quad \phi : \mathbb{R}^m \to [0, \infty) \quad \text{is a continuous function.} \tag{22}$$

Since the function ϕ from (22) is continuous, random variables $\chi_{i,n}$, $i = 1, \ldots, m$, $n \in \mathbb{N}$, are mutually independent, and $\chi_{i,n}$ are identically distributed (for each i), random variables $\nu(n)$ are also mutually independent and identically distributed.

For each $\omega \in \Omega$, the stochastic controls $\alpha_i + \ell_i \chi_{i,n+1}(\omega)$, $i = 1, \ldots, m$, $n \in \mathbb{N}_0$, are just numbers in the corresponding intervals $[\alpha_i - \ell_i, \alpha_i + \ell_i]$. Therefore, if we can choose α_i and ℓ_i in such a way that $\nu(n) \le \nu^* < 1$ on Ω, for some $\nu^* \in (0, 1)$, we are in the same situation as in Section 2, see Propositions 3.4 and 3.5 in Section 3.2. The idea to some extent generalizes the approach in [9] to systems.

In this section, we concentrate on the case when $\nu(n)$ might exceed one on some set with a nonzero probability, assuming instead that

$$\mathbb{E} \ln \nu(n) = -\lambda_1 < 0 \tag{23}$$

and applying Kolmogorov's Law of Large Numbers, see [40, p. 391].

Lemma 3.2 (Kolmogorov's Law of Large Numbers): *Let $(v_n)_{n\in\mathbb{N}}$ be a sequence of independent identically distributed random variables, where $\mathbb{E}|v_n| < \infty$, $n \in \mathbb{N}$, their common mean is $\mu := \mathbb{E}v_n$, and the partial sum is $S_n := \sum_{k=1}^{n} v_k$. Then $\lim_{n\to\infty} \frac{S_n}{n} = \mu$, a.s.*

Even though the result of Theorem 3.3 is new, its proof is quite standard, and we give it here only for completeness of presentation. Since random variables $\chi_{i,n}$, $i = 1, \dots, m$, $n \in \mathbb{N}$, are bounded, we follow the approach recently employed in [10].

Theorem 3.3: *Let Assumption 3.1 and conditions (21), (22), (23) hold. Then, for each $\gamma \in (0, 1)$, there is $\Omega(\gamma) \subseteq \Omega$, $\mathbb{P}(\Omega(\gamma)) > 1 - \gamma$, and $0 < \delta < R$ such that for the initial values $X_0 \in B_R(X^*)$ satisfying $\|X_0 - X^*\| \le \delta$, a solution of (6) converges to X^* on $\Omega(\gamma)$.*

Proof: Condition (21) implies

$$\|(I - U_n)[F(X) - X^*]\| \le \exp\{\ln \nu(n)\}\|X - X^*\|, \quad n \in \mathbb{N}, \quad X \in B_R(X^*).$$

Since $(\ln \nu(n))_{n\in\mathbb{N}}$ is a sequence of independent identically distributed bounded random variables, Lemma 3.2 (Kolmogorov's Law of Large Numbers) can be applied. So, by (23), there is a random integer $\mathcal{N}(\omega)$ such that, for $n \ge \mathcal{N}(\omega)$, a.s.,

$$\sum_{i=1}^{n} \ln \nu(i) < -\frac{\lambda_1}{2} n. \tag{24}$$

Therefore, for each $\gamma \in (0, 1)$, there exists a nonrandom integer $N = N(\gamma)$ and a set $\Omega(\gamma) \subseteq \Omega$ with $\mathbb{P}(\Omega(\gamma)) > 1 - \gamma$ such that (24) holds for $n \ge N(\gamma)$ on $\Omega(\gamma)$.

Since $d_{i,n} \in [0, 1)$, $i = 1, \dots, m$, $n \in \mathbb{N}$, see (20), and by continuity of ϕ, see (22), there exists a nonrandom number $\mathcal{M} > 1$ such that, $\forall n \in \mathbb{N}$, on Ω,

$$\nu(n) < \mathcal{M}.$$

Choose now $\delta = R\mathcal{M}^{-N(\gamma)}$ and let $\|X_0 - X^*\| < \delta$. Then

$$\begin{aligned}
\|X_1 - X^*\| &= \|(I - U_1)[F(X_0) - X^*]\| \le \nu(1)\|X_0 - X^*\| < \mathcal{M}\delta < R, \\
&\quad \text{so } X_1 \in B_R(X^*), \\
\|X_2 - X^*\| &= \|(I - U_2)[F(X_1) - X^*]\| \le \nu(2)\|X_1 - X^*\| < \nu(2)\nu(1)\|X_0 - X^*\| \\
&= \exp\left\{\sum_{i=1}^{2} \ln \nu(i)\right\} \|X_0 - X^*\| < \mathcal{M}^2\delta < R, \quad \text{so } X_2 \in B_R(X^*), \\
&\dots \\
\|X_N - X^*\| &= \|(I - U_N)[F(X_{N-1}) - X^*]\| \le \nu(N)\|X_{N-2} - X^*\| \\
&\le \exp\left\{\sum_{i=1}^{N} \ln \nu(i)\right\} \|X_0 - X^*\| < \mathcal{M}^N\delta < R, \quad \text{so } X_N \in B_R(X^*).
\end{aligned}$$

Then we can keep solution X_n in $B_R(X^*)$ for at least $N = N(\gamma)$ steps until estimate (24) (Kolmogorov's Law of Large Numbers) starts working on $\Omega(\gamma)$. Further, we get, on $\Omega(\gamma)$,

$$\begin{aligned}\|X_{N+1} - X^*\| &= \|(I - U_{N+1})[F(X_N) - X^*]\| \le \nu(N+1)\|X_N - X^*\| \\ &\le \exp\left\{\sum_{i=1}^{N+1} \ln \nu(i)\right\} \|X_0 - X^*\| \le \exp\left\{-\frac{\lambda_1(N+1)}{2}\right\} \|X_0 - X^*\| \\ &\le \|X_0 - X^*\| < \delta, \quad \text{and for any } n > N+1, \\ \|X_n - X^*\| &\le \exp\left\{\sum_{i=0}^{n} \ln \nu(i)\right\} \|X_0 - X^*\| \le \exp\left\{-\frac{\lambda_1 n}{2}\right\} \|X_0 - X^*\| \\ &< \exp\left\{-\frac{\lambda_1 n}{2}\right\} \delta \to 0, \quad \text{as } n \to \infty,\end{aligned}$$

which provides convergence of a solution to X^* on $\Omega(\gamma)$. ■

3.2. Hénon and Lozi maps with a stochastic control

Let $d_{1,n}$ and $d_{2,n}$ be defined as in (20). Following the methods and the notations of Section 2.3, we conclude that stochastic VMTOC of Hénon (14) and Lozi (15) maps can be transformed to the equation

$$(I - U_{n+1})[F(X_n) - X^*] = \mathcal{C}_{n+1}(x_n) \begin{bmatrix} x_n - x^* \\ y_n - y^* \end{bmatrix}.$$

In the case of Hénon map (14), we have

$$\mathcal{C}_{n+1}(x_n) =: \begin{bmatrix} -a\left(1 - \alpha_1 - \ell_1 \chi_{1,n+1}\right)(x_n + x^*) & \left(1 - \alpha_1 - \ell_1 \chi_{1,n+1}\right) \\ \left(1 - \alpha_2 - \ell_2 \chi_{2,n+1}\right) b & 0 \end{bmatrix},$$

while, in the case of the Lozi map (15) and $x_n x^* > 0$,

$$\mathcal{C}_{n+1}(x_n) = \mathcal{C}_{n+1} =: \begin{bmatrix} -a\left(1 - \alpha_1 - \ell_1 \chi_{1,n+1}\right) & \left(1 - \alpha_1 - \ell_1 \chi_{1,n+1}\right) \\ b\left(1 - \alpha_2 - \ell_2 \chi_{2,n+1}\right) & 0 \end{bmatrix}.$$

In general, the norms $\|\mathcal{C}_n\|$ are not less than one on all Ω, and we get stability only on a smaller set $\Omega_\gamma \subset \Omega$. If, however, the coefficients α_i and ℓ_i, $i = 1, 2$, are chosen in such a way that $\|\mathcal{C}_n\| \le \mu^* < 1$ on Ω, for all $n \in \mathbb{N}$, we are in the same situation as in Section 2. For the Hénon map it holds when $\alpha_* < \alpha_1 - \ell_1$ and $\beta_* < \alpha_2 - \ell_2$, since $\alpha_1 - \ell_1 \le \alpha_1 + \ell_1 \chi_{1,n}$ and $\alpha_2 - \ell_2 \le \alpha_1 + \ell_1 \chi_{2,n}$, for each $n \in \mathbb{N}$. Here α_* and β_* are the convergence bounds from Proposition 2.4. Similar arguments are applied to the Lozi map. Therefore, Propositions 2.4 and 2.5 immediately imply the following results.

Proposition 3.4: *Let (x^*, y^*) be an equilibrium of* (1), *$R < |x^*|$, and constants α_* and β_* be the convergence bounds from Proposition 2.4. Let X be a solution to* (14) *with $d_{1,n}$ and $d_{2,n}$ defined as in* (20). *Then for any $\alpha_1 \in (\alpha_*, 1)$, $\alpha_2 \in (\beta_*, 1)$, $\ell_1 < \min\{\alpha_1 - \alpha_*, 1 - \alpha_1\}$ and $\ell_2 < \min\{\alpha_2 - \beta_*, 1 - \alpha_2\}$, X converges to (x^*, y^*) for any initial value $(x_0, y_0) \in B_R(X^*)$.*

Proposition 3.5: *Let (x^*, y^*) be an equilibrium of* (2), $R < |x^*|$, *and constants α_* and β_* be the convergence bounds from Proposition 2.5. Let X be a solution to* (15) *with $d_{1,n}$ and $d_{2,n}$ defined as in* (20). *Then for any $\alpha_1 \in (\alpha_*, 1)$, $\alpha_2 \in (\beta_*, 1)$, $\ell_1 < \min\{\alpha_1 - \alpha_*, 1 - \alpha_1\}$ and $\ell_2 < \min\{\alpha_2 - \beta_*, 1 - \alpha_2\}$, X converges to (x^*, y^*) for any initial value $(x_0, y_0) \in B_R(X^*)$.*

Proceed now to a more general situation, when introduction of noise contributes to reducing the minimum values of control parameters which guarantee stability.

In order to apply Theorem 3.3, we need to find control parameters $\alpha_1, \alpha_2, \ell_1, \ell_2$ and $\lambda_1 > 0$ such that condition (23) holds for some $\nu(n) > \max_{X \in B(X^*,R)} \|C_n(x)\|$. Note that $\nu(n)$ and values of parameters depend on the particular norm $\|\cdot\|$ and distributions of noises χ_1 and χ_2. In Sections 3.2.1 and 3.2.2, we consider $\|\cdot\|_\infty$ and $\|\cdot\|_1$, Bernoulli and uniform continuous distributions for the noise, and illustrate obtained results with computer simulations.

3.2.1. Hénon

(1) *Norm $\|\cdot\|_\infty$, χ_1 is Bernoulli distributed.* Since $|x + x^*| \le 2x^* + R$, we have, for $X \in B_R(X^*)$,

$$\|C_n(x)\|_\infty \le \max\left\{\left(1 - \alpha_1 - \ell_1\chi_{1,n}\right)[a(2x^* + R) + 1], \left(1 - \alpha_2 - \ell_2\chi_{2,n}\right)b\right\}.$$

Theorem 3.3 states only local stability for any given probability; therefore, we can assume $R \approx 7.1 \times 10^{-5}$, so that $a(2x^* + R) \approx 1.7677$ and use local estimates at the equilibrium point for $\|C_n(x)\|_\infty$:

$$\|C_n(x)\|_\infty \le \max\left\{2.7677\left(1 - \alpha_1 - \ell_1\chi_{1,n}\right), 0.3\left(1 - \alpha_2 - \ell_2\chi_{2,n}\right)\right\}.$$

If

$$1 - \alpha_2 + \ell_2 < 9.2256\,(1 - \alpha_1 - \ell_1), \tag{25}$$

then

$$\max\{2.7677(1 - \alpha_1 - \ell_1\chi_{1,n}), 0.3(1 - \alpha_2 - \ell_2\chi_{2,n})\} = 2.7677(1 - \alpha_1 - \ell_1\chi_{1,n}).$$

Choose first α_1, ℓ_1 such that condition (23) holds for $\nu(n) := 2.7677(1 - \alpha_1 - \ell_1\chi_{1,n})$ and for some $\lambda_1 > 0$. Since χ_1 is Bernoulli distributed, we get $\mathbb{E}\ln\nu(n) = \ln 2.7677 + 0.5\ln((1 - \alpha_1)^2 - \ell_1^2)$, which leads to the following estimation

$$\ell_1^2 > (1 - \alpha_1)^2 - 0.1305.$$

To construct the interval for ℓ_1 in (20), we take $\alpha_1^2 > (1 - \alpha_1)^2 - 0.1305$, which gives $\alpha_1 > \frac{0.8695}{2} = 0.4347$, so, for $\alpha_1 = 0.44$ we have $\ell_1 > \sqrt{0.1831} = 0.4279$. The right-hand side of the inequality in condition (25) for $\alpha_1 = 0.44$, $\ell_1 = 0.4279$ gives $1.2186 > 1$, which means that (25) holds for each α_2, ℓ_2 satisfying (20).

For $\alpha_1 = 0.44$ and no control in y ($\alpha_2 = \ell_2 = 0$), Figure 3 confirms that there is stabilization already for $\ell_1 = 0.3$. The series of runs with the same initial conditions illustrate that, without noise in the control, we have a stable two-cycle (Figure 3(a)). As small noise with $\ell_1 = 0.15$ is introduced, a noisy around this two-cycles solution nearly reaches the equilibrium (Figure 3(b)). As noise increases to $\ell_1 = 0.25$, the

x-trajectory already looks as a stochastic perturbation of the equilibrium (Figure 3(c)), and the equilibrium becomes stable for $\ell_1 = 0.3 < 0.4279$ (Figure 3(d)). We can also observe that much smaller noise does not guarantee stabilization.

Two-dimensional Figure 4 with the attracting limit sets also illustrates the way to stabilization for $\alpha = 0.3 < 0.4279$. There is a stable two cycle without noise (not illustrated), and close to this cycle blurred set for $\ell_1 = 0.05$ (Figure 4(a)), which widens with the increase of the noise to $\ell_1 = 0.15$, still not reaching the equilibrium point (Figure 4(b)), getting a blurred equilibrium for $\ell_1 = 0.25$ (Figure 4(c)) and stabilization of the equilibrium point for $\ell_1 = 0.3$ (Figure 4(d)).

(2) *Norm* $\|\cdot\|_1$, χ_1 *is Bernoulli distributed,* $\ell_2 = 0$. Since $2x^*a = 1.7676 > 1$, we have

$$\begin{aligned}\|\mathcal{C}_n(x)\|_1 &\le \max\left\{\left(1-\alpha_1-\ell_1\chi_{1,n}\right)a(2x^*+R)+\left(1-\alpha_2-\ell_2\chi_{2,n}\right)b,\right.\\ &\qquad\left.\left(1-\alpha_1-\ell_1\chi_{1,n}\right)\right\}\\ &\approx 1.7677\left(1-\alpha_1-\ell_1\chi_{1,n}\right)+0.3\left(1-\alpha_2-\ell_2\chi_{2,n}\right) =: \nu(n),\\ \mathbb{E}\ln\nu(n) &\approx 0.5\ln\left[(1.7677(1-\alpha_1)+0.3(1-\alpha_2))^2-(1.7677\ell_1)^2\right].\end{aligned}$$

So $\ell_1^2 > ((1-\alpha_1)+0.1697(1-\alpha_2))^2 - 0.3200 \implies \mathbb{E}\ln\nu(n) < 0$. For $\alpha_2 = 0.8$, $\ell_1 = 0.2862$ and $\alpha_1 = 0.4$, we get $\mathbb{E}\ln\nu(n) = \ln(0.9999) < 0$ along with the fact that condition (20) holds.

(3) *Norm* $\|\cdot\|_1$, χ_1 *has uniform continuous distribution,* $\ell_2 = 0$. We have

$$\mathbb{E}\ln\nu(n) \approx \frac{1}{2}\int_{-1}^{1}\ln\left[(1.7677(1-\alpha_1)+0.3(1-\alpha_2))-1.7677\ell_1 z\right]dz.$$

The set of parameters $\alpha_2 = 0.9, \alpha_1 = 0.44, \ell_1 = 0.2862$ gives $\mathbb{E}\ln\nu(n) \approx -0.0251 < 0$ and, also, condition (20) holds.

Figure 5 illustrates and compares introduction of Bernoulli and uniformly distributed on $[-1, 1]$ noises for the Hénon map, using a part of the bifurcation diagram for x in the range of parameters of α_1 close to the last bifurcation leading to stabilization of the equilibrium, with $\alpha_2 = 0.8$-control in y. In Figure 5(a), for $\ell_1 = 0$, we see a period-halving bifurcation; in (b) and (c) figures (Bernoulli and uniform noise, respectively, both with the amplitude $\ell_1 = 0.2861$), the value of α_1 for this bifurcation is smaller (≈ 0.39 and ≈ 0.432, respectively), with some stabilization advantage of the Bernoulli distribution. Figure 5 also shows that the image for the uniformly distributed perturbation looks 'noisier' ((c) compared to (b)). Note that compared to the above computations, Figure 5(c) illustrates stabilization for $\ell_1 = 0.2861, \alpha_1 \approx 0.432 < 0.44$ and $\alpha_2 = 0.8 < 0.9$. The case $\alpha_2 = 0.9$ is further tested in Figure 6.

The fact that convergence is observed for lower control values than theoretically predicted, is also illustrated in Figure 6 with limit two-dimensional sets for the Hénon map, with $\ell_1 = 0.2861$, and the control in y being $\alpha_2 = 0.9$, starting with $\alpha_1 = 0.3$ and the Bernoulli noise (Figure 6(a)), then for $\alpha_1 = 0.3$ and the uniform noise (Figure 6(b)), next for $\alpha_1 = 0.36$ and the Bernoulli noise (Figure 6(c)) and, finally, for $\alpha_1 = 0.36$ and the uniform noise (Figure 6(d)).

(4) *Norm* $\|\cdot\|_1$, χ_1 *is Bernoulli distributed,* $\ell_2 = 0$, $a = 2$ *and* $b = 0.5$. In this part we take different parameters a and b, which represent a chaotic case. In this case, $x^* \approx 0.593$,

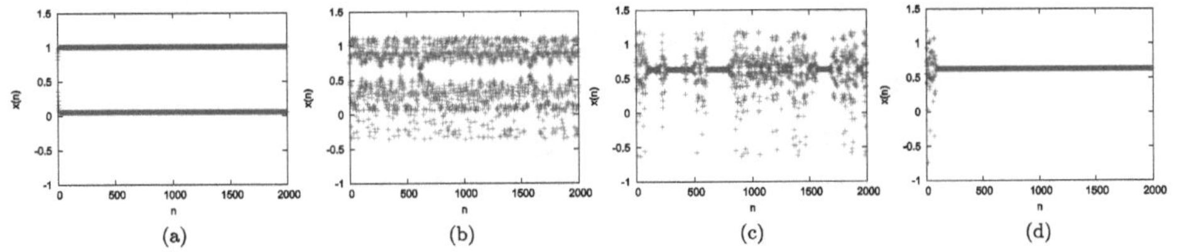

Figure 3. Runs of the Hénon map for x-coordinate only, $\alpha = 0.44$, no control in y, $x_0 = 0.3$, $y_0 = 0.1$ and (a) $\ell = 0$, (b) $\ell = 0.15$, (c) $\ell = 0.25$, (d) $\ell = 0.3$ for the Bernoulli noise.

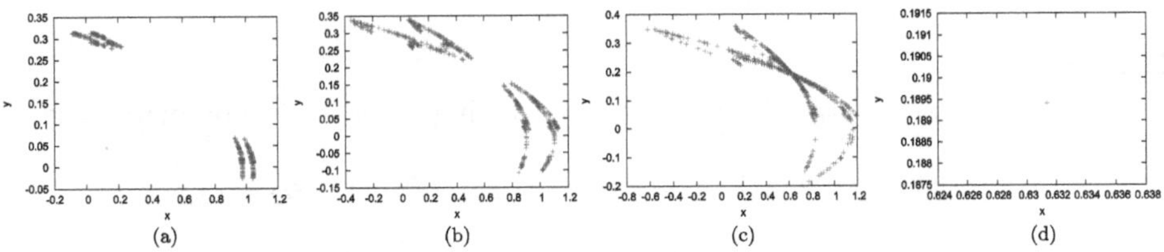

Figure 4. The limit set for the Hénon map for $\alpha = 0.44$, no control in y, $x_0 = 0.3$, $y_0 = 0.1$ and (a) $\ell = 0.05$, (b) $\ell = 0.15$, (c) $\ell = 0.25$, (d) $\ell = 0.3$ for the Bernoulli noise.

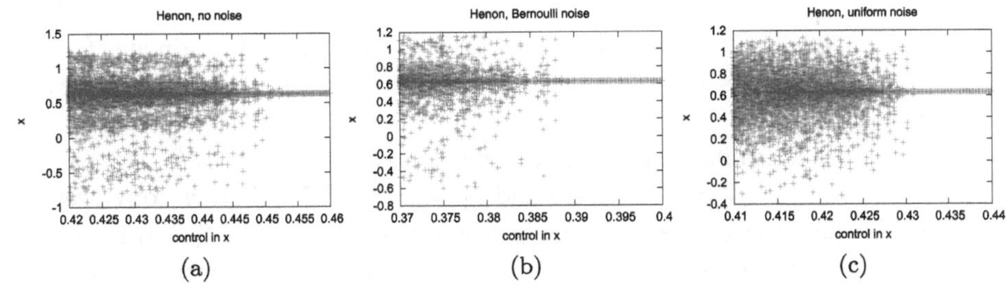

Figure 5. Bifurcation diagram of the Hénon map for x with no control in y and (a) no noise $\ell_1 = 0$, (b) $\ell_1 = 0.2861$, Bernoulli distribution, (c) $\ell_1 = 0.2861$, uniform on $[-1, 1]$ distribution, $\alpha_2 = 0.8$. The range of initial values is $x_0 \in [0.1, 0.8]$ and $y_0 \in [0.1, 0.2]$.

$2ax^* \approx 2.372$ and

$$\begin{aligned}\|C_n(x)\|_\infty &\le \max\left\{\left(1-\alpha_1-\ell_1\chi_{1,n}\right)2ax^* + \left(1-\alpha_2-\ell_2\chi_{2,n}\right)b,\right.\\ &\qquad \left.\left(1-\alpha_1-\ell_1\chi_{1,n}\right)\right\}\\ &\approx 2.372\left(1-\alpha_1-\ell_1\chi_{1,n}\right) + 0.5\left(1-\alpha_2-\ell_2\chi_{2,n}\right) =: \nu(n).\end{aligned}$$

This leads to the estimate $\ell_1^2 > ((1-\alpha_1) + 0.2107(1-\alpha_2))^2 - 0.1777$, which, along with (20), holds for $\alpha_1 = 0.45$, $\alpha_2 = 0.8$ and $\ell_1 = 0.416$.

Note that in the deterministic case, when $\ell_1 = \ell_2 = 0$, we get the best value (using the estimation of eigenvalues) as $\alpha_1 > 0.6518$ for $\alpha_2 = 0$ and $\alpha_1 > 0.5801$ for $\alpha_2 = 0.8$.

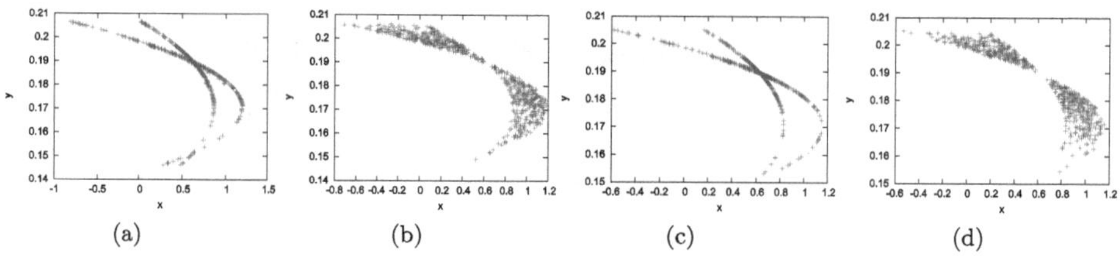

Figure 6. The limit sets for the Hénon map for $\ell_1 = 0.2861$ and $\alpha = 0.3$, (a) for Bernoulli, (b) for the uniform on $[-1, 1]$ noise, then $\alpha = 0.36$, (c) for Bernoulli, (d) for the uniform on $[-1, 1]$ noise. The control in y is 0.9, $x_0 = 0.3$, $y_0 = 0.1$ in all four simulations.

3.2.2. Lozi

(1) *Norm* $\|\cdot\|_\infty$, χ_1 *is Bernoulli distributed.* Reasoning as for the Hénon map, we get

$$\begin{aligned}\|\mathcal{C}_n(x)\|_\infty &\le \max\{2.4\left(1-\alpha_1-\ell_1\chi_{1,n}\right), 0.3\left(1-\alpha_2-\ell_2\chi_{2,n}\right)\}\\ &= 2.4\left(1-\alpha_1-\ell_1\chi_{1,n}\right) =: \nu(n)\end{aligned}$$

if

$$1-\alpha_2+\ell_2 \le 8\left(1-\alpha_1-\ell_1\right). \tag{26}$$

Then $\mathbb{E}\ln\nu(n) = \ln 2.4 + 0.5\ln((1-\alpha_1)^2 - \ell_1^2) < 0$ leads to the estimate $\ell_1^2 > (1-\alpha_1)^2 - 0.1736$. The values $\alpha_1 = 0.414$, $\ell_1 = 0.17$ satisfy (20) as well as (26) (for arbitrary α_2, ℓ_2).

Simulations illustrate convergence for smaller values of α_1 and ℓ_1 for $\alpha_2 = \ell_2 = 0$. In Figure 7(d), for $\alpha_1 = 0.4 < 0.413$ and $\ell_1 = 0.15 < 0.17$, there is convergence to the equilibrium; however, such convergence is achieved for any $\ell_1 \ge 0.15$. According to Figure 7(a), for $\ell_1 = 0$, there is a stable two-cycle, which becomes blurred and approaches the equilibrium for $\ell_1 = 0.05$ (Figure 7(b)), leading to a blurred equilibrium at $\ell_1 = 0.1$ (Figure 7(c)) and a stable equilibrium for $\ell_1 = 0.15$ (Figure 7(d)). In Figure 7, we took the initial point $(-10, -15)$ quite far from the equilibrium; the same stabilization pattern is observed for any initial point. The only difference is that for very large by absolute value $x_0 \approx -100$, up to 70 first iterations can be outside of the interval $[-1; 1.5]$ chosen to illustrate convergence.

In addition to separate runs in x, asymptotic behaviour is illustrated by two-dimensional diagrams of the limit sets in Figure 8. This set forms a blurred two-cycle for $\ell_1 = 0.03$ (Figure 8(a)) which expands and approached the equilibrium point for

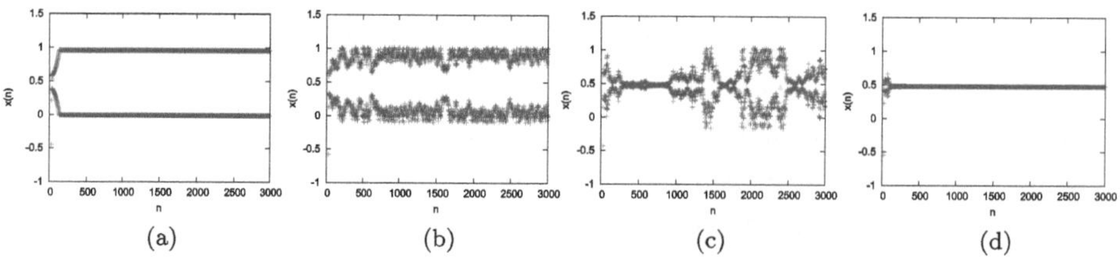

Figure 7. Runs of the Lozi map for x-coordinate only, $\alpha = 0.4$, no control in y, $x_0 = -10$, $y_0 = -15$ and (a) $\ell = 0$, (b) $\ell = 0.05$, (c) $\ell = 0.1$, (d) $\ell = 0.15$ for the Bernoulli noise.

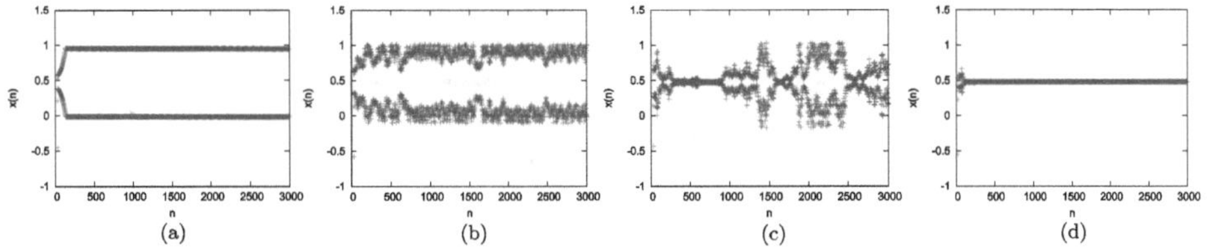

Figure 8. The limit set for the Lozi map for $\alpha = 0.4$, no control in y, $x_0 = -10$, $y_0 = -15$ and (a) $\ell = 0.03$, (b) $\ell = 0.08$, (c) $\ell = 0.135$, (d) $\ell = 0.145$ for the Bernoulli noise.

$\ell_1 = 0.08$ (Figure 8(b)), becomes a blurred equilibrium for $\ell_1 = 0.135$ (Figure 8(c)) and a stable equilibrium for $\ell_1 = 0.145$ (Figure 8(d)) or larger control values.

(2) *Norm* $\|\cdot\|_1$, χ_1 *is Bernoulli distributed,* $\ell_2 = 0$. Now,

$$\|\mathcal{C}_n(x)\|_1 \le 1.4\left(1 - \alpha_1 - \ell_1\chi_{1,n}\right) + 0.3\left(1 - \alpha_2 - \ell_2\chi_{2,n}\right) =: \nu(n),$$
$$\mathbb{E}\ln\nu(n) = 0.5\ln\left[(1.4(1-\alpha_1) + 0.3(1-\alpha_2))^2 - (1.4\ell_1)^2\right].$$

We can show that each set of parameters $\alpha_1 = 0.3$, $\alpha_2 = 0.8$, $\ell_1 = 0.2039$ and $\alpha_1 = 0.27$, $\alpha_2 = 0.9$, $\ell_1 = 0.2332$, satisfies all necessary conditions and leads to $\mathbb{E}\ln \nu(n) < 0$.

(3) *Norm* $\|\cdot\|_1$, χ_1 *and* χ_2 *are Bernoulli distributed.* We have

$$\mathbb{E}\ln\nu(n) = 0.25\left(\ln\left[(1.4(1-\alpha_1) + 0.3(1-\alpha_2))^2 - (1.4\ell_1 + 0.3\ell_2)^2\right] + \ln\left[(1.4(1-\alpha_1) + 0.3(1-\alpha_2))^2 - (1.4\ell_1 - 0.3\ell_2)^2\right]\right).$$

By introduction of the second nonzero noise, we can decrease a little bit the intensity ℓ_1 of the first noise for $\alpha_1 = 0.27, \alpha_2 = 0.9$. In particular, we can take $\ell_1 = 0.2, \ell_1 = 0.55$ and get $\mathbb{E}\ln\nu(n) \approx 0.25\ln 0.9936 \approx -0.25 \cdot 0.0064 < 0$.

Figure 9 compares the influence of Bernoulli and uniformly distributed on $[-1, 1]$ noises for the Lozi map, illustrating a part of the bifurcation diagram for x in the range of parameters of α close to the last bifurcation leading to stabilization of the equilibrium, with the control $\alpha_2 = 0.9$ in y. In Figure 9(a), for $\ell_1 = 0$, we see a period-halving bifurcation; in (b) and (c) (Bernoulli and uniform noise, respectively, both with the

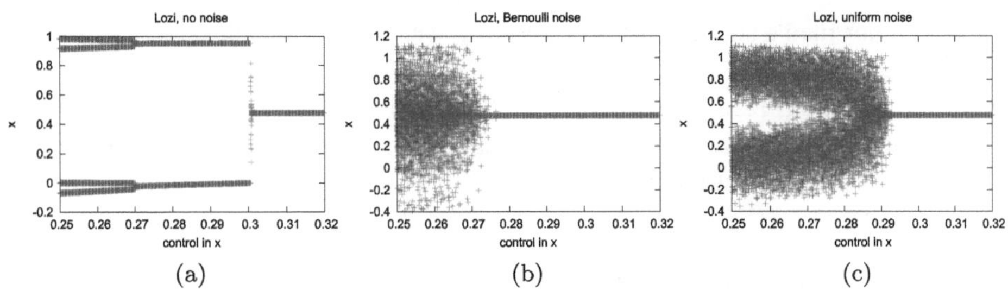

Figure 9. Bifurcation diagram of the Lozi map for x with 0.9-control in y and (a) no noise $\ell_1 = 0$, (b) $\ell_1 = 0.2$, Bernoulli distribution, (c) $\ell_1 = 0.2$, uniform on $[-1, 1]$ noise distribution. The range of initial values is $x_0 \in [0.1, 0.8]$ and $y_0 \in [0.1, 0.2]$.

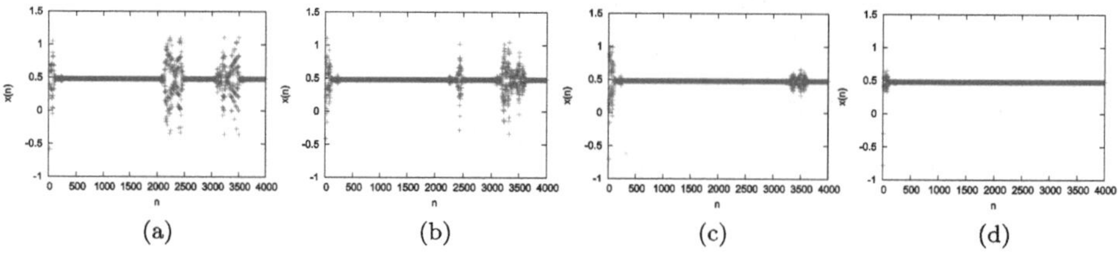

Figure 10. Runs of the Lozi map for x-coordinate only, $\alpha_1 = 0.27, \ell_1 = 0.2$, the control in y is $\alpha_2 = 0.9$, $x_0 = -10, y_0 = -15$ and (a) $\ell_2 = 0$, (b) $\ell_2 = 0.4$, (c) $\ell_2 = 0.5$, (d) $\ell_2 = 0.55$ for the Bernoulli noise for both ℓ_i.

amplitude $\ell_1 = 0.2$), the value of α for this bifurcation is smaller, with some stabilization advantage of the Bernoulli distribution. Figure 9(c) also shows that the image for the uniformly distributed perturbation looks 'noisier' (compared to (b)).

Finally, we illustrate the role of noise ℓ_2 in stabilization. If $\alpha_1 = 0.27$, $\alpha_2 = 0.9$, without noise, stabilization is not observed (Figure 10(a)). Even with $\ell_1 = 0.4$ and $\ell_1 = 0.5$, there is no stabilization, see Figure 10(b) and 10(c). However, if ℓ_2 increases to 0.55, stabilization is observed, as theoretically predicted, see Figure 10(d).

4. Conclusions and discussion

In the present paper, we explored both deterministic and stochastic stabilization of Hénon and Lozi maps with TOC, where an equilibrium is chosen as a target.

(1) We proved that TOC allows stabilizing a chosen equilibrium in a ball centred at it, once controls are close enough to one. For an autonomous model, simulations illustrate, however, that local stabilization parameters work in a reasonably large neighbourhood. This brings us to the problem of whether there is an equivalence of local and global stability observed in certain one-dimensional cases of TOC [18] and PBC [30]. However, for both Hénon and Lozi maps, it is still an open question, even if we reduce the domain to the same half-plane where the equilibrium lies, and consider autonomous cases only.

(2) Investigation of non-autonomous models makes it possible to add noise in certain bounds, still keeping stabilization. Moreover, we demonstrate theoretically and with simulations that stochastic stabilization may be achieved when there is, e.g. a stable cycle rather than a stable equilibrium for the average control values.

Let us compare our results to other types of control.

Prediction-Based Control (PBC) keeps all the original equilibrium points. The general form of PBC applied to a scalar map $x_{n+1} = f(x_n)$ is [41]

$$x_{n+1} = f(x_n) - \alpha \left(f^k(x_n) - x_n\right), \quad x_0 > 0, \ n \in \mathbb{N}_0,$$

which can be treated as a discrete analogue of delayed control for continuous models [37]. Here the control $\alpha \in (0, 1)$, while f^k is the kth iteration of f. The application and the analysis become simpler for $k = 1$ [30]

$$x_{n+1} = f(x_n) - \alpha(f(x_n) - x_n) = (1-\alpha)f(x_n) + \alpha x_n, \quad x_0 > 0, \ n \in \mathbb{N}_0.$$

However, as mentioned in [31], immediate generalization of this scheme to the vector case

$$X_{n+1} = F(X_n) - \alpha(F(X_n) - X_n)$$

does not stabilize the Hénon map for any $\alpha \in (0,1)$. A modification of PBC proposed in [36] was applied to get local stabilization of both equilibrium points for the Hénon map in [31].

Compared to these methods, the advantage of VMTOC is robustness allowing stabilization in a prescribed domain, not just a small neighbourhood of the equilibrium point. The introduction of several parameters allows more adaptivity, controlling, for example, one of two variables only. Also, Liz and Pötzsche [31] and Polyak [36] considered constant control only. The introduction of step-dependent and stochastic control allowed us to evaluate the positive effect of noise on stabilization. On the other hand, PBC-type design allowed to get lower control values ('weak control' in [31]) for local stabilization and could be applied at selected steps only. Note that PBC in principle can provide local stabilization only, as it stabilizes all chosen equilibrium points with the same control equation. In addition, Liz and Pötzsche [31] and Polyak [36] considered cycles stabilization, which is not in the framework of the present research.

While it is generally hard to compare continuous and discrete controls, some ideas are similar. For example, in [19], as in the present paper, the control is a function of some difference, which becomes small as the solution approaches the target orbit. On the other hand, in [19], to some extent similarly to [31,36], this is the difference between the current state and some other state, past in [19] and future, or predicted, in [31,36]. The other common feature of [19] and our research is that the distributed nature of the control can contribute to stabilization. However, in each realization, control distribution in [19] is fixed and, for a given map and prescribed initial values, the results are deterministic. In our models, each realization of the stochastic control, generally, leads to different results. This justifies the necessity to apply Kolmogorov's Law of Large Numbers to verify stabilization.

There are some other control types, which cannot be directly compared to the results of the paper as they stabilize not the equilibria of the original map but some other points. For example, we considered TOC with an equilibrium as a target, but it could be interesting to include other possible targets, for example, the zero target; in this case, we study Proportional Feedback control [14,22,29]

$$x_{n+1} = f(cx_n), \quad c \in [0,1).$$

It is also natural to consider its 'external' modification

$$x_{n+1} = cf(x_n), \quad c \in [0,1).$$

We recall that, once we choose an equilibrium point as the target T in (3), this is also a fixed point of (3). However, PF control shifts equilibrium points (as well as TOC with a target not at an equilibrium point). Moreover, for a scalar PF control existence of a positive equilibrium is not guaranteed; for an unstable equilibrium, there is usually a range of parameters (c_*, c^*) such that a new positive equilibrium exists and is stable for $0 < c_* < c < c^* < 1$ only.

Acknowledgments

The authors are very grateful to anonymous reviewers whose valuable comments significantly contributed to the presentation of our results and the current form of the manuscript.

Disclosure statement

No potential conflict of interest was reported by the author(s).

Funding

The first author was supported by NSERC [grant RGPIN-2020-03934].

ORCID

E. Braverman http://orcid.org/0000-0002-6476-585X
A. Rodkina http://orcid.org/0000-0001-8758-0287

References

[1] J.A.D. Appleby, X. Mao, and A. Rodkina, *On stochastic stabilization of difference equations*, Dyn. Contin. Discrete Syst. 15(3) (2006), pp. 843–857.
[2] L. Arosio, A.M. Benini, J.E. Fornæss, and H. Peters, *Dynamics of transcendental Hénon maps*, Math. Ann. 373(1–2) (2019), pp. 853–894.
[3] K.J. Åström, *Introduction to Stochastic Control Theory, Mathematics in Science and Engineering*, Vol. 70, Academic Press, New York, 1970.
[4] G. Berkolaiko, E. Buckwar, C. Kelly, and A. Rodkina, *Almost sure asymptotic stability analysis of the θ-Maruyama method applied to a test system with a.s. stabilizing and destabilizing stochastic perturbations*, LMS J. Comput. Math. 15 (2012), pp. 71–83.
[5] G. Berkolaiko and A. Rodkina, *Almost sure convergence of solutions to non-homogeneous stochastic difference equation*, J. Differ. Equations Appl. 12 (2006), pp. 535–553.
[6] E. Braverman and B. Chan, *Stabilization of prescribed values and periodic orbits with regular and pulse target oriented control*, Chaos 24(1) (2014), p. 013119.
[7] E. Braverman and D. Franco, *Stabilization with target oriented control for higher order difference equations*, Phys. Lett. A 379(16) (2015), pp. 1102–1109.
[8] E. Braverman and D. Franco, *Stabilization of structured populations via vector target-oriented control*, Bull. Math. Biol. 79 (2017), pp. 1759–1777.
[9] E. Braverman, C. Kelly, and A. Rodkina, *Stabilisation of difference equations with noisy prediction-based control*, Phys. D 326 (2016), pp. 21–31.
[10] E. Braverman, C. Kelly, and A. Rodkina, *Stabilization of cycles with stochastic prediction-based and target-oriented control*, Chaos 30 (2020), p. 15.
[11] E. Braverman and A. Rodkina, *Stabilization of difference equations with noisy proportional feedback control*, Discrete Contin. Dyn. Syst. Ser. B 22 (2017), pp. 2067–2088.
[12] E. Braverman and A. Rodkina, *Stochastic control stabilizing unstable or chaotic maps*, J. Differ. Equ. Appl. 25 (2019), pp. 151–178.
[13] E. Buckwar and C. Kelly, *Towards of systematic linear stability analysis of numerical methods for systems of stochastic differential equations*, SIAM J. Numer. Anal. 48(1) (2010), pp. 298–321.
[14] P. Carmona and D. Franco, *Control of chaotic behavior and prevention of extinction using constant proportional feedback*, Nonlinear Anal. Real World Appl. 12 (2011), pp. 3719–3726.
[15] P.L. Chesson, *The stabilizing effect of a random environment*, J. Math. Biol. 15(1) (1982), pp. 1–36.
[16] J. Dattani, J.C. Blake, and F.M. Hilker, *Target-oriented chaos control*, Phys. Lett. A 375(45) (2011), pp. 3986–3992.
[17] S.N. Elaydi, *An Introduction to Difference Equations*, 2nd ed. Springer, Berlin, 1999.

[18] D. Franco and E. Liz, *A two-parameter method for chaos control and targeting in one-dimensional maps*, Int. J. Bifur. Chaos Appl. Sci. Eng. Int. 23(1) (2013), p. 1350003.
[19] A. Gjurchinovski, T. Jungling, V. Urumov, and E. Schöll, *Delayed feedback control of unstable steady states with high-frequency modulation of the delay*, Phys. Rev. E. 88(3) (2013), p. 032912.
[20] A. Gjurchinovski and V. Urumov, *Stabilization of unstable steady states and unstable periodic orbits by feedback with variable delay*, Romanian J. Phys. 58(1–2) (2013), pp. 36–49.
[21] A. Gjurchinovski, A. Zakharova, and E. Schöll, *Amplitude death in oscillator networks with variable-delay coupling*, Phys. Rev. E. 89(3) (2014), p. 032915.
[22] J. Güémez and M.A. Matías, *Control of chaos in unidimensional maps*, Phys. Lett. A181 (1993), pp. 29–32.
[23] M. Hénon, *A two-dimensional mapping with a strange attractor*, Commun. Math. Phys. 50(1) (1976), pp. 69–77.
[24] P. Hitczenko and G. Medvedev, *Stability of equilibria of randomly perturbed maps*, Discrete Contin. Dyn. Syst. Ser. B 22(2) (2017), pp. 269–281.
[25] R. Khasminskii, *Stochastic Stability of Differential Equations*, 2nd ed., Stochastic Modelling and Applied Probability, Vol. 66, Springer, Heidelberg, 2012.
[26] B. Leal and S. Muñoz, *Hénon-Devaney like maps*, Nonlinearity 34(5) (2021), pp. 2878–2896.
[27] C.G. Lee and J. Silverman, *GIT stability of Hénon maps*, Proc. Amer. Math. Soc. 148(10) (2020), pp. 4263–4272.
[28] H. Li, K. Li, M. Chen, B. Mo, and B. Bao, *Coexisting infinite orbits in an area-preserving Lozi map*, Entropy 22(10) (2020), 14 pp.
[29] E. Liz, *How to control chaotic behavior and population size with proportional feedback*, Phys. Lett. A 374 (2010), pp. 725–728.
[30] E. Liz and D. Franco, *Global stabilization of fixed points using predictive control*, Chaos 20 (2010), p. 023124.
[31] E. Liz and C. Pötzsche, *PBC-based pulse stabilization of periodic orbits*, Phys. D 272 (2014), pp. 26–38.
[32] R. Lozi, *Un attracteur étrange (?) du type attracteur de Hénon*, J. Physique (Paris) 39(8) (1978), pp. 9–10.
[33] M. Lyubich and H. Peters, *Structure of partially hyperbolic Hénon maps*, J. Eur. Math. Soc. (JEMS) 23(9) (2021), pp. 3075–3128.
[34] M. Misiurewicz and S. Štimac, *Symbolic dynamics for Lozi maps*, Nonlinearity 29(10) (2016), pp. 3031–3046.
[35] E. Ott, C. Grebogi, and J.A. Yorke, *Controlling chaos*, Phys. Rev. Lett. 64 (1990), pp. 1196–1199.
[36] B.T. Polyak, *Chaos stabilization by predictive control*, Autom. Remote Control 66 (2005), pp. 1791–1804.
[37] K. Pyragas, *Continuous control of chaos by self-controlling feedback*, Phys. Lett. A 170 (1992), pp. 421–428.
[38] E. Schöll and H.G. Schuster (Eds), *Handbook of Chaos Control*. Second edition, Wiley-VCH Verlag GmbH & Co, Weinheim, 2008.
[39] S.J. Schreiber, *Persistence for stochastic difference equations: A mini-review*, J. Differ. Equations Appl. 18(8) (2012), pp. 1381–1403.
[40] A.N. Shiryaev, *Probability*, 2nd ed. Springer, Berlin, 1996.
[41] T. Ushio and S. Yamamoto, *Prediction-based control of chaos*, Phys. Lett. A 264 (1999), pp. 30–35.
[42] M. Yampolsky and J. Yang, *Structural instability of semi-Siegel Hénon maps*, Adv. Math. 389 (2021), 36 pp.

Controlling chimera and solitary states by additive noise in networks of chaotic maps

Elena Rybalova, Eckehard Schöll and Galina Strelkova

ABSTRACT

We study numerically the spatio-temporal dynamics of ring networks of coupled discrete-time systems in the presence of additive noise. The robustness of chimera states with respect to noise perturbations is explored for two ensembles in which the individual elements are described by either logistic maps or Henon maps in the chaotic regime. The influence of noise on the behavior of solitary states is investigated for a ring of nonlocally coupled Lozi maps. Numerical simulations are performed for a set of different noise realizations and random initial conditions to provide reliable statistical data. The type of dynamics of the considered networks is quantified by using a cross-correlation coefficient. We find that there is a finite and sufficiently wide region with respect to the coupling strength and the noise intensity where the probability of observing the chimeras and solitary states is high.

1. Introduction

Random fluctuations are unavoidable in real-world systems. They can appear as intrinsic noise or may come as external random perturbations. Studying system responses to noise is very important from a viewpoint of stability of operating modes, predicting and controlling the functioning of a majority of life-significant systems, such as infrastructure systems, power grids, transport networks, communication and information systems, biological, epidemiological, economic and social networks [4, 13, 14, 25, 38], etc. Numerical studies have demonstrated that besides a degrading function, noise can play a constructive role and gives rise to new dynamical behaviour and induces novel spatio-temporal patterns which do not exist in networks of coupled dynamical systems without noise [2, 6, 19, 25, 31, 43, 51, 73]. The noise influence can also effectively control the operating modes and essentially improve certain properties of the system dynamics [2, 7]. Recently, the interest in the study of noise impact in real-world systems has significantly increased with the discovery of special partial synchronization patterns, such as chimera states [1, 10, 29, 45, 46, 49, 61, 63, 81] and solitary states [8, 26, 34, 66]. A chimera state was reported for the first time for a network of nonlocally coupled identical phase oscillators [1, 29]. This peculiar spatio-temporal pattern represents an intermediate stage at a transition from complete coherence (synchronization) to fully incoherent (spatio-temporal chaos) dynamics and denotes the coexistence of spatially localized domains with coherent (synchronized) and incoherent (desynchronized) dynamics of network oscillators. Recent theoretical and numerical studies have demonstrated that chimera states can emerge in networks with various types of local dynamics and a variety of topologies [1, 10, 29, 45, 48, 49, 58, 60, 67, 72, 76, 82]. These partial synchronization patterns were also observed in a wide range of experimental systems [18, 21, 28, 30, 37, 53, 59, 75, 79]. Besides, chimera states can be linked to various processes occurring in real-world systems, for example, in neuroscience [5, 35, 62], in power-grids networks [39, 41, 78], in social systems [20, 24], etc.

Solitary states represent another important type of partial synchronization patterns and are defined as network states for which single or several elements behave differently compared with neighbouring elements which can behave coherently or can be completely synchronized. As a rule, solitary nodes are typically randomly distributed over the network space and their number increases when the coupling strength between the network elements decreases. It was uncovered that the emergence of solitary states is related to the appearance of bistable individual dynamics due to nonlocal interaction [26, 34]. Recently, solitary states have been found in networks of the Kuramoto-Sakaguchi models and the Kuramoto oscillators with inertia [8, 26, 27, 34, 80], the discrete-time systems [54, 67, 69, 71], the FitzHugh–Nagumo systems [40, 55, 64, 65], models of power grids [9, 22, 74], and even in experimental setups of coupled pendula [28].

The robustness of chimera states with respect to noise was investigated in nonlocally coupled networks of discrete-time maps with Gaussian and uniformly distributed parameters [36, 42] and in the presence of additive and multiplicative noise [12, 44, 56], as well as in networks of continuous-time systems [32, 47, 68, 83]. In particular, it has been demonstrated that noise can induce the appearance of novel chimera patterns, such as coherence-resonance chimeras in a ring network of FitzHugh-Nagumo neurons [68], or new spatio-temporal patterns, e.g. solitary states and solitary state chimeras in a network of nonlocally coupled Henon maps [11, 57]. The dependence of the chimera lifetime on the noise was studied in nonlocally coupled networks of periodic and chaotic oscillators [32, 56, 70, 83]. It was shown that amplitude chimeras are long-living transients and can be controlled by noise influence. The noise permits to decrease the lifetime of amplitude chimeras in a ring network of Stuart-Landau oscillators [32, 83] and to increase it to infinity in ensembles of nonlocally coupled chaotic oscillators [56, 70]. Nevertheless, it is still not completely understood how the probability of observing chimera states in networks of nonlocally coupled chaotic oscillators depends on the additive noise intensity, the coupling strength, and random initial conditions.

In contrast to the analysis of chimera states in the presence of noise, the robustness of solitary states with respect to noise has not been elucidated at all until very recently, except [17], where it was shown that the presence of noise in a ring of nonlocally coupled FitzHugh-Nagumo oscillators leads to a transition from solitary states to patched synchrony. Thus, a more detailed study is needed to analyze the behaviour of solitary states in the presence of external random fluctuations.

In the present paper we investigate the influence of additive noise on chimera states in networks of nonlocally coupled logistic and Henon maps and on solitary states in a network of nonlocally coupled Lozi maps. We explore how the existence of these peculiar partial synchronization patterns depends on the coupling strength and random initial conditions in the presence of external noise. In order to provide good statistical data we use different noise realizations for each of 50 different realizations of randomly distributed initial conditions. The cross-correlation coefficient between network elements is calculated to characterize the observed spatio-temporal structures and to automatically detect and count chimera states and solitary nodes in the considered networks both without and in the presence of noise.

2. Networks under study

2.1. Model equations

In this paper we investigate numerically the spatiotemporal dynamics of a network which is represented by a ring of nonlocally coupled chaotic maps and is subjected to additive noise. The mathematical model of the network, in its general form, reads:

$$\begin{aligned} x_i(n+1) &= F_i(n) + \frac{\sigma}{2R}\sum_{j=i-R}^{i+R}[F_j(n) - F_i(n)] + D\xi_i(n), \\ y_i(n+1) &= G_i(n), \end{aligned} \tag{1}$$

where $x_i(n)$ and $y_i(n)$ are dynamical variables, $i = 1, 2, 3, \ldots, N$, $N = 1000$ is the total number of elements in the ensemble, n denotes the discrete time. Functions $F_i(n)$ and $G_i(n)$ are defined by the right-hand sides of the equations of the respective map (they will be described below). The elements within the ring are coupled in a nonlocal way, i.e. each ith element is linked with R neighbours on the left and right, respectively. Parameter R denotes the coupling range and σ is the coupling strength between the elements. We choose and fix $R = 320$ for all our numerical simulations. The last term in the first equation of (1) corresponds to the influence of additive noise which is uniformly distributed within the interval $[-1, 1]$, where $\xi_i(n)$ is a noise generator, and D is the noise intensity. We have also studied the effect of additive Gaussian noise sources on the network of coupled maps. However, the obtained results have shown a qualitative similarity to the case of uniformly distributed noise.

As individual elements in the network (1), we have used the logistic map [15, 16], the Henon [23] and the Lozi [33] maps. The logistic map is the simplest one-dimensional map demonstrating chaotic dynamics with a nonhyperbolic attractor and multistability [3]. The logistic map is given as follows:

$$x^l(n+1) = F^l(n) = \alpha^l x^l(n)(1 - x^l(n)), \tag{2}$$

where $x^l(t)$ is the dynamical variable, α^l is the control (bifurcation) parameter. We fix $\alpha^l = 3.8$ which corresponds to the chaotic dynamics of individual nodes when they are uncoupled.

The two-dimensional Henon map is described by the following equations:

$$\begin{aligned} x^H(n+1) &= F^H(n) = 1 - \alpha^H(x^H(n))^2 + y^H(n), \\ y^H(n+1) &= G^H(n) = \beta^H x^H(n), \end{aligned} \tag{3}$$

where $x^H(n)$ and $y^H(n)$ are the dynamical variables, and $\alpha^H > 0$ and $\beta^H > 0$ are the control parameters. The Henon map under high-ratio compression ($\beta^H \to 0$) reduces to the logistic map. When the parameters are changed, the Henon map undergoes a period-doubling bifurcation cascade which results in the emergence of a nonhyperbolic chaotic attractor [3]. The Henon map is also characterized by the property of multistability [3]. In our calculations, the parameters of the Henon map are fixed at $\alpha^H = 1.4$ and $\beta^H = 0.2$ which corresponds to the chaotic dynamics in each uncoupled element and prevents the trajectories from divergence to infinity when the maps are coupled in a ring. It was found [45, 46, 54, 67] that when logistic maps or Henon maps are nonlocally coupled within a ring, the transition from complete chaotic synchronization to spatio-temporal chaos is accompanied by the appearance of amplitude and phase chimera states when the coupling strength decreases.

The two-dimensional Lozi map is defined by the following system of equations:

$$\begin{aligned} x^L(n+1) &= F^L(n) = 1 - \alpha^L|x^L(n)| + y^L(n)), \\ y^L(n+1) &= G^L(n) = \beta^L x^L(n), \end{aligned} \tag{4}$$

where $x^L(n)$, $y^L(n)$ are the dynamical variables, and $\alpha^L > 0$ and $\beta^L > 0$ are the control parameters. In contrast to the two aforementioned maps, the Lozi map belongs to the class of chaotic maps which can be obtained as the Poincaré section of the hyperbolic class of Lorenz-type attractors. The Lozi map exhibits a quasihyperbolic chaotic attractor [50], and there is no multistability in the system phase space. We also fix $\alpha^L = 1.4$ and $\beta^L = 0.3$ which provides the chaotic behaviour of the individual maps in (1) without coupling. As was shown in Ref. [67], the ring of nonlocally coupled Lozi maps demonstrates the transition from complete coherence to incoherence through the emergence of solitary states when the coupling strength between the elements decreases. The nonlocal coupling causes the birth of bistability in the dynamics of individual Lozi maps. It was found that it is precisely the coexistence of two attracting sets (bistability) in the phase space that is the mechanism for the emergence of solitary nodes in this network [71].

Due to multistability of the considered networks not all of the initial conditions provide chimera states or solitary states with the equal number of solitary nodes. The spatio-temporal dynamics of each network is studied for 50 different realizations of randomly distributed initial states of the dynamical variables $(x_i(0), y_i(0))$, each having its own additive noise realization. The noise realizations used do not change when the coupling strength and the noise intensity are varied.

2.2. *Cross-correlation coefficient*

In addition to the graphical representations for tracking the changes in the network dynamics (snapshots and spatio-temporal diagrams), we calculate the cross-correlation coefficient for each element of the network [77] and then plot the spatial profile (spatial distribution) of its values along the ring. The cross-correlation coefficient is defined as follows:

$$C_{1,i} = \frac{\langle \tilde{x}_1(n)\tilde{x}_i(n)\rangle}{\sqrt{\langle(\tilde{x}_1(n))^2\rangle\langle(\tilde{x}_i(n))^2\rangle}}, \quad i = 2, 3, \ldots, N, \tag{5}$$

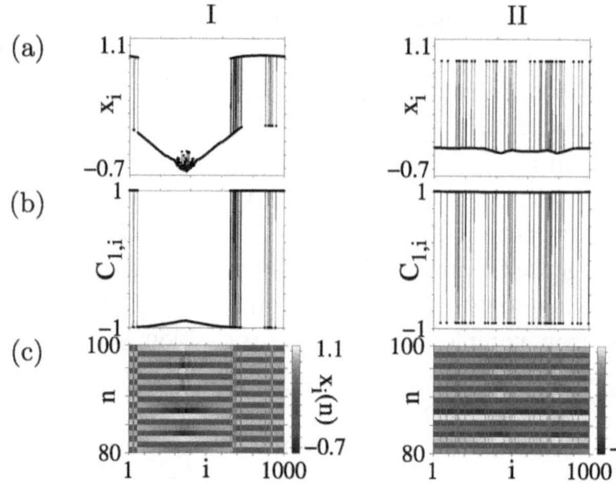

Figure 1. Examples of a combined chimera state in the Henon map ring at $\sigma = 0.275$ (panel I) and a solitary state mode in the Lozi map ring at $\sigma = 0.18$ (panel II) without noise. Snapshots of the x_i variables (a), spatial distributions of the cross-correlation coefficient (b), and spatio-temporal diagrams of $x_i(n)$ (c). Parameters: $\alpha^H = 1.4$, $\beta^H = 0.2$, $\alpha^L = 1.4$, $\beta^L = 0.3$, $R = 320$, $D = 0$, $N = 1000$

where $\tilde{x}(n) = x(n) - \langle x \rangle$, $\langle x \rangle$ is the averaging of the x variable over time $T = 50,000$ iterations. This measure shows the degree of correlation between the first element of the ensemble and all the others, and accordingly illustrates the synchronous or correlated behaviour of the ring elements. The values of $C_{1,i}$ range from -1 to $+1$, where $+1$ relates to full in-phase synchronization, and -1 corresponds to full anti-phase synchronization. $C_{1,i} = 0$ indicates the absence of correlation between the network elements. Calculation of the cross-correlation coefficient enables one to detect automatically phase chimera states and solitary states in the ensembles under study. It was found earlier [77] that an incoherent cluster of a phase chimera is characterized by irregular switchings of cross-correlation coefficient values between ≈ -1 and $\approx +1$, that indicates the anti-phase nature of the oscillations of elements in time. In the case of solitary state mode, the correlation coefficient takes the values $\approx +1$ and ≈ -0.9.

2.3. *Network dynamics without noise*

We first briefly overview the peculiar spatiotemporal structures (chimeras and solitary states) in the networks of Henon maps and Lozi maps without additive noise. Figure 1 shows exemplary snapshots of the x_i variables (Figure 1(a)), spatial distributions of $C_{1,i}$ (Figure 1(b)), and spatio-temporal plots of $x_i(n)$ (Figure 1(c)) for the chimera state in the ensemble of nonlocally coupled Henon maps (Figure 1, left column) and for the solitary state in the ensemble of nonlocally coupled Lozi maps (Figure 1, right column). A chimera state observed in the ring of nonlocally coupled logistic maps is similar to that in the Henon map ring (Figure 1, left column).

The chimera state chosen in our simulation and shown in Figure 1(a–c),I represents the coexistence of a phase chimera (two incoherent clusters with elements $25 < i < 55$ and $652 < i < 727$), an amplitude chimera (the incoherent cluster includes elements $300 < i < 420$), and a solitary state chimera (the incoherent cluster with nodes $877 < i < 950$). As can be seen from Figure 1(b)I, the cross-correlation coefficient values tend to 1 or -1 for the oscillators belonging to the coherent clusters, the clusters of the phase chimera and of the solitary state chimera, while they lie within the interval $C_{1,i} \in [-1, -0.8]$ for the amplitude chimera cluster (or $C_{1,i} \in [0.8, 1]$ if the first element relative to which the cross-correlation coefficients are calculated belongs to the amplitude chimera cluster, for example $i = 100$). To automatically detect phase chimeras we take into account the fact that there are clusters where the cross-correlation coefficient changes from -1 to 1 several times (inside the incoherent cluster of phase chimeras), and clusters in which $C_{1,i}$ approaches only either -1 or 1 for all the oscillators in the coherent clusters. However, introducing the additive noise violates the ideal picture that $C_{1,i}$ is confined to values 1 and -1. Therefore, we must define the confidence intervals $[-1, -0.8]$ and $[0.8, 1]$ instead of -1 and 1, respectively. For the same reason, since it is rather difficult to automatically detect the amplitude chimera in a noisy ensemble, we do not determine its presence. Here, we did not distinguish between the phase chimeras and the solitary state chimeras, since they are similar in the spatial distributions of $C_{1,i}$.

In the case of solitary states (Figure 1(a–c),II), the snapshot consists of a coherent plateau (typical states) and solitary nodes evenly distributed over the entire ensemble (Figure 1(a),II). The cross-correlation coefficients for oscillators in the typical state are equal to 1 (or -1 if the first oscillator is a solitary node), and for solitary nodes $C_{1,i} \approx -0.9$ (or 0.9, if the first oscillator is a solitary node) (see Figure 1(b),II). In the presence of additive noise, the absolute value of the cross-correlation coefficient changes, as in the case of chimera states. If the first oscillator is in the typical state, the $C_{1,i}$ values lie within the interval $[-1, -0.7]$ for the solitary nodes. If the first oscillator is a solitary node, the cross-correlation coefficient for the solitary nodes takes the values from the interval $[0.7, 1]$. We have used these values to automatically find the solitary nodes.

3. Effect of additive noise on chimera states

In this section, we consider the effect of external additive noise on the spatio-temporal dynamics in the networks of nonlocally coupled chaotic maps with a nonhyperbolic chaotic attractor. As individual elements we choose the logistic map (Section 3.1) and the Henon map (Section 3.2). These ensembles can demonstrate different types of chimera states and we now explore their robustness to additive uniformly distributed noise sources with different intensities.

3.1. Network of nonlocally coupled logistic maps

We start with considering the dynamics of the logistic map network (1) without noise and with increasing coupling strength σ between the elements. The initial conditions for the dynamical variables $x_i(0)$ are chosen to be randomly distributed in the interval $[0, 1)$. The changes in the spatio-temporal structures of the logistic map ring are illustrated in Figure 2 by snapshots, spatial distributions of cross-correlation coefficients and spatio-temporal diagrams. At a weak coupling strength ($0 < \sigma < 0.15$), the network shows the regime of complete spatio-temporal incoherence, which turns into spatial incoherence with periodic dynamics of the elements in time when the coupling strength increases (Figure 2,I). When

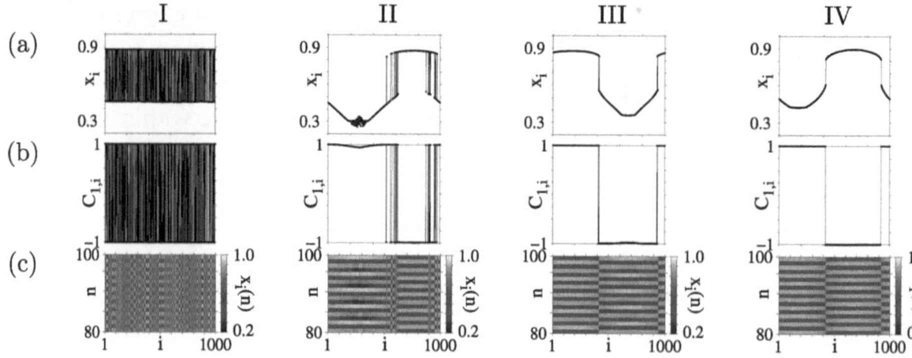

Figure 2. Dynamics of the noise-free network of nonlocally coupled logistic maps for different values of the coupling strength σ: 0.24 (panel I), 0.295 (panel II), 0.345 (panel III), 0.38 (panel IV). Snapshots of the x_i variables (a), spatial distributions of cross-correlation coefficient (b), and spatio-temporal diagrams of $x_i(n)$(c). Parameters: $\alpha^l = 3.8$, $R = 320$, $D = 0$, $N = 1000$

$0.25 < \sigma < 0.4$, we have some probability to obtain either the coexistence of amplitude and phase chimeras (Figure 2,II) or only phase chimeras (Figure 2,III). For some realizations of the initial conditions, only amplitude chimeras occur in the ensemble, which are formed on the snapshot with profile discontinuities. As the coupling strength increases further, the incoherent clusters of phase and amplitude chimeras become narrower, and the probability of emerging amplitude chimeras decreases. It was analytically shown in Ref. [46] that chimera states in the ring of nonlocally coupled logistic maps with the control parameter $\alpha = 3.8$ occur up to the critical coupling strength $\sigma \approx 0.41$ where the spatial wave profile has two turning points with infinite slope. For larger values of σ, the spatial profile in the limit $N \to \infty$ is continuous, while for lower values it breaks at these turning points into two discontinuous branches, and the incoherent clusters are characterized by spatially uncorrelated sequences of nodes alternating randomly between these two branches. Due to the finite number of nodes in our simulations, chimera states are observed in our studies only up to $\sigma \approx 0.4$. However, for some realizations of the initial conditions, profiles with two discontinuities and without any chimeras appear at $\sigma < 0.4$ (See Figure 2,IV). These profiles show the discontinuities but not the two coexisting stable branches, since they may be confined to a very small range of nodes, and their basins of attraction may have quite different sizes, so that for random initial conditions only one of the two branches may be predominantly reached by each node, and hence the incoherent bistable profile typical of chimeras is not observed. At $\sigma > 0.62$, the logistic map ensemble shifts to the regime of complete synchronization in space, which is characterized by chaotic behaviour of the elements in time [46].

Now we add a uniformly distributed noise source to the first equation of the logistic map network (1) and analyze its effect on the network dynamics. Figure 3 presents a two-parameter diagram of regimes which shows the probability P of observing chimera states depending on the coupling strength σ and the noise intensity D. The quantity P denotes the normalized number of initial realizations $P = K/50$, where K is the number of initial sets for which chimeras arise in the network. Qualitatively, one may distinguish between the intervals of weak ($D \in [0, 0.005]$) and strong noise ($D > 0.005$).

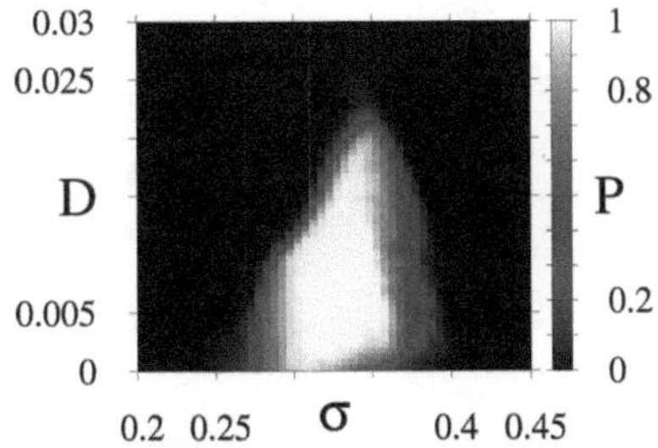

Figure 3. 2D diagram in the (σ, D) parameter plane showing the probability P of observing chimera states in the logistic map ring. Calculations were carried out for 50 different realizations of randomly distributed initial conditions, each having its own noise realization. Parameters: $\alpha^l = 3.8$, $R = 320$, $N = 1000$

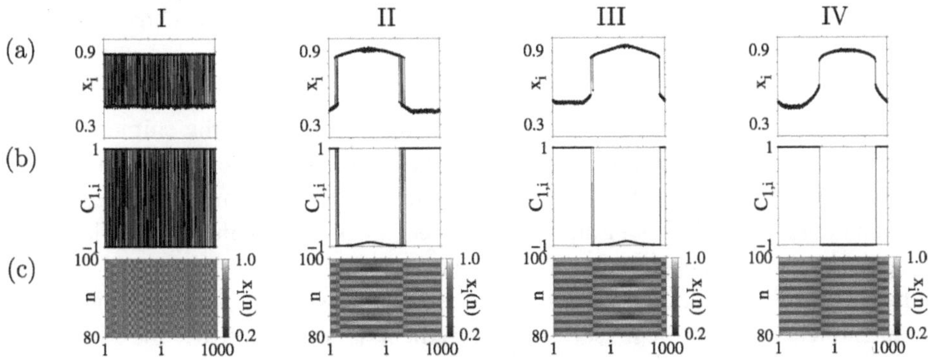

Figure 4. Dynamics of the network of nonlocally coupled logistic maps in the presence of weak noise $D = 0.0045$ and for different values of the coupling strength σ: 0.24 (panel I), 0.295 (panel II), 0.345 (panel III), 0.38 (panel IV). Snapshots of the x_i variables (a), spatial distributions of cross-correlation coefficient (b), and spatio-temporal diagrams of $x_i(n)$ (c). Parameters: $\alpha^l = 3.8$, $R = 320$, $N = 1000$

As follows from the diagram (Figure 3), for weak noise ($D < 0.005$), there is still a wide interval of the coupling strength σ where the chimera states are observed with a rather high probability (yellow and orange colours). With increasing noise, the probability of chimeras near the right boundary of this interval even increases because noise induces transitions between the basins of attraction of the two branches of the profile, and thus the incoherence clusters of the phase chimera in the vicinity of the discontinuity points of the spatial profile become better visible. Figures 4,I–IV show the dynamics of the logistic map network for weak noise intensity and at the same values of σ as in the noise-free case (Figure 2). It is seen that the snapshots of the noisy network dynamics (Figure 4,I–III) are rather similar to those without noise, and only minor noise-induced fluctuations are observed on the spatial profiles. Figure 4,IV illustrates the case when in the presence of noise the incoherence clusters of the phase chimera near the discontinuity points of the snapshot are broadened (compare Figure 2,IV without chimeras and Figure 4,IV with the phase chimera with the incoherence cluster $895 < i < 899$). The reason for this is the noise-induced transitions between the two basins of attraction of the two branches, which broadens the incoherent clusters. This constitutes a constructive influence of noise.

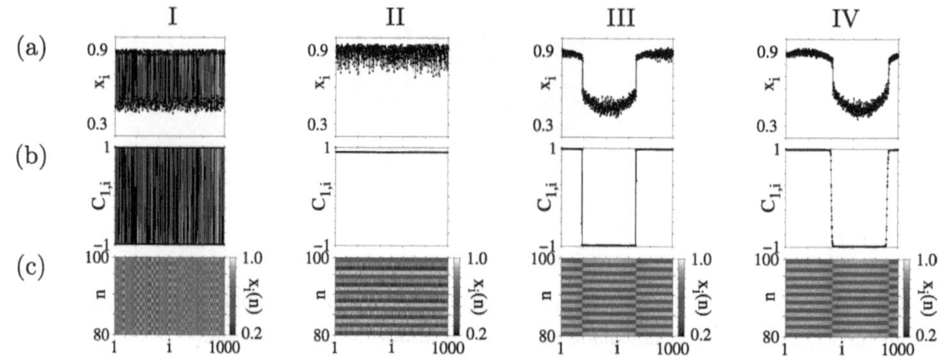

Figure 5. Dynamics of the network of nonlocally coupled logistic maps in the presence of strong noise $D = 0.02$ and for different values of the coupling strength σ: 0.24 (panel I), 0.295 (panel II), 0.345 (panel III), 0.38 (panel IV). Snapshots of the x_i varables (a), spatial distributions of cross-correlation coefficient (b), and spatio-temporal diagrams of $x_i(n)$ (c). Parameters: $\alpha^l = 3.8$, $R = 320$, $N = 1000$

In the case of strong noise influence, $D > 0.005$, the width of the chimera region is reduced markedly with respect to σ (Figure 3). The changes in the network dynamics in the presence of strong noise are illustrated in Figures 5,I–IV for the same values of σ as in the previous cases. When the coupling strength is very weak, the incoherent profile does not undergo qualitative changes, only small fluctuations occur in the system profile (compare Figure 2,I and Figure 5,I). At a weak coupling strength which is related to the observation of chimera states in the noise-free system, the noise effect destroys the whole spatial structure in the network (Figure 5,II). Now a strongly incoherent snapshot of spatial profile corresponds to the network dynamics (Figure 5(a),II). However, the elements still oscillate periodically in time (Figure 5(c),II) and the cross-correlation coefficients take rather large values $C_{1,i} > 0.89$ (Figure 5(b),II). With increasing coupling strength, a phase chimera with noisy coherent clusters is clearly observed in the logistic map ring (Figure 5,III). Finally, when the coupling strength corresponds to a spatial profile with a discontinuity or narrow clusters of the phase chimera, additive noise smears out and 'smooths' the profile (Figure 5(a),IV), which is especially apparent on the distribution of cross-correlation coefficient values (Figure 5(b),IV).

In summary, from Figure 3 it follows that there exists an optimum non-zero noise intensity D where the σ-interval of chimera observation is maximum. This counter-intuitive constructive effect of noise is in some sense can be treated as a resonance-like effect since the constructive and destructive action of noise are balanced for some intermediate value of the noise strength.

3.2. Network of nonlocally coupled henon maps

The ring of nonlocally coupled Henon maps can demonstrate phase and amplitude chimeras within a certain range of the coupling strength [10, 67]. The introduction of multiplicative noise into this ensemble can induce the appearance of a solitary state chimera [57], which is characterized by the presence of an incoherent cluster containing only solitary nodes. This type of chimera can also be observed in the Henon map ensemble without external noise, but its basin of attraction is small and is unlikely observed for random initial conditions.

As before, we first briefly consider the spatio-temporal dynamics of the noise-free Henon map ring when the coupling strength increases. The initial conditions of the dynamical variables for all the elements $x_i(0)$, $y_i(0)$ are randomly distributed in the intervals $x_i(0) \in [-0.5, 0.5]$ and $y_i(0) \in [-0.15, 0.15]$. Figure 6 illustrates the changes in the spatio-temporal dynamics of the ring of nonlocally coupled Henon maps for selected increasing values of σ. In each of the eight panels, the upper row (a) includes snapshots of the variables x_i, the middle one (b) shows spatial distributions of cross-correlation coefficient (5), and spatio-temporal diagrams of $x_i(n)$ are given in the bottom row (c). In the case of weak coupling, a dynamical regime is observed when the network elements behave incoherently in space and non-periodically in time (Figure 6,I). Amplitude and phase chimera states may appear in the network when σ is varied within the range $[0.17, 0.385]$. This interval of nonlocal coupling strength was established on the base of the analysis of 50 different initial distributions of the dynamical variables. For some realizations of the initial conditions, chimeras are observed at a rather weak coupling strength, and for a number of other realizations, at a stronger one. To illustrate the dynamics of the Henon map ring, one realization of the initial conditions was chosen, which leads to chimera states only at the coupling strengths corresponding to Figures. 6,II,IV,V. Figure 6,II shows the coexistence of a phase chimera (incoherent clusters with elements $1 < i < 131$, $196 < i < 428$, $717 < i < 1000$) and an amplitude chimera (the incoherent cluster includes elements $480 < i < 630$). As the coupling strength grows, it is possible to observe the coexistence of a phase chimera (two incoherent clusters: $315 < i < 417$ and $836 < i < 1000$) and a solitary states chimera (the incoherent cluster consists of elements $216 < i < 224$) (Figure 6,IV) or only a phase chimera (two incoherent clusters: $24 < i < 45$, $583 < i < 587$) (Figure 6,V). For some initial distributions of the dynamical variables, the σ-interval of chimera state existence can be interrupted, and in this case, the dynamics of the Henon map network corresponds to a completely incoherent spatial profile with periodic oscillations of the elements in time (Figure 6,III). An increase in the coupling strength within the range of chimera existence leads to a decrease in the probability of observing amplitude chimeras and solitary state chimeras. At the same time, the width of the incoherent clusters of the phase chimeras decreases. When $\sigma > 0.36$ for the chosen initial conditions, the phase chimeras are not visible and one can observe an x_i profile with discontinuity (Figure 6,VI), where the discontinuity decreases smoothly (branches must approach each other) with increasing coupling strength and, as a result, the profile must represent a smooth function. This behaviour is similar as for the logistic map, and a similar analytical argument has been derived for the Henon map in [67]. However, at certain values of σ from the range corresponding to the establishment of a profile with discontinuity, smooth spatial profiles may appear, as shown in Figure 6,VII, which can be replaced again by discontinuous profiles as the coupling strength increases (Figure 6,VIII). For $\sigma > 0.41$, the profile discontinuity gradually decreases and, ultimately, the system goes into a chaotic synchronization mode (complete synchronization in space and non-periodic dynamics of elements in time). It should be noted that the jumps from discontinuous profiles to smooth ones and vice versa observed within the parameter range $\sigma \in [0.365, 0.410]$ are caused by the peculiarities of the phase space structure and are not related to using different initial conditions. This means that all

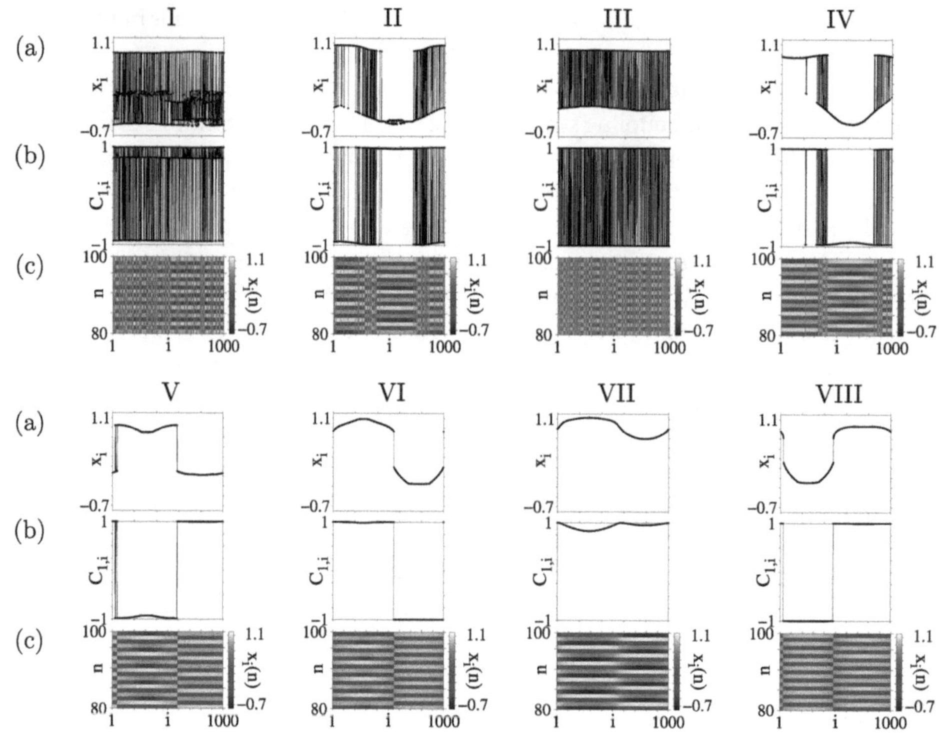

Figure 6. Dynamics of the noise-free network of nonlocally coupled Henon maps for different values of the coupling strength σ: 0.135 (panel I), 0.190 (panel II), 0.230 (panel III), 0.255 (panel IV), 0.290 (panel V), 0.380 (panel VI), 0.395 (panel VII), 0.405 (panel VIII). Snapshots of the x_i varables (a), spatial distributions of cross-correlation coefficient (b), and spatio-temporal diagrams of $x_i(n)$ (c). Parameters: $\alpha^H = 1.4$, $\beta^H = 0.2$, $R = 320$, $D = 0$, $N = 1000$

the used initial condition do not lead to a gradual transition from the discontinuous profiles to the smooth ones with increasing the coupling strength unlike the case of the ring of nonlocally coupled logistic maps.

Now let us turn to the analysis of the effect of additive noise on chimera states in the ring of nonlocally coupled Henon maps. Figure 7 demonstrates the dependence of the probability of establishing chimera states from random initial distributions of the dynamical variables on the coupling strength σ and the noise intensity D. As can be seen, when the noise intensity is low, there is a high probability of chimera state observation within a rather wide interval of the coupling strength. The maximum probability P occurs in the interval $\sigma \in [0.3, 0.35]$. However, at $D > 0.01$, a second maximum and a second region where the chimeras can be observed appear for strong coupling. These two regions are separated by a blue area, which corresponds to a low probability of observing chimeras. When $D > 0.03$, the region of chimera existence in the weak coupling range disappears, and only the second region remains. From the analysis of Figure 7 it can be concluded that the right boundary of the chimera region even shifts towards a stronger coupling with increasing noise intensity. However, we assume that this effect is rather similar to that observed in the logistic map ring. In the noise-free case when $\sigma \in [0.35, 0.42]$, the probability of chimera state observation vanishes but it can be increased by applying noise influence.

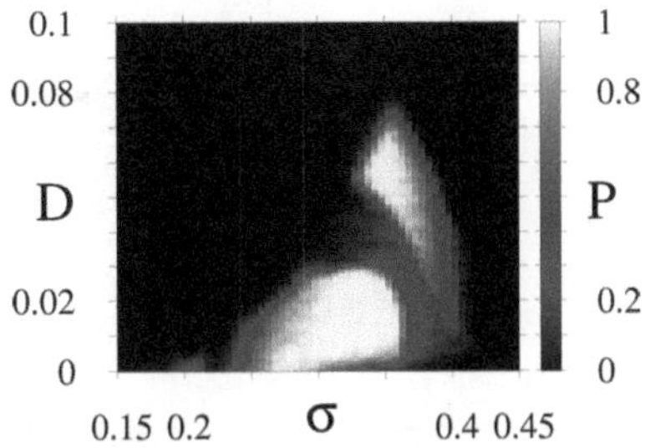

Figure 7. 2D diagram in the (σ, D) parameter plane showing the probability P of observing chimera states in the Henon map ring. Calculations were carried out for 50 different realizations of randomly distributed initial conditions, each having its own noise realization. Parameters: $\alpha^H = 1.4$, $\beta^H = 0.2$, $R = 320$, $N = 1000$

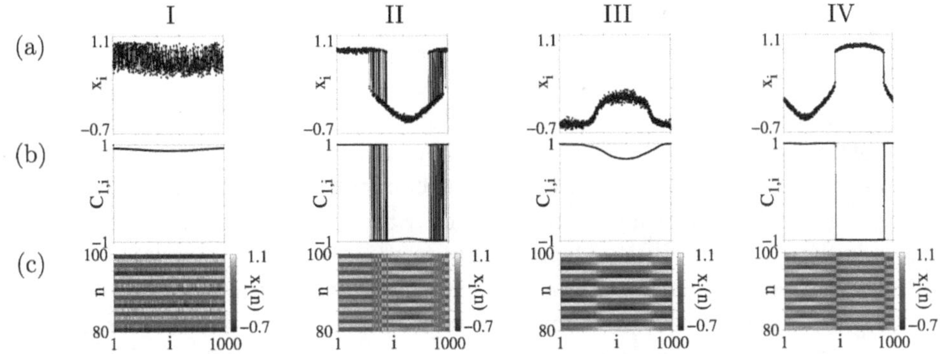

Figure 8. Snapshots of the x_i variables (a), spatial distributions of cross-correlation coefficient (b), and spatio-temporal diagrams of $x_i(n)$ (c) for the Henon map network for a fixed noise intensity $D = 0.025$ and for different values of the coupling strength σ: 0.255 (panel I), 0.285 (panel II), 0.36 (panel III), and 0.38 (panel IV). Other parameters: $\alpha^H = 1.4$, $\beta^H = 0.2$, $R = 320$, $N = 1000$

We now analyze the network behaviour at a fixed noise intensity $D = 0.025$ and for different values of the coupling strength (Figure 8). As shown above, the boundary of the chimera state region shifts to stronger coupling strength, chimera states are no longer observed at weak coupling strength, and the ensemble passes into the regime of complete spatial incoherence illustrated in Figure 8,I. However, the cross-correlation coefficient does not vanish, i.e. the network elements demonstrate nearly periodic dynamics in time (Figure 8(b),I). With stronger coupling, a chimera state can appear in the network even in the presence of noise with $D = 0.025$, but the width of the incoherent domain is larger than in the noise-free case (Figure 8,II). Upon further increase of σ, the probability of observing chimera states decreases, and the network dynamics is typically characterized by a spatial wave profile with fluctuations due to noise perturbations (Figure 8,III). For even larger values of the coupling strength, we again find a region of chimera existence. In the neighbourhood of the discontinuity narrow regions of incoherence occur which correspond to incoherent clusters of the phase chimera (Figure 8,IV).

We now fix the coupling strength $\sigma = 0.355$ and increase the noise intensity. The corresponding changes in the network dynamics are shown in Figure 9. The chosen value of

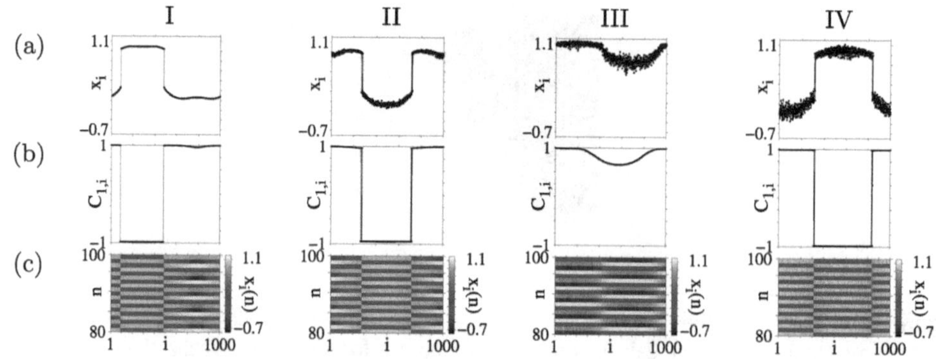

Figure 9. Snapshots of the x_i variables (a), spatial distributions of cross-correlation coefficient (b), and spatio-temporal diagrams of $x_i(n)$ (c) for the Henon map network for a fixed coupling strength $\sigma = 0.355$ and for different noise intensities D: 0 (panel I), 0.02 (panel II), 0.04 (panel III), 0.06 (panel IV). Other parameters: $\alpha^H = 1.4$, $\beta^H = 0.2$, $R = 320$, $N = 1000$

the coupling strength corresponds to a snapshot with two discontinuities in the noise-free case (Figure 9,I), and the probability of observing a chimera state is extremely small (see Figure 7). When the noise intensity is low, a phase chimera with narrow incoherent clusters appears (Figure 9,II). Increasing the noise intensity leads to the disappearance of discontinuities in the spatial profile, and, accordingly, to the absence of chimera states (Figure 9,III). This case corresponds to the region with a low probability of detecting chimera states (Figure 7). When the noise intensity increases, we enter into the second region with a high probability of chimera existence, and the spatial profile contains narrow clusters of incoherence of the phase chimera (Figure 9,IV).

4. Influence of additive noise on solitary states in the ring of nonlocally coupled Lozi maps

In this section we investigate the robustness of solitary states in a ring of nonlocally coupled Lozi maps with respect to additive noise and analyze their existence depending on the noise intensity and the coupling strength. The initial conditions of all the dynamical variables in the Lozi map ring are randomly distributed in the intervals $x_i(0) \in [-0.5, 0.5]$ and $y_i(0) \in [-0.5, 0.5]$.

The ring of nonlocally coupled Lozi maps without noise can demonstrate solitary states when the coupling strength σ between the elements is varied within the range $[0, 0.22]$. Figure 10 shows how the noise-free network dynamics changes with increasing coupling strength. As can be seen from Figure 10,I, for weak coupling strength, an incoherent mode of solitary nodes is observed in the network and is characterized by a snapshot with an incoherent plateau. Increasing the coupling strength changes the spatial profile which consists now of a coherent basis with solitary nodes (Figure 10,II,III). The number of solitary nodes gradually decreases as σ grows. Finally, for $\sigma > 0.22$, all solitary nodes disappear, and the network dynamics corresponds to a coherent spatial profile, while the elements behave irregularly in time (Figure 10,IV). When the coupling is sufficiently strong, the network elements become completely synchronized.

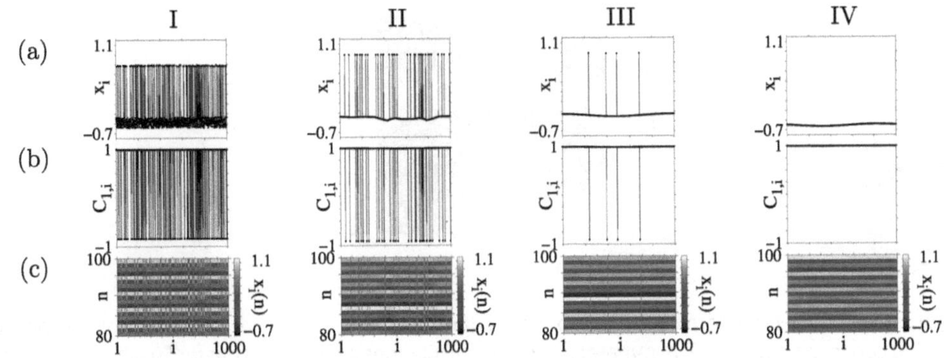

Figure 10. Dynamics of the noise-free network of nonlocally coupled Lozi maps for different values of the coupling strength σ: 0.1 (panel I), 0.18 (panel II), 0.22 (panel III), 0.24 (panel IV). Snapshots of the x_i variables (a), spatial distributions of cross-correlation coefficient (b), and spatio-temporal diagrams of $x_i(n)$ (c). Other parameters: $\alpha^L = 1.4, \beta^L = 0.3, R = 320, D = 0, N = 1000$

We now consider the influence of a uniformly distributed additive noise source on the formation and evolution of solitary states in the Lozi map ring. The obtained numerical results are summarized in a two-parameter diagram of regimes (Figure 11) where the number of solitary states is plotted in dependence of the coupling strength σ and the noise intensity D. The quantity N_s is calculated by averaging the normalized number of solitary nodes S over all 50 different random initial distributions, where S is the number of solitary nodes obtained for each initial realization normalized by the total number of nodes $N = 1000$. As can be seen, the distribution of S forms an arc-shaped domain with a maximum at $D \approx 0.08$ and $\sigma \approx 0.11$. Qualitatively, one may distinguish between the intervals of weak and strong noise. Weak noise corresponds to $D < 0.02$ for which the number of solitary nodes remains almost unchanged in the presence of additive noise. If $0.02 < D < 0.04$, only the right boundary of the region of solitary node existence changes in dependence on the coupling strength. The number of solitary nodes is the largest for weak coupling strength, $0 < \sigma < 0.11$, and is gradually reduced as σ becomes stronger. When $D > 0.04$, the left boundary of the solitary state region also changes (in the region of weak coupling). In this case the additive noise is equivalent to the introduction of an inhomogeneity into the Lozi map network. This leads to the destruction of the spatio-temporal structure and to a strong inhomogeneity in the network when the cross-correlation coefficient vanishes. When the noise intensity increases, the region of solitary node existence is smoothly narrowing, while the boundaries lying both in the regions of weak and strong coupling smoothly shift to the value of $\sigma \approx 0.1$. At $D > 0.085$, the solitary states no longer exist in the Lozi map ring.

The evolution of the Lozi map dynamics is illustrated in Figure 12 for a fixed noise intensity $D = 0.04$ and with increasing coupling strength. It is seen that at a weak coupling strength, the additive noise almost does not change the spatio-temporal structure in the ring (compare Figures 10,I and 12,I). At a stronger coupling, the coherent basis of the spatial profile becomes noisy but the number of solitary nodes remains the same (also compare Figs. 10,II and 12,II). If the value of σ corresponds to the existence of a few solitary nodes only, then the noise intensity $D = 0.04$ is quite sufficient to reduce the basin of attraction of a solitary node attractor, and the solitary nodes are no longer observed in the network

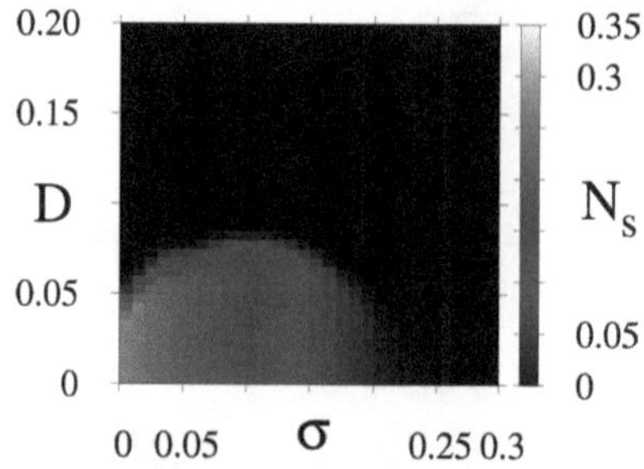

Figure 11. The average normalized number of solitary nodes N_s observed in the ring of nonlocally coupled Lozi maps in the (σ, D) parameter plane. The calculations were carried out for 50 different realizations of randomly distributed initial conditions of the dynamical variables and noise. Other parameters: $\alpha^L = 1.4, \beta^L = 0.3, R = 320, N = 1000$.

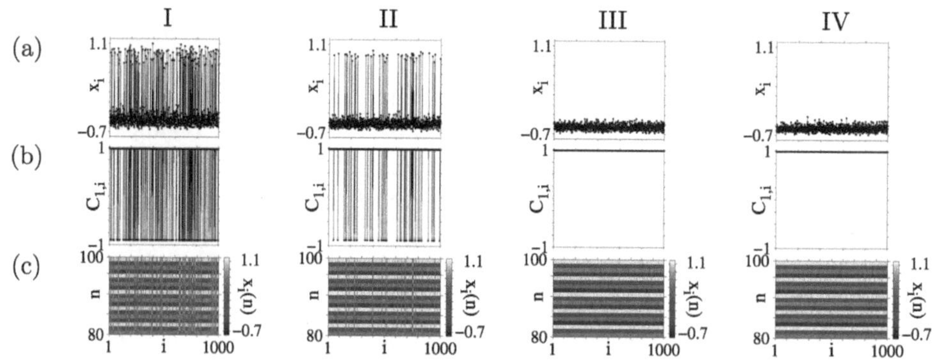

Figure 12. Snapshots of the x_i variables (a), spatial distributions of cross-correlation coefficient (b), and spatio-temporal diagrams of $x_i(n)$ (c) for the Lozi map network for a fixed noise intensity $D = 0.04$ and for different values of the coupling strength σ: 0.1 (panel I), 0.18 (panel II), 0.22 (panel III), and 0.24 (panel IV). Other parameters: $\alpha^L = 1.4, \beta^L = 0.3, R = 320, N = 1000$

(Figure 12,III). A further increase in the coupling strength does not cause the existing dynamical regime to qualitatively change, but leads to the increase of the cross-correlation coefficient (compare Figure 12,III and IV).

We now fix the coupling strength at $\sigma = 0.16$ and analyze the changes in the Lozi map ring dynamics as the noise intensity increases. The corresponding snapshots of the x_i variables (upper row), spatial distributions of cross-correlation coefficient values (middle row), and spatio-temporal diagrams (lower row) are shown in Figure 13. The value of σ is chosen in such a way to observe a sufficiently large amount of solitary nodes in the network (Figure 13,I). As noted above, the introduction of a relatively low noise intensity leads only to a noisy plateau of the spatial profile, while the solitary node number does not change in any way as compared with the noise-free case (Figure 13,II). When approaching the solitary state region with increasing noise intensity, the number of solitary nodes is markedly reduced since some elements jump on the attractor of typical states under the noise influence. Their cross-correlation coefficients lie within the interval $C_{1,i} \in [-0.7, 0.7]$ (Figure 13,III). Finally, when the noise intensity is sufficiently high, the solitary state mode is destroyed and spatially incoherent dynamics is observed in the Lozi map ring (Figure

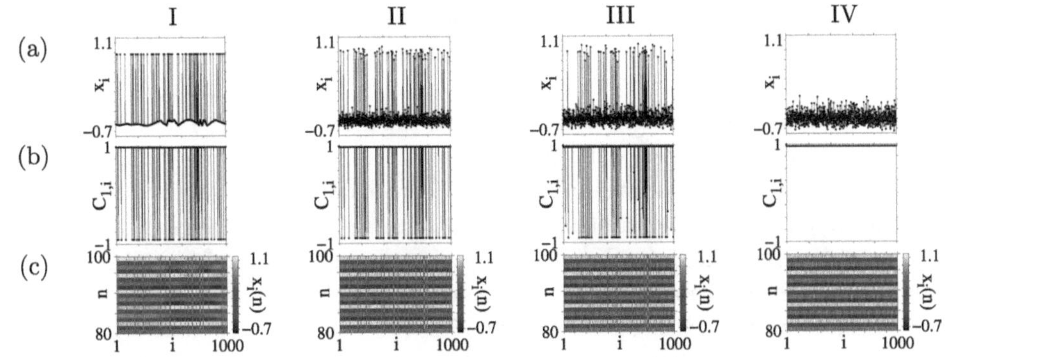

Figure 13. Snapshots of the x_i variables (a), spatial distributions of cross-correlation coefficient (b), and spatio-temporal diagrams of $x_i(n)$ (c) for the Lozi map network for a fixed coupling strength $\sigma = 0.16$ and for different noise intensities D: 0 (panel I), 0.04 (panel II), 0.065 (panel III), 0.075 (panel IV). Other parameters: $\alpha^L = 1.4, \beta^L = 0.3, R = 320, N = 1000$

13,IV). It should be noted that there are differences between incoherent regimes which exist for weak and strong coupling. In the first case the values of the cross-correlation coefficients of all the network elements are close to zero, while $C_{1,i} > 0.9$ in the second case.

Let us return to Figure 13,III and pay attention to the oscillators, the cross-correlation coefficient values of which lie in the interval $C_{1,i} \in [-0.7, 0.7]$. As mentioned above, the

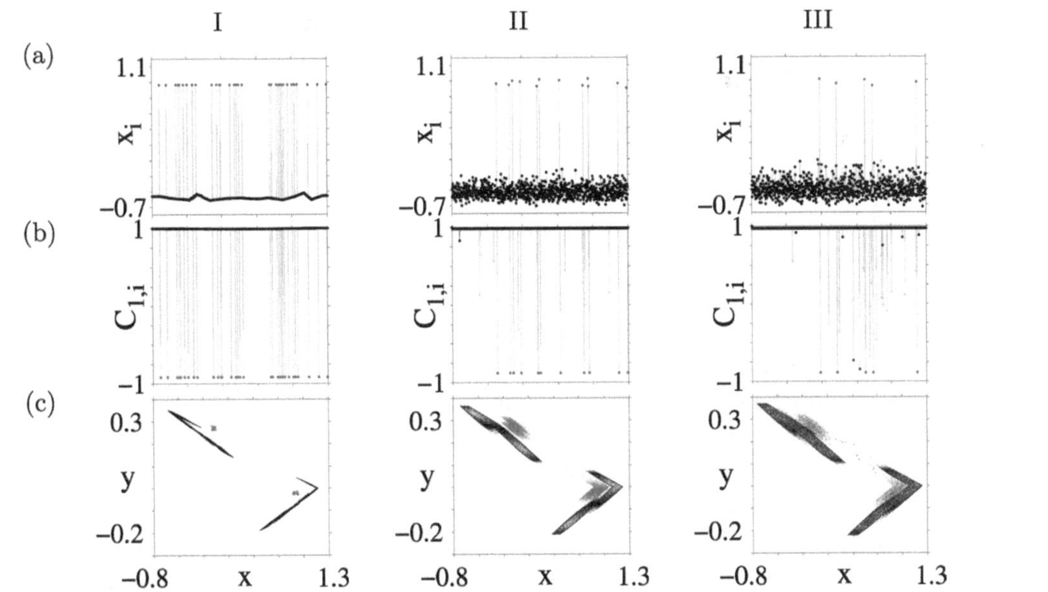

Figure 14. Snapshots of the x_i variables (a), spatial distributions of cross-correlation coefficient $C_{1,i}$ (b), and phase portraits of all elements (c) of the Lozi map for different values of the coupling strength and the noise intensity: $D = 0, \sigma = 0.19$ (panel I), $D = 0.05, \sigma = 0.19$(panel II), $D = 0.65, \sigma = 0.17$ (panel III). Black points correspond to the typical states, red ones to the solitary nodes, green ones to the typical states with $C_{1,i} \in [-0.7, 0.7]$, and blue ones to oscillators which switch between the typical state attractor and the solitary state one. Other parameters: $\alpha^L = 1.4, \beta^L = 0.3, R = 320, N = 1000$.

solitary state regime is characterized by the bistable dynamics of the individual elements of the network. This means that there are two attracting sets in the phase space, one of which corresponds to the typical states or coherent dynamics (black dots in Figure 14(c),I–III) and the other one – to the solitary nodes (red dots in Figure 14(c),I–III). The introduction of additive noise causes some of the network oscillators to switch from the solitary node attractor to the typical state attractor (Figure 14,II, green dots). However, their behaviour is not completely synchronous with that of the typical states and thus their cross-correlation coefficient takes the values from the interval $C_{1,i} \in [-0.7, 0.7]$ (Figure 14(b),II). For a larger noise intensity, the probability of switching between the two coexisting metastable states increases, and one can observe the solitary nodes with $C_{1,i} < -0.7$ in the network. However, they can switch between the typical state attractor and the solitary state one (blue dots in Figure 14(c),III). Despite the fact that these oscillators were taken into account when calculating the solitary nodes, one can consider them as 'imaginary' solitary nodes which belong to the typical state attractor for most of the time, but their oscillations are not synchronous with those of the coherent part $C_{1,i} < -0.7$.

5. Conclusion

In this study we have demonstrated how additive uniformly distributed noise influences the emergence and existence of special partial synchronization patterns, such as chimera states and solitary states, in networks of nonlocally coupled chaotic maps. We have studied the robustness of chimera states with respect to noise using ring networks of either logistic maps or Henon maps. It is known that these networks demonstrate a transition from complete chaotic synchronization to complete incoherence through the appearance of phase and amplitude chimeras. The effect of noise sources on solitary states has been investigated in a ring of Lozi maps with nonlocal interaction. The latter typically exhibits the existence of solitary states along the transition to spatio-temporal chaos when the coupling strength decreases.

For both cases (chimeras and solitary states) we have investigated how regions of existence of these patterns change when the nonlocal coupling strength σ and the additive noise intensity D are varied. To this purpose we have calculated the cross-correlation coefficients between network elements. Using this correlation measure we can diagnoze the type of spatio-temporal dynamics of the considered networks, as well as detect phase chimeras and solitary states. In order to provide reliable statistical data we have performed numerical simulations for 50 sets of randomly distributed initial conditions each possessing its own independent noise realizations. Furthermore, we have illustrated the effects of noise on chimera and solitary states by snapshots, spatial distributions of cross-correlation coefficient values and space-time plots.

We have computed two-parameter diagrams of regimes in the (σ, D) parameter plane, which show the probability of observing chimera states in networks of coupled logistic maps and Henon maps. It has been shown that in the case of the logistic map, the phase chimeras can exist within a rather wide interval of the noise intensity and a finite range of the coupling strength, and a high probability of observing chimera states is detected for rather strong noise. In the case of the Henon map, it has been found that there are two clearly distinguished regions in the (σ, D) parameter plane, within which the phase chimera regimes persist even for rather large values of the noise intensity. However, in both networks, already very weak noise can completely suppress the chimera states in the case of weak and strong nonlocal coupling. Interestingly, we have observed a counter-intuitive non-monotonic dependence of the chimera existence upon noise intensity in both networks, which is reminiscent of the constructive influence of noise known from coherence resonance.

To analyze the robustness of solitary states in a ring network of nonlocally coupled Lozi maps with respect to additive noise we have plotted the normalized number of solitary nodes in the (σ, D) parameter plane. The main result is that even weak noise can sustain the solitary state regime in the range of weak nonlocal coupling. Stronger noise causes the solitary nodes to disappear for any values of the coupling strength. It can be assumed that this effect is related to noise-induced preference of attractors [52]. In the presence of noise, a small basin of attraction of solitary nodes becomes smeared and thus most of the system elements go to a typical attractor with a larger basin. The similar results were reported in Ref. [17] for the network of coupled FitzHugh-Nagumo neurons in the presence additive noise.

Our findings provide a detailed understanding of the stability of chimera and solitary states and open the possibility of controlling them using additive noise sources.

Disclosure statement

The authors declare that there is no potential conflict of interest.

Funding

E.R. and G.S. acknowledge financial support from the Russian Science Foundation (project No. 20-12-00119) (numerical simulation of the dynamics of the Henon and Lozi map networks). E.S. acknowledges financial support from the German Science Foundation (DFG-Projektnummer 429685422 and 440145547) (numerical simulation of the dynamics of the logistic map network).

ORCID

Elena Rybalova http://orcid.org/0000-0003-3008-1078
Eckehard Schöll http://orcid.org/0000-0002-7318-2672
Galina Strelkova http://orcid.org/0000-0002-8667-2742

References

[1] D.M. Abrams and S.H. Strogatz, Chimera states for coupled oscillators, Phys. Rev. Lett. 93 (2004), p. Aritcle ID 174102.
[2] V.S. Anishchenko, A.B. Neiman, F. Moss, and L. Schimansky-Geier, Stochastic resonance: Noise-enhanced order, Phys. Uspekhi 42 (1999), pp. 7.
[3] V.S. Anishchenko, T.E. Vadivasova, and G.I. Strelkova, Deterministic Nonlinear Systems, Springer Series in Synergetics, Springer, 2014.
[4] L. Arnold, *Random dynamical systems*, in *Dynamical Systems. Lecture Notes in Mathematics*, 1609 Springer, Berlin, Heidelberg, 1995. pp. 1–43.
[5] K. Bansal, J.O. Garcia, S.H. Tompson, T. Verstynen, J.M. Vettel, and S.F. Muldoon, Cognitive chimera states in human brain networks, Sci. Adv. 5 (2019), p. Article ID eaau8535.
[6] I. Bashkirtseva, L. Ryashko, and H. Schurz, Analysis of noise-induced transitions for Hopf system with additive and multiplicative random disturbances, Chaos Soliton Fract. 39 (2009), pp. 72–82.

[7] R. Benzi, A. Sutera, and A. Vulpiani, The mechanism of stochastic resonance, J. Phys. A Math. Gen.14 (1981), p. L453.
[8] R. Berner, A. Polanska, E. Schöll, and S. Yanchuk, Solitary states in adaptive nonlocal oscillator networks, Eur. Phys. J. Spec. Top. 229 (2020), pp. 2183–2203.
[9] R. Berner, S. Yanchuk, and E. Schöll, What adaptive neuronal networks teach us about power grids, Phys. Rev. E. 103 (2021), p. Article ID 042315.
[10] S.A. Bogomolov, A.V. Slepnev, G.I. Strelkova, E. Schöll, and V.S. Anishchenko, Mechanisms of appearance of amplitude and phase chimera states in ensembles of nonlocally coupled chaotic systems, Commun. Nonlinear Sci. Numer. Simul. 43 (2017), pp. 25–36.
[11] A. Bukh, E. Rybalova, N. Semenova, G. Strelkova, and V. Anishchenko, New type of chimera and mutual synchronization of spatiotemporal structures in two coupled ensembles of nonlocally interacting chaotic maps, Chaos Interdisci. J. Nonlinear Sci. 27 (2017), p. Article ID 111102.
[12] A.V. Bukh, A.V. Slepnev, V.S. Anishchenko, and T.E. Vadivasova, Stability and noise-induced transitions in an ensemble of nonlocally coupled chaotic maps, Regul. Chaotic Dyn. 23 (2018), pp. 325–338.
[13] M.J. Chacron, A. Longtin, and L. Maler, The effects of spontaneous activity, background noise, and the stimulus ensemble on information transfer in neurons, Netw. Comput. Neural Syst. 14 (2003), pp. 803–824.
[14] A. Destexhe and M. Rudolph-Lilith, Neuronal Noise, Vol. 8, Springer Science & Business Media, 2012.
[15] M.J. Feigenbaum, Quantitative universality for a class of nonlinear transformations, J. Stat. Phys. 19 (1978), pp. 25–52.
[16] M.J. Feigenbaum, The universal metric properties of nonlinear transformations, J. Stat. Phys. 21 (1979), pp. 669–706.
[17] I. Franović, S. Eydam, N. Semenova, and A. Zakharova, Unbalanced clustering and solitary states in coupled excitable systems, Chaos Interdisci. J. Nonlinear Sci. 32 (2022), p. Article ID 011104.
[18] L.V. Gambuzza, A. Buscarino, S. Chessari, L. Fortuna, R. Meucci, and M. Frasca, Experimental investigation of chimera states with quiescent and synchronous domains in coupled electronic oscillators, Phys. Rev. E. 90 (2014), p. Article ID 032905.
[19] L. Gammaitoni, F. Marchesoni, E. Menichella-Saetta, and S. Santucci, Stochastic resonance in bistable systems, Phys. Rev. Lett. 62 (1989), p. 349.
[20] J.C. González-Avella, M.G. Cosenza, and M. San Miguel, Localized coherence in two interacting populations of social agents, Phys. A Stat. Mech. Appl. 399 (2014), pp. 24–30.
[21] A.M. Hagerstrom, T.E. Murphy, R. Roy, P. Hövel, I. Omelchenko, and E. Schöll, Experimental observation of chimeras in coupled-map lattices, Nat. Phys. 8 (2012), pp. 658–661.
[22] F. Hellmann, P. Schultz, P. Jaros, R. Levchenko, T. Kapitaniak, J. Kurths, and Y. Maistrenko, Network-induced multistability through lossy coupling and exotic solitary states, Nat. Commun. 11 (2020), pp. 1–9.
[23] M. Henon, Numerical study of quadratic area-preserving mappings, Quart. Appl. Math. 27(3) (1969), pp. 291–312.
[24] S. Hong and Y. Chun, Efficiency and stability in a model of wireless communication networks, Soc. Choice. Welfare. 34 (2010), pp. 441–454.
[25] W. Horsthemke and R. Lefever, *Noise-Induced Transitions in Physics, Chemistry, and Biology*, in *Noise-Induced Transitions. Springer Series in Synergetics*.15.Springer, Berlin, Heidelberg, 1984. pp. 164–200.
[26] P. Jaros, Y. Maistrenko, and T. Kapitaniak, Chimera states on the route from coherence to rotating waves, Phys. Rev. E. 91 (2015), p. Article ID 022907.
[27] P. Jaros, S. Brezetsky, R. Levchenko, D. Dudkowski, T. Kapitaniak, and Y. Maistrenko, Solitary states for coupled oscillators with inertia, Chaos Interdisci. J. Nonlinear Sci. 28 (2018), p. Article ID 011103.
[28] T. Kapitaniak, P. Kuzma, J. Wojewoda, K. Czolczynski, and Y. Maistrenko, Imperfect chimera states for coupled pendula, Sci. Rep. 4 (2014), p. 6379.

[29] Y. Kuramoto and D. Battogtokh, Coexistence of coherence and incoherence in nonlocally coupled phase oscillators, Nonlinear Phenom. Complex Syst. 5 (2002), pp. 380–385.
[30] L. Larger, B. Penkovsky, and Y. Maistrenko, Virtual chimera states for delayed-feedback systems, Phys. Rev. Lett. 111 (2013), p. Article ID 054103.
[31] B. Lindner and L. Schimansky-Geier, Analytical approach to the stochastic FitzHugh-Nagumo system and coherence resonance, Phys. Rev. E. 60 (1999), p. 7270.
[32] S.A. Loos, J.C. Claussen, E. Schöll, and A. Zakharova, Chimera patterns under the impact of noise, Phys. Rev. E. 93 (2016), p. Article ID 012209.
[33] R. Lozi, Un attracteur étrange (?) du type attracteur de Hénon, J. Phys. Colloques 39 (1978), pp. C5–9.
[34] Y. Maistrenko, B. Penkovsky, and M. Rosenblum, Solitary state at the edge of synchrony in ensembles with attractive and repulsive interactions, Phys. Rev. E. 89 (2014), p. Article ID 060901.
[35] S. Majhi, B.K. Bera, D. Ghosh, and M. Perc, Chimera states in neuronal networks: A review, Phys. Life. Rev. 28 (2019), pp. 100–121.
[36] A.K. Malchow, I. Omelchenko, E. Schöll, and P. Hövel, Robustness of chimera states in nonlocally coupled networks of nonidentical logistic maps, Phys. Rev. E. 98 (2018), p. 012217.
[37] E.A. Martens, S. Thutupalli, A. Fourriere, and O. Hallatschek, Chimera states in mechanical oscillator networks, Proc. Nat. Acad. Sci. 110 (2013), pp. 10563–10567.
[38] M.D. McDonnell and L.M. Ward, The benefits of noise in neural systems: Bridging theory and experiment, Nat. Rev. Neurosci. 12 (2011), pp. 415–425.
[39] P.J. Menck, J. Heitzig, J. Kurths, and H.J. Schellnhuber, How dead ends undermine power grid stability, Nat. Commun. 5 (2014), pp. 1–8.
[40] M. Mikhaylenko, L. Ramlow, S. Jalan, and A. Zakharova, Weak multiplexing in neural networks: Switching between chimera and solitary states, Chaos Interdisci. J. Nonlinear Sci. 29 (2019), p. Article ID 023122.
[41] A.E. Motter, S.A. Myers, M. Anghel, and T. Nishikawa, Spontaneous synchrony in power-grid networks, Nat. Phys. 9 (2013), pp. 191–197.
[42] V.A. Nechaev, E.V. Rybalova, and G.I. Strelkova, Influence of parameters inhomogeneity on the existence of chimera states in a ring of nonlocally coupled maps, Izvestiya VUZ. Appl. Nonlinear Dyn.29 (2021), pp. 943–952.
[43] A. Neiman, Synchronizationlike phenomena in coupled stochastic bistable systems, Phys. Rev. E. 49 (1994), pp. 3484.
[44] N.N. Nikishina, E.V. Rybalova, G.I. Strelkova, and T.E. Vadivasova, Destruction of cluster structures in an ensemble of chaotic maps with noise-modulated nonlocal coupling, Reg. Chaotic Dyn. 27 (2022), pp. 242–251. (accepted).
[45] I. Omelchenko, Y. Maistrenko, P. Hövel, and E. Schöll, Loss of coherence in dynamical networks: Spatial chaos and chimera states, Phys. Rev. Lett. 106 (2011), p. Article ID 234102.
[46] I. Omelchenko, B. Riemenschneider, P. Hövel, Y. Maistrenko, and E. Schöll, Transition from spatial coherence to incoherence in coupled chaotic systems, Phys. Rev. E. 85 (2012), p. Article ID 026212.
[47] I. Omelchenko, A. Provata, J. Hizanidis, E. Schöll, and P. Hövel, Robustness of chimera states for coupled FitzHugh-Nagumo oscillators, Phys. Rev. E. 91 (2015), p. Article ID 022917.
[48] M.J. Panaggio and D.M. Abrams, Chimera states on a flat torus, Phys. Rev. Lett. 110 (2013), p. Article ID 094102.
[49] M.J. Panaggio and D.M. Abrams, Chimera states: Coexistence of coherence and incoherence in networks of coupled oscillators, Nonlinearity 28 (2015), p. R67.
[50] Y.B. Pesin, Dynamical systems with generalized hyperbolic attractors: Hyperbolic, ergodic and topological properties, Ergodic Theory Dyn. Syst. 12 (1992), pp. 123–151.
[51] A.S. Pikovsky and J. Kurths, Coherence resonance in a noise-driven excitable system, Phys. Rev. Lett. 78 (1997), p. 775.
[52] A.N. Pisarchik and U. Feudel, Control of multistability, Phys. Rep. 540 (2014), pp. 167–218.

[53] D.P. Rosin, D. Rontani, N.D. Haynes, E. Schöll, and D.J. Gauthier, Transient scaling and resurgence of chimera states in networks of Boolean phase oscillators, Phys. Rev. E.90 (2014), p. Article ID 030902.

[54] E. Rybalova, N. Semenova, G. Strelkova, and V. Anishchenko, Transition from complete synchronization to spatio-temporal chaos in coupled chaotic systems with nonhyperbolic and hyperbolic attractors, Euro. Phys. J. Spec. Top. 226 (2017), pp. 1857–1866.

[55] E. Rybalova, V. Anishchenko, G. Strelkova, and A. Zakharova, Solitary states and solitary state chimera in neural networks, Chaos Interdisci. J. Nonlinear Sci. 29 (2019), p. Article ID 071106.

[56] E.V. Rybalova, D.Y. Klyushina, V.S. Anishchenko, and G.I. Strelkova, Impact of noise on the amplitude chimera lifetime in an ensemble of nonlocally coupled chaotic maps, Regul. Chaotic Dyn.24 (2019), pp. 432–445.

[57] E. Rybalova, G. Strelkova, and V. Anishchenko, Mechanism of realizing a solitary state chimera in a ring of nonlocally coupled chaotic maps, Chaos Soliton Fract. 115 (2018), pp. 300–305.

[58] J. Sawicki, I. Omelchenko, A. Zakharova, and E. Schöll, Chimera states in complex networks: Interplay of fractal topology and delay, Euro. Phys. J. Spec. Top. 226 (2017), pp. 1883–1892.

[59] L. Schmidt, K. Schönleber, K. Krischer, and V. García-Morales, Coexistence of synchrony and incoherence in oscillatory media under nonlinear global coupling, Chaos Interdisci. J. Nonlinear Sci. 24 (2014), p. Article ID 013102.

[60] E. Schöll, Synchronization patterns and chimera states in complex networks: Interplay of topology and dynamics, Euro. Phys. J. Spec. Top. 225 (2016), pp. 891–919.

[61] E. Schöll, Chimeras in physics and biology: Synchronization and desynchronization of rhythms, Nova. Acta. Leopold. 425 (2021), pp. 67–95.

[62] E. Schöll, Partial synchronization patterns in brain networks, Europhys. Lett.136 (2021), p. Article ID 18001.

[63] E. Schöll, A. Zakharova, and R.G. Andrzejak, Chimera States in Complex Networks, Frontiers Media SA, 2020.

[64] L. Schülen, S. Ghosh, A.D. Kachhvah, A. Zakharova, and S. Jalan, Delay engineered solitary states in complex networks, Chaos Soliton Fract. 128 (2019), pp. 290–296.

[65] L. Schülen, D.A. Janzen, E.S. Medeiros, and A. Zakharova, Solitary states in multiplex neural networks: Onset and vulnerability, Chaos Soliton Fract. 145 (2021), p. Article ID 110670.

[66] V. Semenov, A. Zakharova, Y. Maistrenko, and E. Schöll, Delayed-feedback chimera states: Forced multiclusters and stochastic resonance, EPL (Euro. Lett.) 115 (2016), p. Article ID 10005.

[67] N. Semenova, A. Zakharova, E. Schöll, and V. Anishchenko, Does hyperbolicity impede emergence of chimera states in networks of nonlocally coupled chaotic oscillators?, EPL (Euro. Lett.) 112 (2015), p. Article ID 40002.

[68] N. Semenova, A. Zakharova, V. Anishchenko, and E. Schöll, Coherence-resonance chimeras in a network of excitable elements, Phys. Rev. Lett. 117 (2016), p. Article ID 014102.

[69] N.I. Semenova, E.V. Rybalova, G.I. Strelkova, and V.S. Anishchenko, 'Coherence–incoherence' transition in ensembles of nonlocally coupled chaotic oscillators with nonhyperbolic and hyperbolic attractors, Reg. Chaotic Dyn. 22 (2017), pp. 148–162.

[70] N. Semenova, G. Strelkova, V. Anishchenko, and A. Zakharova, Temporal intermittency and the lifetime of chimera states in ensembles of nonlocally coupled chaotic oscillators, Chaos Interdisci. J. Nonlinear Sci. 27 (2017), p. Article ID 061102.

[71] N. Semenova, T. Vadivasova, and V. Anishchenko, Mechanism of solitary state appearance in an ensemble of nonlocally coupled Lozi maps, Euro. Phys. J. Spec. Top. 227 (2018), pp. 1173–1183.

[72] S. Shima and Y. Kuramoto, Rotating spiral waves with phase-randomized core in nonlocally coupled oscillators, Phys. Rev. E. 69 (2004), p. Article ID 036213.

[73] B. Shulgin, A. Neiman, and V. Anishchenko, Mean switching frequency locking in stochastic bistable systems driven by a periodic force, Phys. Rev. Lett. 75 (1995), pp. 4157.

[74] H. Taher, S. Olmi, and E. Schöll, Enhancing power grid synchronization and stability through time-delayed feedback control, Phys. Rev. E. 100 (2019), p. Article ID 062306.

[75] M.R. Tinsley, S. Nkomo, and K. Showalter, Chimera and phase-cluster states in populations of coupled chemical oscillators, Nat. Phys. 8 (2012), pp. 662–665.

[76] S. Ulonska, I. Omelchenko, A. Zakharova, and E. Schöll, Chimera states in networks of Van der Pol oscillators with hierarchical connectivities, Chaos Interdisci. J. Nonlinear Sci. 26 (2016), p. Article ID 094825.
[77] T.E. Vadivasova, G.I. Strelkova, S.A. Bogomolov, and V.S. Anishchenko, Correlation analysis of the coherence-incoherence transition in a ring of nonlocally coupled logistic maps, Chaos Interdisci. J. Nonlinear Sci. 26 (2016), p. Article ID 093108.
[78] B. Wang, H. Suzuki, and K. Aihara, Enhancing synchronization stability in a multi-area power grid, Sci. Rep. 6 (2016), pp. 1–11.
[79] M. Wickramasinghe and I.Z. Kiss, Spatially organized dynamical states in chemical oscillator networks: Synchronization, dynamical differentiation, and chimera patterns, PLoS. One. 8 (2013), p. Article ID e80586.
[80] H. Wu and M. Dhamala, Dynamics of Kuramoto oscillators with time-delayed positive and negative couplings, Phys. Rev. E. 98 (2018), p. Article ID 032221.
[81] A. Zakharova, Chimera Patterns in Networks: Interplay Between Dynamics, Structure, Noise, and Delay, Springer Nature, 2020.
[82] A. Zakharova, M. Kapeller, and E. Schöll, Chimera death: Symmetry breaking in dynamical networks, Phys. Rev. Lett. 112 (2014), p. Article ID 154101.
[83] A. Zakharova, S. Loos, J. Siebert, A. Gjurchinovski, and E. Schöll, Chimera patterns: Influence of time delay and noise, IFAC-PapersOnLine 48 (2015), pp. 7–12.

Statistical self-similarity in Lozi and Hénon's strange attractors

Alexandre M. de Paula Viveiros

ABSTRACT

Non-classical characterizations of strange attractors assign occupancy times and numbers to each of the hypercubes that cover them, local information, and generate statistical distributions over such quantities, global information, at the expense of losing all the local ones. The confrontation between local and global information would allow statistical quantification, important for the characterization of the self-similarity of the involved attractor: correspondence between global and local properties are expected as manifestations of the self-similarity that characterizes that fractal set (statistical self-similarity). Hence, the importance of clarifying what the local distributions would be. The core of this work is the presentation of statistical distributions of the mentioned times and numbers, and quantities derived from them, linked to each hypercube of a set that corresponds to a small fraction of their total, that confirm the correspondence between local and global statistical properties.

1. Introduction

Strange attractors are multifractal sets. They do not exhibit, as with many other fractal objects, strict self-similarity: there is no strict equality between the spatial distribution of points of the whole set and that linked to smaller portions of it at various scales of reduction. Statistical self-similarity is that expected for such attractors: correspondence between statistical properties derived from the time-dependent spatial distribution of the entire set and the equivalents linked to smaller portions. The article [3] reveals that, with respect to ten strange attractors immersed in two-dimensional spaces and linked to maps, including those of Lozi and Hénon, statistical distributions of occupancy times and occupancy numbers associated with each set as a whole are given by power laws:

$$\varrho_{ocn}(Q) \propto Q^{-\gamma_{ocn}} \quad \varrho_{oct}(\tau_o) \propto \tau_o^{-\gamma_{oct}}, \tag{1}$$

where Q and τ_o are respectively the occupancy numbers and times, ϱ_{ocn} and ϱ_{oct} are the corresponding probability densities, $1.71 \leq \gamma_{ocn} \leq 2.45$ and $1.92 \leq \gamma_{oct} \leq 2.76$. In [3], although local information regarding each hypercube is considered (local information), it is used for generations of information regarding each attractor taken with a single object (global information), so that only the latter are present . All local information is lost. Detailed information about the procedure employed and its origin are present in the work [3]. Statistical distributions of the same times and numbers, associated with 18 hypercubes that cover small portions of Lozi's strange attractor and 16 that cover small portions of Hénon's strange attractor, are considered here. As for the Lozi map, each of the two probability densities associated with each hypercube is given by power law. With respect to the Hénon map, only two hypercubes do not correspond to pairs of densities given by such laws. Therefore, when considering the two attractors, there are 32 hypercubes:

$$\varrho_{ocn}^{(q)}(Q_q) \propto Q_q^{-\gamma_{ocn}^{(q)}} \quad \varrho_{oct}^{(q)}(\tau_{o_q}) \propto \tau_{o_q}^{-\gamma_{oct}^{(q)}} \quad \text{Lozi and Hénon,} \tag{2}$$

where Q_q are the occupancy numbers coming from the hypercubes connected to the strange attractors of Lozi and Hénon, τ_{o_q} the corresponding occupancy times, $\varrho_{ocn}^{(q)}$ the probability densities relative to occupancy numbers and $\varrho_{oct}^{(q)}$ those relating to occupancy times. Values of q in the interval $1 \leq q \leq 18$ refer to the strange attractor of Lozi while values in interval $19 \leq q \leq 32$ refer to the strange attractor of Hénon. Finally, for $1 \leq q \leq 18$, $1.667 \leq \gamma_{ocn}^{(q)} \leq 1.79$ and $1.86 \leq \gamma_{oct}^{(q)} \leq 2.02$ and, for $19 \leq q \leq 32$, $1.53 \leq \gamma_{ocn}^{(q)} \leq 1.68$ and $1.898 \leq \gamma_{oct}^{(q)} \leq 1.94$.

Each set of hypercubes covering some attractor forms a mesh. These are known meshes that could be used to determine the singularity spectra, $f(\alpha)$ and the generalized dimensions, D_q, in each case [1, 2, 4–6]. I will refer to them here as 'classic characterizations'. However, the 'non-classical' characterizations performed in [3] are based on statistical distributions of occupancy times and numbers, associated with each of the hypercubes, generated by the chaotic advection of a large amount of phase points in each case. This approach is quite different from the one underlying $f(\alpha)$ and D_q. In that work I considered four statistical properties of a random variable z with probability density given by $\varrho(z) = Az^{-\gamma}$, $\gamma > 0$ and normalization constant A: the effective length, Δ_σ, the effective ratio, f_σ, the effective cumulative probability, P_σ, and the effective variability, M_σ. The first represents the length of the interval with most likely values of z, the effective values of z. The second represents the fraction of the interval with all possible values of z corresponding to Δ_σ. The third represents the probability of occurrence of some value pertinent to the effective interval. Finally, the last one concerns the concentration of effective values around the mean. With respect to the ten attractors in [3], four results were obtained:

(a) The effective intervals of both occupancy numbers and times, Δ_{σ_n} and Δ_{σ_t} respectively [1], are much smaller than the intervals with all possible corresponding values. Facts represented by the small values of the effective ratios, $f_{\sigma_n} \ll 1$ and $f_{\sigma_t} \ll 1$ respectively.
(b) The probabilities of occurrence of the effective values of occupancy numbers and times are high, $P_{\sigma_n} \geq 0.95$ and $P_{\sigma_t} \geq 0.95$ respectively.
(c) From the two previous results it can be inferred that effectively only a small fraction of all possible values occur, effectively the vast majority of possible values do not occur.

(d) The value of the effective variability linked to numbers, M_{σ_n}, is a measure of the multifractality of the strange attractor. Such values indicate that one of the attractors is almost (mono) fractal, M_{σ_n} remarkably smaller than the others. The remaining values indicate multifractal attractors with distinct multifractalities. As for the values of M_{σ_t}, numerical simulations currently in progress, relating to the 32 hypercubes considered here, aim to verify whether such values represent multifractalities of the one-dimensional sets of points formed by the occupation times associated with each hypercube.

Although effective lengths, ratios, cumulative probabilities and variability have been employed for global characterizations of strange attractors in [3], the aforementioned revelation of local probability distributions of occupancy times and numbers given by power laws makes it possible to calculate corresponding quantities, $\Delta_{\sigma_t}^{(q)}, f_{\sigma_t}^{(q)}, P_{\sigma_t}^{(q)}, M_{\sigma_t}^{(q)}$ and $\Delta_{\sigma_n}^{(q)}, f_{\sigma_n}^{(q)}, P_{\sigma_n}^{(q)}, M_{\sigma_n}^{(q)\ 2}$, associated with each of the 32 hypercubes to which power laws are associated. These distributions in local and global scopes are consistent with the self-similarity exhibited by strange attractors:properties associated with the entire set of points can also be associated with small portions of it. Thus, from the values of these eight quantities, can be obtained 32 new local results $a^{(q)}, b^{(q)}, c^{(q)}$ and $d^{(q)}$ comparable to the global ones a, b, c and d. Furthermore, local results relating to a certain hypercube can be compared to those relating to any other. These possibilities of confrontations of the strange attractor with itself are the foundations of a new way of characterizing it.: – if the attractor is (mono) fractal, there must be equality both between global and local results, $(a, b, c, d) = (a^{(q)}, b^{(q)}, c^{(q)}, d^{(q)})$, as between locals, $(a^{(q)}, b^{(q)}, c^{(q)}, d^{(q)}) = (a'^{(q)}, b'^{(q)}, c'^{(q)}, d'^{(q)})$ – if the attractor is multifractal the equalities will not hold $(a, b, c, d) \neq (a^{(q)}, b^{(q)}, c^{(q)}, d^{(q)})$, and $(a^{(q)}, b^{(q)}, c^{(q)}, d^{(q)}) \neq (a'^{(q)}, b'^{(q)}, c'^{(q)}, d'^{(q)})$.

The quantities relative to each of the attractors taken as a whole are those present in [3]. These are the distributions given by Equation (1) and the values of the quantities $\Delta_{\sigma_t}, f_{\sigma_t}, P_{\sigma_t}, M_{\sigma_t}$ and $\Delta_{\sigma_n}, f_{\sigma_n}, P_{\sigma_n}, M_{\sigma_n}$. While the distributions related to each of the 32 hypercubes are presented here. These are the distributions given by Equation (2), the values of $\Delta_{\sigma_t}^{(q)}, f_{\sigma_t}^{(q)}, P_{\sigma_t}^{(q)}, M_{\sigma_t}^{(q)}$ and $\Delta_{\sigma_n}^{(q)}, f_{\sigma_n}^{(q)}, P_{\sigma_n}^{(q)}, M_{\sigma_n}^{(q)}$. This work is dedicated to the comparison between quantities linked to the global description and the corresponding ones from the local descriptions. Confrontations which will allow, at least partially, the establishment of the aforementioned statistical self-similarity in specific quantitative terms. Comparisons between local quantities will be the subject of work published in a later article.

2. Method

The statistics involved consist of simple derivations from those used in [3]. In that article, ten strange attractors, immersed in two-dimensional spaces, associated with one-dimensional, two-dimensional and three-dimensional maps, are considered. Each one of them is covered by a specific hypercube mesh arranged in rows and columns like the squares in a game of chess. Here only two attractors are considered and each of them is not covered by a network of small squares. Only small portions are covered as the purpose is to bring local information.

2.1. Maps, strange attractors, hypercubes and normalization of density $\varrho(z) \propto z^{-\gamma}$

The maps used are given by the systems of Equations (3) and (4) and the positions of the hypercubes are shown in Figures 1 and 2. The vertices of the latter were highlighted by circles. There are 16 and 18 hypercubes respectively for the maps of Hénon and Lozi. Each hypercube is a square with a side 0.025. The coordinates of the centres of the hypercubes are shown in Tables 1 and 2. I determined them using the 'Grace' chartplotter: I positioned the course on the point that would be the centre and checked which coordinates were displayed by the software. In turn, the centres were chosen so as to meet two criteria: – number of hypercubes not greater than 20 with respect to each of the attractors, given the limitations of time and computational resources involved – small portions of attractors with higher density of points verified by visual inspection.

Hénon map:

$$\begin{cases} x_{n+1} = a - x_n^2 + by_n, \\ y_{n+1} = x_n, \end{cases} \tag{3}$$

where $a = 1.4$ e $b = 0.3$.

Lozi map :

$$\begin{cases} x_{n+1} = 1 - \alpha|x_n| + \beta y_n, \\ y_{n+1} = x_n, \end{cases} \tag{4}$$

where $\alpha = 1.7$ e $\beta = 0.5$.

Hypercubes and initial conditions boxes: The hypercubes are small squares with side $\varepsilon = 0.025$. The initial conditions are regularly arranged inside small rectangles, the initial

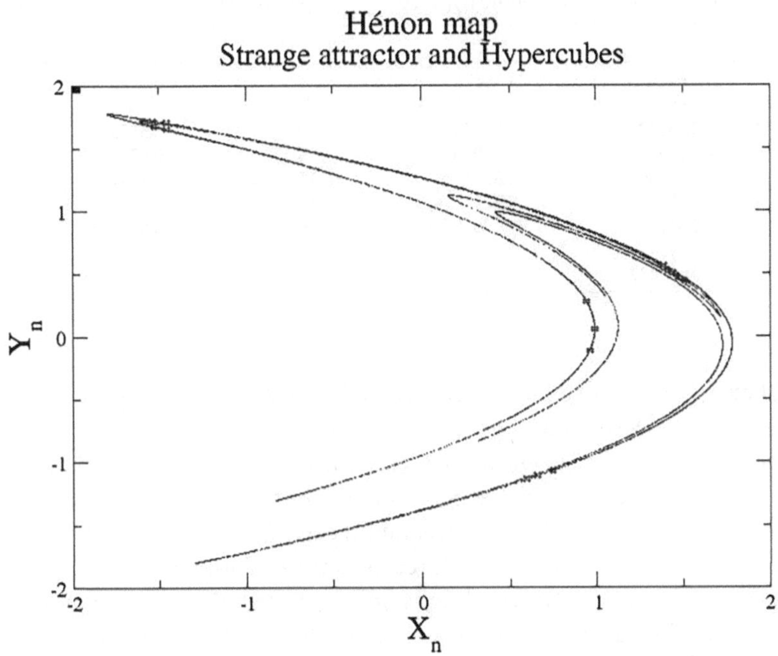

Figure 1. The strange attractor of Hénon and the vertices of the 16 hypercubes used to determine the local distributions of occupancy times and occupancy numbers.

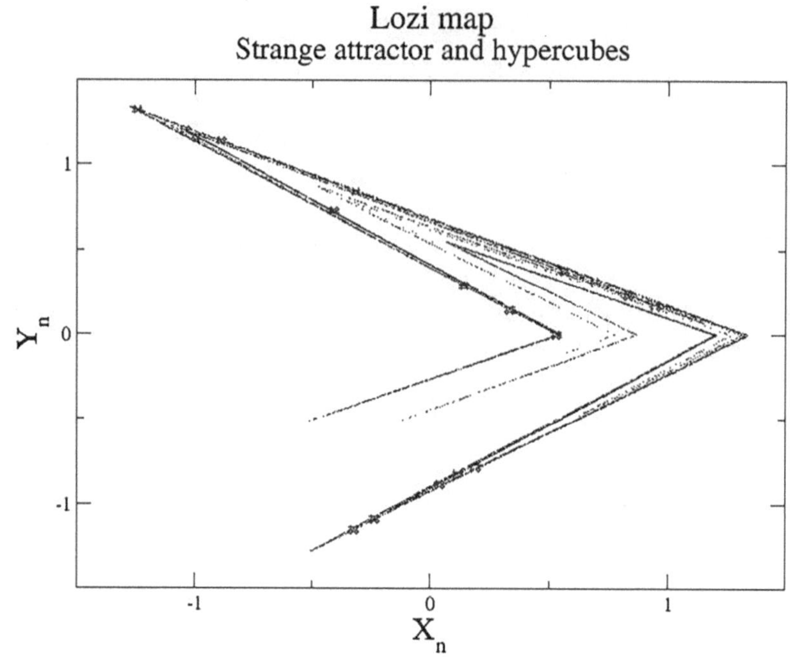

Figure 2. Lozi's strange attractor and the vertices of the 18 hypercubes used to determine the local distributions of occupancy times and occupancy numbers.

Table 1. Coordinates of the centres of the hypercubes linked to the Lozi map.

Number	X	Y	Number	X	Y
01	−1.0425	1.19876	10	0.948088	0.168172
02	−0.994755	1.160192	11	0.145147	0.281618
03	−0.889608	1.13745	12	0.337304	0.144578
04	−1.24488	1.31958	13	0.527745	−0.001996
05	−0.312892	0.83451	14	0.193186	−0.78666
06	−0.40848	0.72271	15	0.119412	−0.824475
07	0.562059	0.376155	16	0.042206	−0.885924
08	0.69245	0.309979	17	−0.233235	−1.08697
09	0.833137	0.229622	18	−0.320613	−1.1507

condition boxes. The size of each of them is equal to or greater than the size of each corresponding hypercube. Several of them are considered according to appropriate sequences determined in such a way as to ensure that each initial condition is considered only once.

Temporal evolution: Time units are iterations, the number of initial conditions is always 10,000. Each evolution occurs over an observation period, $T_{obs} = t_f - t_i$, where t_f and t_i are the final and initial instants. When T_{obs} is reached, the evolution in progress is interrupted and another, under a new initial conditions box, is started. Each evolution in a sequence of them consists of an 'achievement'. The value of the observation period common to all systems is $T_{obs} = 15,000$ ($t_i = 0$ and $t_f = 15,000$).

Occupations: For a certain hypercube: – the occupancy time is given by counting the successive units of time during which there was at least one phase point in its interior or

Table 2. Coordinates of the centres of the hypercubes linked to the Hénon map.

Number	X	Y	Number	X	Y
19	1.44048	0.503401	26	1.452450	1.706410
20	1.41035	0.524883	27	0.989704	0.052273
21	1.38523	0.560680	28	0.963963	−0.117616
22	−1.58313	1.720730	29	0.943872	0.272646
23	−1.52299	1.682090	30	0.597589	−1.13749
24	−1.5329	1.72157	31	0.655499	−1.10526
25	−1.45285	1.70541	32	0.747112	−1.06767

its border – the occupancy number is given by the sum between the number of points in its interior, or boundary, verified in each unit of time of occupancy time.

Maintained transients: Detailed discussion in [3] makes clear the need for maintenance of transients in that work. Here they remain to ensure that transient contributions are present in both works so that comparisons between the results of one with the other are legitimate.

Histograms, intervals, and classes: Histograms of absolute frequencies relative to occupancy numbers and times are attributed to each hypercube. The values of each time or number are organized according to 63 classes that cover the range $1 \leq z < 10^7$. There are seven intervals, each consisting of nine classes. The lengths of the intervals grow exponentially with their positions, the ith interval is given by

$$10^{i-1} \leq z < 10^i, \quad i = 1, .., 7.$$

In turn, the jth class contained therein is given by

$$j \cdot 10^{i-1} \leq z < (j+1) \cdot 10^{i-1}, \quad j = 1, .., 9.$$

Each of the 63 classes is specified by its centre, the arithmetic average between its extremes

$$c_{ij} = \frac{(2j+1) \cdot 10^{i-1}}{2}, \quad i = 1, .., 7, j = 1, .., 9.$$

Probability distributions: Absolute frequencies are accumulated over successive achievements. Finally, the absolute frequency in each class of general histogram, N_{ij}, is replaced by n_{ij}:

$$n_{ij} = \frac{N_{ij}}{L_i}, \tag{5}$$

where L_i is the width of the class involved. Thus, each n_{ij} consists of a density (number of events per unit width of the class). Each set of n_{ij} values linked to occupancy times and occupancy numbers, with i and j assuming all possible values, determines the probability densities. All of them are given by power laws with the general form $\varrho(z) \propto z^{-\gamma}$, with $z > 0$ and $\gamma > 0$. Finally, density is properly normalized:the substitution $z^{-\gamma} \rightarrow Az^{-\gamma}$ is performed and then the value of the normalization constant A is determined:

$$1 = \int_{z_i}^{z_f} Az^{-\gamma}\,\mathrm{d}z \tag{6}$$

so the new density becomes

$$\varrho(z) = Az^{-\gamma} \tag{7}$$

where

$$A = \frac{1-\gamma}{z_f^{1-\gamma} - z_i^{1-\gamma}}. \tag{8}$$

2.2. Statistical quantities derived from density $\varrho(z) = Az^{-\gamma}$

Given a distribution $\varrho(z) = Az^{-\gamma}$, $\gamma > 0$ and $z > 0$, valid for values of z in the range $z_i < z < z_f$, the following definitions apply:

$$\begin{aligned} &\overline{z} = \int_{z_i}^{z_f} \varrho(z)z\,\mathrm{d}z \quad \sigma^2 = \int_{z_i}^{z_f} \varrho(z)(\overline{z}-z)^2\,\mathrm{d}z, \\ &z_\sigma \equiv \overline{z} + \sigma \quad \Delta_\sigma \equiv z_\sigma - z_i \quad M_\sigma \equiv \Delta_\sigma/\overline{z} \; f_\sigma \equiv \Delta_\sigma/\Delta_{tot}, \\ &\Delta_{tot} \equiv z_f - z_i \quad (\Delta_{tot})_{dec} \equiv \log_{10}(z_f) - \log_{10}(z_i), \end{aligned} \tag{9}$$

$$\begin{aligned} &P_\sigma \equiv P(z_i \le z \le z_\sigma) = \int_{z_i}^{z_\sigma} \varrho(z)\,\mathrm{d}z \quad P_{\overline{\sigma}} \equiv P(z_\sigma \le z \le z_f) = \int_{z_\sigma}^{z_f} \varrho(z)\,\mathrm{d}z, \\ &I_\sigma \equiv [z_i, z_\sigma] \quad I_{\overline{\sigma}} \equiv [z_\sigma, z_f] \quad I_{tot} \equiv [z_i, z_f] = I_\sigma \cup I_{\overline{\sigma}}, \\ &\qquad P_\sigma + P_{\overline{\sigma}} = 1. \end{aligned} \tag{10}$$

Effective cumulative probability, effective ratio and effective variability: The curves in Figures 3, 4 and 5 were relevant to obtain the results (*a*), (*b*), (*c*) and (*d*) in the article [3], mentioned in the Section 1. Here they will be important for generations of analogous results linked to hypercubes, in addition to other unprecedented results.

General features of $P_\sigma \times \gamma$: The curves in Figure 3 refer to the effective cumulative probability, P_σ. The numerical simulations indicated that the behaviour of $P_\sigma \times \gamma$ depends on the values of $(\Delta_{tot})_{dec}$, and not on the individual values of x_i and x_f: different values of the pair (x_i, x_f) corresponding to a certain value of $(\Delta_{tot})_{dec}$ correspond to a single value P_σ according to Table 3. The curves associated with $3 \le (\Delta_{tot})_{dec} \le 6$ exhibit plateaus, or almost plateaus, whose extensions grow with $(\Delta_{tot})_{dec}$ for which $P_\sigma \ge 0.95$, corresponding to intervals around $\gamma = 2$ with extensions between one and two units. All the intervals in the table are contained in another defined in Section 2.1, extensively used in the simulations, composed of 63 classes, whose centres are given by the numbers $c_{ij} = (2j+1)\cdot 10^{i-1}/2, i = 1,\ldots,7$ and $j = 1,\ldots,9$. *General features of $f_\sigma \times \gamma$:* The curves in Figure 4 show the values of f_σ for various values of $(\Delta_{tot})_{dec}$. The curves with $3 \le (\Delta_{tot})_{dec} \le 6$ present the most pronounced decays among those presented. From $\gamma \simeq 3$, the rate of decrease assumes an intermediate value, depending on $(\Delta_{tot})_{dec}$, between those at $0 < \gamma < 1$ and at $1 < \gamma < 3$.

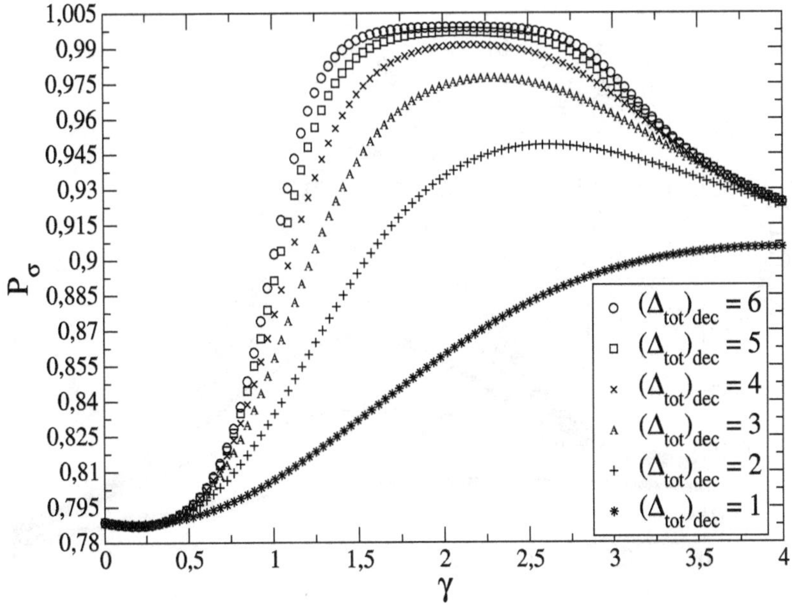

Figure 3. Curves $P_\sigma \times \gamma$ for the different values of $(\Delta_{tot})_{dec}$ according to Table 3.

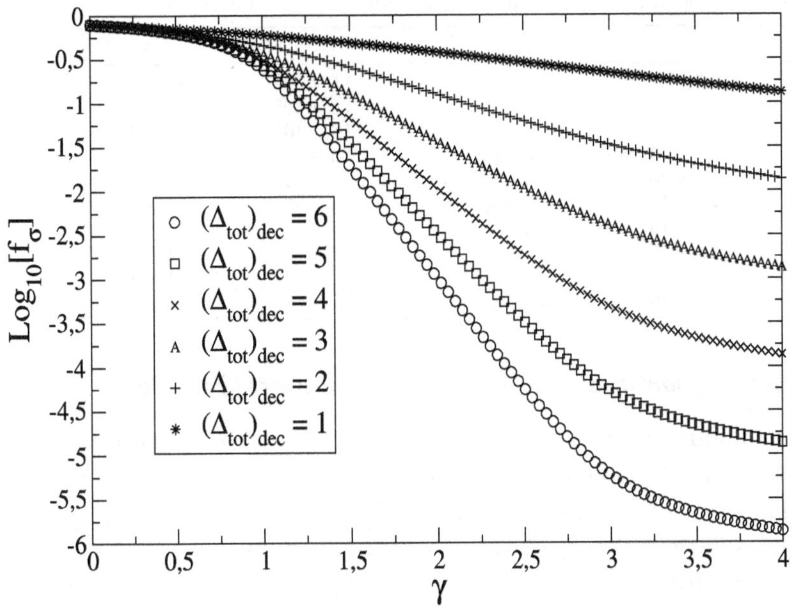

Figure 4. Curves $\log[f_\sigma] \times \gamma$ for the values of $(\Delta_{tot})_{dec}$ used in Figure 1.

General features of $M_\sigma \times \gamma$: The curves in Figure 5 show the values of M_σ for various values of $(\Delta_{tot})_{dec}$. The maximum highlighted in $\gamma = 2$ is notable regardless of the extent in decades involved. The increasingly acute aspect of each curve corresponding to the elevation of $(\Delta_{tot})_{dec}$ from $(\Delta_{tot})_{dec} = 2$ consists of another aspect to highlight.

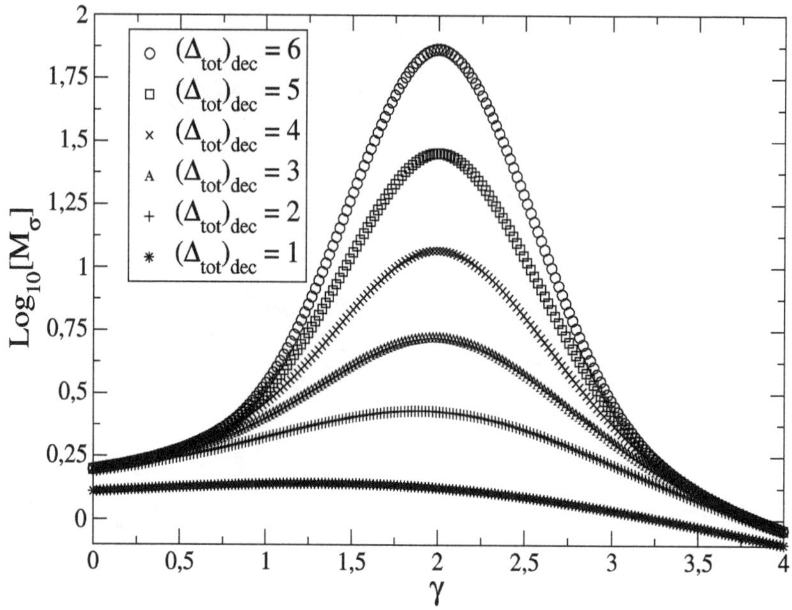

Figure 5. Curves $\log_{10}[M_\sigma] \times \gamma$ for the values $(\Delta_{tot})_{dec}$ used in Figure 1.

Table 3. Relationship between the values of z_i, z_f, and $(\Delta_{tot})_{dec}$ considered.

z_i	z_f	$(\Delta_{tot})_{dec}$	z_i	z_f	$(\Delta_{tot})_{dec}$
1.5×10^0	1.5×10^6	$(\Delta_{tot})_{dec} = 6$	1.5×10^4	1.5×10^6	$(\Delta_{tot})_{dec} = 2$
1.5×10^1	1.5×10^6	$(\Delta_{tot})_{dec} = 5$	1.5×10^3	1.5×10^5	$(\Delta_{tot})_{dec} = 2$
1.5×10^0	1.5×10^5	$(\Delta_{tot})_{dec} = 5$	1.5×10^2	1.5×10^4	$(\Delta_{tot})_{dec} = 2$
1.5×10^2	1.5×10^6	$(\Delta_{tot})_{dec} = 4$	1.5×10^1	1.5×10^3	$(\Delta_{tot})_{dec} = 2$
1.5×10^1	1.5×10^5	$(\Delta_{tot})_{dec} = 4$	1.5×10^0	1.5×10^2	$(\Delta_{tot})_{dec} = 2$
1.5×10^0	1.5×10^4	$(\Delta_{tot})_{dec} = 4$	1.5×10^5	1.5×10^6	$(\Delta_{tot})_{dec} = 1$
1.5×10^3	1.5×10^6	$(\Delta_{tot})_{dec} = 3$	1.5×10^4	1.5×10^5	$(\Delta_{tot})_{dec} = 1$
1.5×10^2	1.5×10^5	$(\Delta_{tot})_{dec} = 3$	1.5×10^3	1.5×10^4	$(\Delta_{tot})_{dec} = 1$
1.5×10^1	1.5×10^4	$(\Delta_{tot})_{dec} = 3$	1.5×10^2	1.5×10^3	$(\Delta_{tot})_{dec} = 1$
1.5×10^0	1.5×10^3	$(\Delta_{tot})_{dec} = 3$	1.5×10^1	1.5×10^2	$(\Delta_{tot})_{dec} = 1$

2.3. Statistical quantities derived from global and local densities

From the definitions in Section 2.2, valid for $\varrho(z) = Az^{-\gamma}$, the inference on how the global quantities, $\Delta_{\sigma_t}, f_{\sigma_t}, P_{\sigma_t}, M_{\sigma_t}$ and $\Delta_{\sigma_n}, f_{\sigma_n}, P_{\sigma_n}, M_{\sigma_n}$, would be defined from the densities given by Equation (1) as for local quantities, $\Delta_{\sigma_t}^{(q)}, f_{\sigma_t}^{(q)}, P_{\sigma_t}^{(q)}, M_{\sigma_t}^{(q)}$ and $\Delta_{\sigma_n}^{(q)}, f_{\sigma_n}^{(q)}, P_{\sigma_n}^{(q)}, M_{\sigma_n}^{(q)}$, from the densities given by Equation (2), becomes quick and direct. After normalizing the probability densities given by (1) and (2), according to Equations (6), (7) and (8), only the new normalized densities, under the same names as the previous ones, are then used

$$\varrho_{oct}^{(q)}(\tau_{o_q}) = A_{t,q}\tau_{o_q}^{-\gamma_{oct}^{(q)}} \qquad \varrho_{oct}^{(q)}(Q_q) = A_{n,q}Q_q^{-\gamma_{ocn}^{(q)}} \tag{11}$$

The numbers $A_{n,q}$ and $A_{t,q}$ are normalization constants. Substituting the two new densities originally linked to Equation (1) into Equations (9) and (10), we get Equations (12) and (13) resulting from the replacement of $\varrho(z) \to \varrho_{ocn}(Q)$ and the Equations (14) and (15) from the replacement $\varrho(z) \to \varrho_{oct}(\tau_o)$. Equations (16) and (17) come from the substitution

$\varrho(z) \to \varrho^{(q)}_{ocn}(Q_q)$ and Equations (18) and (19), from the substitution $\varrho(z) \to \varrho^{(q)}_{oct}(\tau_{o_q})$, both derived from the probability densities originally linked to Equation (2)

$$\begin{aligned}
&\overline{Q} = \int_{Q_i}^{Q_f} \varrho_{ocn}(Q) Q \,\mathrm{d}Q \quad \sigma_n^2 = \int_{Q_i}^{Q_f} \varrho_{ocn}(Q)(\overline{Q} - Q)^2 \,\mathrm{d}Q, \\
&Q_\sigma \equiv \overline{Q} + \sigma_n \quad \Delta_{\sigma_n} \equiv Q_\sigma - Q_i \quad M_{\sigma_n} \equiv \Delta_{\sigma_n}/\overline{Q} \quad f_{\sigma_n} \equiv \Delta_{\sigma_n}/\Delta_{tot,n}, \\
&\Delta_{tot,n} \equiv Q_f - Q_i \quad (\Delta_{tot,n})_{dec} \equiv \log_{10}(Q_f) - \log_{10}(Q_i),
\end{aligned} \tag{12}$$

$$\begin{aligned}
&P_{\sigma_n} \equiv P(Q_i \le Q \le Q_\sigma) = \int_{Q_i}^{Q_\sigma} \varrho_{ocn}(Q) \,\mathrm{d}Q \quad I_{\sigma_n} \equiv [Q_i, Q_\sigma], \\
&P_{\overline{\sigma}_n} \equiv P(Q_{\sigma_n} \le Q \le Q_f) = \int_{Q_\sigma}^{Q_f} \varrho_{ocn}(Q) \,\mathrm{d}Q \quad I_{\overline{\sigma}_n} \equiv [Q_\sigma, Q_f], \\
&I_{tot} \equiv [Q_i, Q_f] = I_{\sigma_n} \cup I_{\overline{\sigma}_n} \quad P_{\sigma_n} + P_{\overline{\sigma}_n} = 1,
\end{aligned} \tag{13}$$

$$\begin{aligned}
&\overline{\tau_o} = \int_{\tau_{o,i}}^{\tau_{o,f}} \varrho_{oct}(\tau_o) \tau_o \,\mathrm{d}\tau_o \quad \sigma_t^2 = \int_{\tau_{o,i}}^{\tau_{o,f}} \varrho_{oct}(\tau_o)(\overline{\tau_o} - \tau_o)^2 \,\mathrm{d}\tau_o, \\
&\tau_{o,\sigma} \equiv \overline{\tau_o} + \sigma_t \quad \Delta_{\sigma_t} \equiv \tau_{o,\sigma} - \tau_{o,i} \quad M_{\sigma_t} \equiv \Delta_{\sigma_t}/\overline{\tau_o} \quad f_{\sigma_t} \equiv \Delta_{\sigma_t}/\Delta_{tot,t}, \\
&\Delta_{tot,t} \equiv \tau_{o,f} - \tau_{o,i} \quad (\Delta_{tot,t})_{dec} \equiv \log_{10}(\tau_{o,f}) - \log_{10}(\tau_{o,i}),
\end{aligned} \tag{14}$$

$$\begin{aligned}
&P_{\sigma_t} \equiv P(\tau_{o,i} \le \tau_o \le \tau_{o,\sigma}) = \int_{\tau_{o,i}}^{\tau_{o,\sigma}} \varrho_{oct}(\tau_o) \,\mathrm{d}\tau_o \quad I_{\sigma_t} \equiv [\tau_{o,i}, \tau_{o,\sigma}], \\
&P_{\overline{\sigma}_t} \equiv P(\tau_{o,\sigma} \le \tau_o \le \tau_{o,f}) = \int_{\tau_{o,\sigma}}^{\tau_{o,f}} \varrho_{oct}(\tau_o) \,\mathrm{d}\tau_o \quad I_{\overline{\sigma}_t} \equiv [\tau_{o,\sigma}, \tau_{o,f}], \\
&I_{tot} \equiv [\tau_{o,i}, \tau_{o,f}] = I_{\sigma_t} \cup I_{\overline{\sigma}_t} \quad P_{\sigma_t} + P_{\overline{\sigma}_t} = 1,
\end{aligned} \tag{15}$$

$$\begin{aligned}
&\overline{Q_q} = \int_{Q_{q,i}}^{Q_{q,f}} \varrho^{(q)}_{ocn}(Q_q) Q_q \,\mathrm{d}Q_q \quad \sigma_{q,n}^2 = \int_{Q_{q,i}}^{Q_{q,f}} \varrho^{(q)}_{ocn}(Q_q)(\overline{Q_q} - Q_q)^2 \,\mathrm{d}Q_q, \\
&Q_{q,\sigma} \equiv \overline{Q_q} + \sigma_{q,n} \quad \Delta^{(q)}_{\sigma_n} \equiv Q_{q,\sigma} - Q_{q,i} \quad M^{(q)}_{\sigma_n} \equiv \Delta^{(q)}_{\sigma_n}/\overline{Q_q} \quad f^{(q)}_{\sigma_n} \equiv \Delta^{(q)}_{\sigma_n}/\Delta^{(q)}_{tot,n}, \\
&(\Delta^{(q)}_{tot,n})_{dec} \equiv \log_{10}(Q_{q,f}) - \log_{10}(Q_{q,i}) \quad \Delta^{(q)}_{tot,n} \equiv Q_{q,f} - Q_{q,i},
\end{aligned} \tag{16}$$

$$\begin{aligned}
&P^{(q)}_{\sigma_n} \equiv P(Q_{q,i} \le Q_q \le Q_{q,\sigma}) = \int_{Q_{q,i}}^{Q_{q,\sigma}} \varrho^{(q)}_{ocn}(Q_q) \,\mathrm{d}Q_q \quad I^{(q)}_{\sigma_n} \equiv [Q_{q,i}, Q_{q,\sigma}], \\
&P^{(q)}_{\overline{\sigma}_n} \equiv P(Q_{q,\sigma} \le Q_q \le Q_{q,f}) = \int_{Q_{q,\sigma}}^{Q_{q,f}} \varrho^{(q)}_{ocn}(Q_q) \,\mathrm{d}Q_q \quad I^{(q)}_{\overline{\sigma}_n} \equiv [Q_{q,\sigma}, Q_{q,f}], \\
&I^{(q)}_{tot,n} \equiv [Q_{q,i}, Q_{q,f}] = I^{(q)}_{\sigma_n} \cup I^{(q)}_{\overline{\sigma}_n} \quad P^{(q)}_{\sigma_n} + P^{(q)}_{\overline{\sigma}_n} = 1,
\end{aligned} \tag{17}$$

$$\begin{aligned}
&\overline{\tau_{o_q}} = \int_{\tau_{o_q,i}}^{\tau_{o_q,f}} \varrho^{(q)}_{oct}(\tau_{o_q}) \tau_{o_q} \,\mathrm{d}\tau_{o_q} \quad \sigma_{q,t}^2 = \int_{\tau_{o_q,i}}^{\tau_{o,f}} \varrho^{(q)}_{oct}(\tau_{o_q})(\overline{\tau_{o_q}} - \tau_{o_q})^2 \,\mathrm{d}\tau_{o_q}, \\
&\tau_{o_q,\sigma} \equiv \overline{\tau_{o_q}} + \sigma_{q,t} \quad \Delta^{(q)}_{\sigma_t} \equiv \tau_{o_q,\sigma} - \tau_{o_q,i} \quad M^{(q)}_{\sigma_t} \equiv \Delta^{(q)}_{\sigma_t}/\overline{\tau_{o_q}} \quad f^{(q)}_{\sigma_t} \equiv \Delta^{(q)}_{\sigma_t}/\Delta_{tot,t}, \\
&(\Delta^{(q)}_{tot,t})_{dec} \equiv \log_{10}(\tau_{o_q,f}) - \log_{10}(\tau_{o_q,i}) \quad \Delta^{(q)}_{tot,t} \equiv \tau_{o_q,f} - \tau_{o_q,i},
\end{aligned} \tag{18}$$

$$P_{\sigma_t}^{(q)} \equiv P(\tau_{o_q,i} \leq \tau_{o_q} \leq \tau_{o_q,\sigma}) = \int_{\tau_{o_q,i}}^{\tau_{o_q,\sigma}} \varrho_{oct}^{(q)}(\tau_{o_q})\, d\tau_{o_q} \quad I_{\sigma_t}^{(q)} \equiv [\tau_{o_q,i}, \tau_{o_q,\sigma}],$$

$$P_{\overline{\sigma}_t}^{(q)} \equiv P(\tau_{o_q,\sigma} \leq \tau_{o_q} \leq \tau_{o_q,f}) = \int_{\tau_{o_q,\sigma}}^{\tau_{o_q,f}} \varrho_{oct}^{(q)}(\tau_{o_q})\, d\tau_{o_q} \quad I_{\overline{\sigma}_t}^{(q)} \equiv [\tau_{o_q,\sigma}, \tau_{o_q,f}], \tag{19}$$

$$I_{tot,t}^{(q)} \equiv [\tau_{o_q,i}, \tau_{o_q,f}] = I_{\sigma_t}^{(q)} \cup I_{\overline{\sigma}_t}^{(q)} \quad P_{\sigma_t}^{(q)} + P_{\overline{\sigma}_t}^{(q)} = 1.$$

3. Results

The generations of the densities $\varrho_{ocn}^{(q)}(Q)$ and $\varrho_{oct}^{(q)}(\tau_{o_q})$, as well as the other statistical quantities associated with them considered in Section 2.3, are described in detail in the work [3] and will not be addressed in this work (the method used is exactly the same here). All information about probability distributions

$$\varrho_{ocn}^{(q)}(Q_q) \times Q_q \quad \varrho_{oct}^{(q)}(\tau_{o_q}) \times \tau_{o_q}$$

from the simulations relating to the two dynamic systems discussed, given or not by power laws, were extracted from curves with the forms

$$\log_{10}[\varrho_{ocn}^{(q)}(Q_q)] \times \log_{10}[Q_q] \quad \log_{10}[\varrho_{oct}^{(q)}(\tau_{o_q})] \times \log_{10}[\tau_{o_q}].$$

The values of $\gamma_{ocn}^{(q)}$ and $\gamma_{oct}^{(q)}$ were generated using the least squares method. In each case a quantitative indicator of the quality of the linear fit, the latter supported by the relationship

$$\log_{10}(\varrho_{ocn}^{(q)}(Q_q) = \log_{10}(A_{n,q}) - \gamma_{ocn}^{(q)} \log_{10}(Q_q),$$

$$\log_{10}(\varrho_{oct}^{(q)}(\tau_{o_q}) = \log_{10}(A_{t,q}) - \gamma_{oct}^{(q)} \log_{10}(\tau_{o_q}) \tag{20}$$

is the squared value of the correlation coefficient, r^2. Qualitative indication of the same quality is based on inspection of the positioning of the line generated by the method in relation to the points to which it was fitted. The lengths of the Q_q and τ_{o_q} value ranges over which the adjustments were made, expressed in decades, $(\Delta_{tot,n}^{(q)})_{dec}$ e $(\Delta_{tot,t}^{(q)})_{dec}$ are important. This aspect will be taken up again in Section 4.

Tables 4 and 5 display the values related to the fits, quality and extent, in addition to the absolute values of the exponents found. The first refers exclusively to the Lozi map and the other refers to the Hénon map.

The values of the effective lengths, $\Delta_{\sigma_t}^{(q)}$ and $\Delta_{\sigma_n}^{(q)}$, of the effective ratios, $f_{\sigma_t}^{(q)}$ and $f_{\sigma_{q,n}}^{(q)}$, and the effective variabilities, $M_{\sigma_t}^{(q)}$ and $M_{\sigma_n}^{(q)}$, are presented in Table 6 (Lozi) and in Table 7 (Hénon).

4. Discussions

4.1. Power laws: the legitimacy of adjustment

As mentioned in the Introduction, the purpose is to establish, at least partially, the statistical similarity between each of the attractors, taken as a whole, and the small portions of

Table 4. Lozi attractor. Values of the extremes of the intervals in which the adjustments are made, initials $\tau_{o_q,i}$ and $Q_{q,i}$, finals $\tau_{o_q,f}$ and $Q_{q,f}$, their widths in decades, $(\Delta_t^{(q)})_{dec}$ and $(\Delta_n^{(q)})_{dec}$. Value of the linear fit quality indicator, r^2, and the absolute values of the exponents relative to each of the distributions given by laws of power, $\gamma_{oct}^{(q)}$ and $\gamma_{ocn}^{(q)}$. ($(\Delta_t^{(q)})_{dec} \equiv (\Delta_{tot,t}^{(q)})_{dec}$ and $(\Delta_n^{(q)})_{dec} \equiv (\Delta_{tot,n}^{(q)})_{dec}$).

q	Att.	γ	Val.	r^2	$\tau_{o_q,i} \vert Q_{q,i}$	$\tau_{o_q,f} \vert Q_{q,f}$	$(\Delta_t^{(q)})_{dec} \vert (\Delta_n^{(q)})_{dec}$
1	Loz	$\gamma_{oct}^{(q)}$	1.946(8)	0.9995	1.5	25, 000	4.2
		$\gamma_{ocn}^{(q)}$	1.730(6)	0.9995	1.5	75, 000	5.2
2	Loz	$\gamma_{oct}^{(q)}$	1.96(1)	0.9991	1.5	3, 500	3.4
		$\gamma_{ocn}^{(q)}$	1.744(6)	0.9995	1.5	45, 000	4.5
3	Loz	$\gamma_{oct}^{(q)}$	1.926(8)	0.9995	1.5	4, 500	3.5
		$\gamma_{ocn}^{(q)}$	1.714(9)	0.9989	1.5	95, 000	4.8
4	Loz	$\gamma_{oct}^{(q)}$	1.97(1)	0.9992	1.5	2, 500	3.2
		$\gamma_{ocn}^{(q)}$	1.736(7)	0.9994	1.5	25, 000	4.2
5	Loz	$\gamma_{oct}^{(q)}$	1.944(8)	0.9995	1.5	3, 500	3.4
		$\gamma_{ocn}^{(q)}$	1.741(7)	0.9994	1.5	35, 000	4.4
6	Loz	$\gamma_{oct}^{(q)}$	1.98(1)	0.9990	1.5	2, 500	3.2
		$\gamma_{ocn}^{(q)}$	1.758(7)	0.9993	1.5	35, 000	4.4
7	Loz	$\gamma_{oct}^{(q)}$	1.86(1)	0.9992	1.5	4, 500	3.5
		$\gamma_{ocn}^{(q)}$	1.678(6)	0.9996	1.5	35, 000	4.4
8	Loz	$\gamma_{oct}^{(q)}$	1.896(8)	0.9994	1.5	4, 500	3.5
		$\gamma_{ocn}^{(q)}$	1.692(6)	0.9995	1.5	35, 000	4.4
9	Loz	$\gamma_{oct}^{(q)}$	1.91(1)	0.9989	1.5	5, 500	3.6
		$\gamma_{ocn}^{(q)}$	1.713(7)	0.9993	1.5	45, 000	4.5
10	Loz	$\gamma_{oct}^{(q)}$	1.912(7)	0.9996	1.5	3, 500	3.6
		$\gamma_{ocn}^{(q)}$	1.704(6)	0.9995	1.5	45, 000	4.5
11	Loz	$\gamma_{oct}^{(q)}$	1.860(7)	0.9997	1.5	3, 500	3.6
		$\gamma_{ocn}^{(q)}$	1.667(5)	0.9996	1.5	35, 000	4.4
12	Loz	$\gamma_{oct}^{(q)}$	1.94(1)	0.9990	1.5	5, 500	3.6
		$\gamma_{ocn}^{(q)}$	1.736(6)	0.9995	1.5	25, 000	4.2
13	Loz	$\gamma_{oct}^{(q)}$	2.02(2)	0.9982	1.5	2, 500	3.2
		$\gamma_{ocn}^{(q)}$	1.79(1)	0.9983	1.5	45, 000	4.5
14	Loz	$\gamma_{oct}^{(q)}$	1.908(8)	0.9993	1.5	3, 500	3.4
		$\gamma_{ocn}^{(q)}$	1.731(8)	0.9992	1.5	55, 000	4.6
15	Loz	$\gamma_{oct}^{(q)}$	1.97(1)	0.9992	1.5	3, 500	3.4
		$\gamma_{ocn}^{(q)}$	1.755(5)	0.9996	1.5	35, 000	4.4
16	Loz	$\gamma_{oct}^{(q)}$	1.99(1)	0.9988	1.5	3, 500	3.4
		$\gamma_{ocn}^{(q)}$	1.773(7)	0.9994	1.5	35, 000	4.4
17	Loz	$\gamma_{oct}^{(q)}$	1.910(7)	0.9996	1.5	3, 500	3.4
		$\gamma_{ocn}^{(q)}$	1.704(6)	0.9995	1.5	55, 000	5.0
18	Loz	$\gamma_{oct}^{(q)}$	1.936(9)	0.9994	1.5	3, 500	3.4
		$\gamma_{ocn}^{(q)}$	1.719(7)	0.9994	1.5	35, 000	4.4

Table 5. Hénon attractor. Values of the extremes of the intervals in which the adjustments are made, initials $\tau_{o_q,i}$ and $Q_{q,i}$, finals $\tau_{o_q,f}$ and $Q_{q,f}$, their widths in decades, $(\Delta_t^{(q)})_{dec}$ and $(\Delta_n^{(q)})_{dec}$. Value of the linear fit quality indicator, r^2, and the absolute values of the exponents relative to each of the distributions given by laws of power, $\gamma_{oct}^{(q)}$ and $\gamma_{ocn}^{(q)}$. ($(\Delta_t^{(q)})_{dec} \equiv (\Delta_{tot,t}^{(q)})_{dec}$ and $(\Delta_n^{(q)})_{dec} \equiv (\Delta_{tot,n}^{(q)})_{dec}$).

q	Att.	γ	Val.	r^2	$\tau_{o_q,i}\vert Q_{q,i}$	$\tau_{o_q,f}\vert Q_{q,f}$	$(\Delta_t^{(q)})_{dec}\vert(\Delta_n^{(q)})_{dec}$
19	Hen	$\gamma_{oct}^{(q)}$	1.93(1)	0.9992	1.5	7,500	3.7
		$\gamma_{ocn}^{(q)}$	1.712(9)	0.9988	1.5	45,000	4.5
20	Hen	$\gamma_{oct}^{(q)}$	1.93(1)	0.9991	1.5	4,500	3.5
		$\gamma_{ocn}^{(q)}$	1.70(2)	0.9967	1.5	25,000	4.2
21	Hen	$\gamma_{oct}^{(q)}$	1.94(1)	0.9991	1.5	4,500	3.5
		$\gamma_{ocn}^{(q)}$	1.712(9)	0.9988	1.5	45,000	4.5
22	Hen	$\gamma_{oct}^{(q)}$	1.94(1)	0.9992	1.5	4,500	3.5
		$\gamma_{ocn}^{(q)}$	1.70(2)	0.9966	1.5	25,000	4.2
23	Hen	$\gamma_{oct}^{(q)}$	1.926(9)	0.9972	1.5	2,500	3.2
		$\gamma_{ocn}^{(q)}$	1.71(1)	0.9982	1.5	35,000	4.4
24	Hen	$\gamma_{oct}^{(q)}$	1.90(1)	0.9994	1.5	450	2.5
		$\gamma_{ocn}^{(q)}$	1.71(1)	0.9982	1.5	35,000	4.4
25	Hen	$\gamma_{oct}^{(q)}$	1.93(1)	0.9991	1.5	3,500	3.4
		$\gamma_{ocn}^{(q)}$	1.72(1)	0.9976	1.5	25,000	4.2
26	Hen	$\gamma_{oct}^{(q)}$	1.930(9)	0.9994	1.5	4,500	3.5
		$\gamma_{ocn}^{(q)}$	1.71(1)	0.9976	1.5	25,000	4.2
27	Hen	$\gamma_{oct}^{(q)}$	1.898(9)	0.9994	1.5	950	2.8
		$\gamma_{ocn}^{(q)}$	1.68(1)	0.9975	1.5	5,500	3.6
28	Hen	$\gamma_{oct}^{(q)}$	1.92(1)	0.9991	1.5	950	2.8
		$\gamma_{ocn}^{(q)}$	1.68(2)	0.9974	1.5	5,500	3.6
29	Hen	$\gamma_{oct}^{(q)}$	1.91(1)	0.9989	1.5	650	2.8
		$\gamma_{ocn}^{(q)}$	1.68(2)	0.9974	1.5	5,500	3.6
30	Hen	$\gamma_{oct}^{(q)}$	1.83(1)	0.9989	1.5	150	2.0
		$\gamma_{ocn}^{(q)}$	1.62(1)	0.9979	1.5	950	2.8
31	Hen	$\gamma_{oct}^{(q)}$	1.91(1)	0.9992	1.5	65	1.6
		$\gamma_{ocn}^{(q)}$	1.69(2)	0.9975	1.5	6,500	3.6
32	Hen	$\gamma_{oct}^{(q)}$	1.75	0.9992	1.5	65	1.6
		$\gamma_{ocn}^{(q)}$	1.53(1)	0.9983	1.5	450	2.5

them enclosed by the corresponding hypercubes. In other words, clarify the existence of self-similarity associated with each of the two strange attractors involved, based on considering only a small fraction of the locally available information about the two. For that, it is necessary to verify if the assignment of the probability distributions $\varrho_{oct}^{(q)}(\tau_{o_q}) = A_{t,q}\tau_{o_q}^{-\gamma_{oct}^{(q)}}$ and $\varrho_{ocn}^{(q)}(Q_q) = A_{n,q}Q_q^{-\gamma_{ocn}^{(q)}}$, Equation (11), to the data coming from the corresponding hypercubes is a good idea. Exactly as in the article [3], two criteria were adopted: (*i*) value of the correlation coefficient not lower than a certain high minimum value, $r^2 \geq 0.999$ (*ii*) value of length of the total range, in decades, not less than three, $(\Delta_{tot}) \geq 3$. The latter is an indicator of the occurrence of 'scale invariance' and the first indicator of the quality of

Table 6. Lozi attractor. Values of the effective lengths, $\Delta_{\sigma t}^{(q)}$ and $\Delta_{\sigma n}^{(q)}$, of the effective ratios, $f_{\sigma t}^{(q)}$ and $f_{\sigma n}^{(q)}$, of the effective variabilities, $M_{\sigma t}^{(q)}$ and $M_{\sigma n}^{(q)}$, and of the effective accumulated probabilities $P_{\sigma t}^{(q)}$ and $P_{\sigma n}^{(q)}$.

q	Att.	Oct|Ocn	$\Delta_{\sigma t}^{(q)} \mid \Delta_{\sigma n}^{(q)}$	$f_{\sigma t}^{(q)} \mid f_{\sigma n}^{(q)}$	$M_{\sigma t}^{(q)} \mid M_{\sigma n}^{(q)}$	$P_{\sigma t}^{(q)} \mid P_{\sigma n}^{(q)}$
1	Loz	Oct	81.16	0.0324	6.26	0.978344
		Ocn	1,163	0.0155	16.32	0.992603
2	Loz	Oct	91.84	0.0262	7.01	0.981617
		Ocn	801.42	0.0178	14.13	0.991131
3	Loz	Oct	114.81	0.0255	7.59	0.982848
		Ocn	1,447.9	0.0152	17.09	0.992980
4	Loz	Oct	75.92	0.0303	6.28	0.978925
		Ocn	579.80	0.0232	11.53	0.988327
5	Loz	Oct	96.28	0.0275	7.00	0.981265
		Ocn	696.84	0.0199	12.95	0.990029
6	Loz	Oct	73.84	0.0296	6.28	0.979149
		Ocn	650.05	0.0186	13.29	0.990472
7	Loz	Oct	141.15	0.0314	7.40	0.981107
		Ocn	901.34	0.0258	11.64	0.988033
8	Loz	Oct	126.31	0.0281	7.52	0.982119
		Ocn	851.29	0.0243	11.93	0.988530
9	Loz	Oct	134.03	0.0244	8.08	0.983962
		Ocn	398.50	0.00886	17.18	0.990304
10	Loz	Oct	106.45	0.0304	6.95	0.980479
		Ocn	913.10	0.0203	13.40	0.990039
11	Loz	Oct	123.44	0.0353	6.82	0.978944
		Ocn	196.06	0.0560	6.13	0.969624
12	Loz	Oct	125.79	0.0229	8.13	0.984597
		Ocn	579.80	0.0232	11.53	0.988327
13	Loz	Oct	66.09	0.0265	6.26	0.979941
		Ocn	660.28	0.0147	15.17	0.992146
14	Loz	Oct	107.40	0.0307	6.94	0.980373
		Ocn	957.27	0.0174	14.78	0.991563
15	Loz	Oct	144.37	0.0152	9.91	0.981823
		Ocn	658.07	0.0188	13.23	0.990397
16	Loz	Oct	84.09	0.0240	7.02	0.982205
		Ocn	611.37	0.0175	13.59	0.990834
17	Loz	Oct	106.45	0.0304	6.95	0.980426
		Ocn	1,075.2	0.0196	14.05	0.990848
18	Loz	Oct	98.58	0.0282	6.99	0.981080
		Ocn	762.40	0.0218	12.50	0.989400

the fit of the line to the linearized relation according to Equation (20). Table 4 shows that all 36 distributions, associated with the corresponding 18 hypercubes, display values of $(\Delta_{tot,t}^{(q)})_{dec}$ and $(\Delta_{tot,n}^{(q)})_{dec}$ greater than 3. Furthermore, all of them are also associated with values of r^2 greater than 0.999. Therefore, the legitimacy of attributing power laws to the data coming from Lozi's strange attractor is verified. Table 5 shows that although all values of r^2 are greater than 0.999, there are values of $(\Delta_{tot,t}^{(q)})_{dec}$ and $(\Delta_{tot,n}^{(q)})_{dec}$ less than three: – occupancy times linked to hypercubes 24, 27, 28, 29, 30, 31, and 32 – Occupancy numbers linked to the 28, 29, 30 and 32 hypercubes. In this way, the legitimacy of attributing the power laws to the two distributions occurs only for the following hypercubes: 19, 20, 21, 22, 23, 25, 26 and 27. Therefore, only the two distributions associated with the last eight hypercubes linked to the strange attractor of Hénon will be considered from this point on. In this way, the values of q considered will be those relevant to the range I_{con}:

$$I_{con} \equiv 1 \leq q \leq 23 \cup 25 \leq q \leq 27. \tag{21}$$

Table 7. Hénon attractor. Values of the effective lengths, $\Delta_{\sigma_t}^{(q)}$ and $\Delta_{\sigma_n}^{(q)}$, of the effective ratios, $f_{\sigma_t}^{(q)}$ and $f_{\sigma_n}^{(q)}$, and the effective variabilities, $M_{\sigma_t}^{(q)}$ and $M_{\sigma_n}^{(q)}$, and of the effective accumulated probabilities $P_{\sigma_t}^{(q)}$ and $P_{\sigma_n}^{(q)}$.

q	Att.	Oct|Ocn	$\Delta_{\sigma_t}^{(q)}\|\Delta_{\sigma_n}^{(q)}$	$f_{\sigma_t}^{(q)}\|f_{\sigma_n}^{(q)}$	$M_{\sigma_t}^{(q)}\|M_{\sigma_n}^{(q)}$	$P_{\sigma_t}^{(q)}\|P_{\sigma_n}^{(q)}$
19	Hen	Oct	147.00	0.0197	9.05	0.986425
		Ocn	916.95	0.0204	13.38	0.990275
20	Hen	Oct	113.76	0.0253	7.59	0.982938
		Ocn	667.75	0.0267	10.91	0.987119
21	Hen	Oct	110.32	0.0245	7.61	0.983154
		Ocn	916.95	0.0204	13.38	0.990275
22	Hen	Oct	110.32	0.0245	7.61	0.983154
		Ocn	667.75	0.0267	10.91	0.987119
23	Hen	Oct	85.82	0.0343	6.24	0.977810
		Ocn	790.96	0.0226	12.31	0.989122
24	Hen	Oct	38.46	0.0858	3.68	0.953511
		Ocn	790.96	0.0226	12.31	0.989122
25	Hen	Oct	100.34	0.0287	6.98	0.980936
		Ocn	617.37	0.0247	11.25	0.987815
26	Hen	Oct	113.76	0.0253	7.59	0.982938
		Ocn	642.07	0.0257	11.08	0.987475
27	Hen	Oct	56.60	0.0597	4.59	0.965457
		Ocn	278.86	0.0507	6.80	0.975146
28	Hen	Oct	53.78	0.0567	4.60	0.966346
		Ocn	278.86	0.0507	6.80	0.975146
29	Hen	Oct	45.45	0.0701	4.10	0.960267
		Ocn	278.86	0.0507	6.80	0.975146
30	Hen	Oct	20.77	0.140	2.70	0.931501
		Ocn	107.81	0.114	4.08	0.947340
31	Hen	Oct	13.55	0.213	2.14	0.906740
		Ocn	299.58	0.0461	7.21	0.977248
32	Hen	Oct	16.09	0.253	2.15	0.895195
		Ocn	78.79	0.176	3.26	0.923639

4.2. Confrontation between local and global derived statistical quantities

Again in line with the Introduction, the comparison between the results (a), (b), (c) and (d) and the corresponding $(a^{(q)})$, $(b^{(q)})$, $(c^{(q)})$, $(d^{(q)})$, consists of a procedure to establish the specific statistical self-similarity proposed here. Therefore, it is necessary to consider the quantities $\Delta_{\sigma_t}^{(q)}$, $\Delta_{\sigma_n}^{(q)}$, $f_{\sigma_t}^{(q)}$, $f_{\sigma_n}^{(q)}$, $M_{\sigma_t}^{(q)}$ and $M_{\sigma_n}^{(q)}$ displayed in Tables 6 and 7, as the quantities Δ_{σ_t}, Δ_{σ_n}, f_{σ_t}, f_{σ_n}, M_{σ_t} and M_{σ_n} coming from [3] and displayed in Table 8. The same table also displays values of P_{σ_t} and P_{σ_n} not shown in [3]. The comparison will be based 16 histograms presented in four Figures 6, 7, 8 and 9. Each figure contains four histograms. The absolute frequency indicated in each of the histograms is exactly the number of hypercubes for which some value pertinent to the corresponding horizontal interval occurred. In each

Table 8. Lozi and Hénon attractors. Values of the effective lengths, Δ_{σ_t} and Δ_{σ_n}, of the effective ratios, f_{σ_t} and f_{σ_n}, of the effective variabilities, M_{σ_t} and M_{σ_n}, and of the effective accumulated probabilities P_{σ_t} and P_{σ_n}.

Att.	Oct|Ocn	$\Delta_{\sigma_t}\|\Delta_{\sigma_n}$	$f_{\sigma_t}\|f_{\sigma_n}$	$M_{\sigma_t}\|M_{\sigma_n}$	$P_{\sigma_t}\|P_{\sigma_n}$
Loz	Oct	110.2	0.0116	9.84	0.989234
	Ocn	803.4	0.0095	20.09	0.994655
Hen	Oct	171.4	0.0181	9.79	0.987631
	Ocn	1,228	0.0129	18.36	0.993722

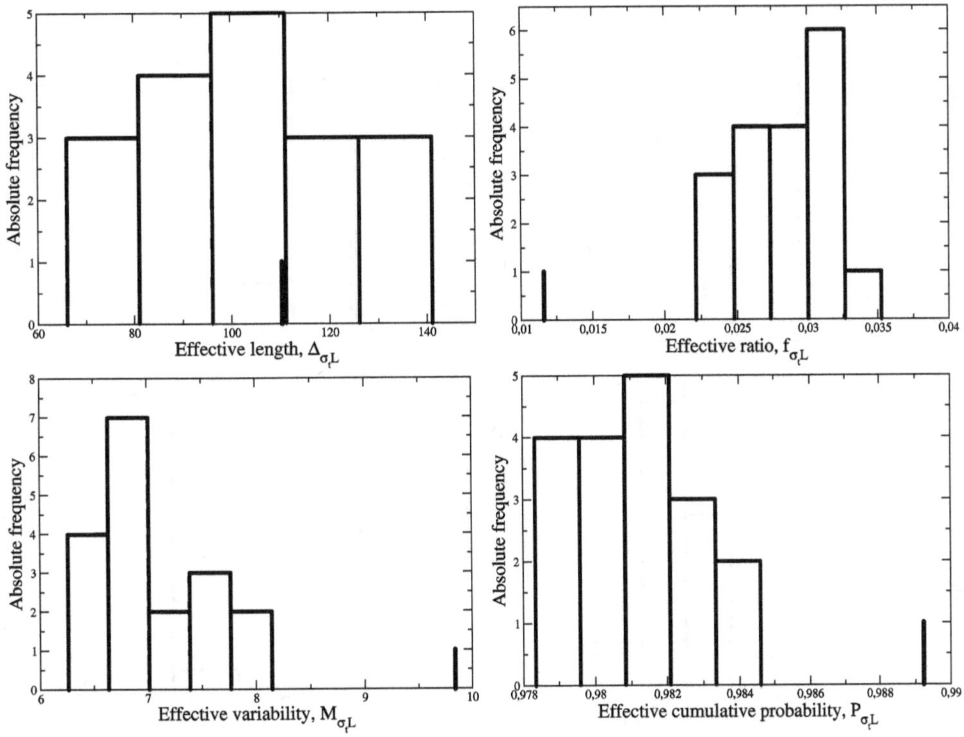

Figure 6. Occupancy times: histograms associated with the 18 hypercubes in the phase space in which the Lozi's strange attractor is immersed. Top left: the effective lengths, $\Delta_{\sigma_t}^{(q)}$, beyond the corresponding length, Δ_{σ_t}, linked to the same attractor taken as a whole (black vertical trace with unit length). Each of the values is designated here by $\Delta_{\sigma_t L}$. Top right: the effective ratios, $f_{\sigma_t}^{(q)}$, beyond the corresponding ratio, f_{σ_t}, linked to the same attractor taken as a whole (black vertical trace with unit length). Each of the values is designated here by $f_{\sigma_t L}$. Botton left: the effective variability, $M_{\sigma_t}^{(q)}$, beyond the corresponding variability, M_{σ_t}, linked to the same attractor taken as a whole (black vertical trace with unit length). Each of the values is designated here by $M_{\sigma_t L}$. Botton right: the effective cumulative probability, $P_{\sigma_t}^{(q)}$, beyond the corresponding probability, P_{σ_t}, linked to the same attractor taken as a whole (black vertical trace with unit length). Each of the values is designated here by $P_{\sigma_t L}$.

of them, the value corresponding to the attractor taken as a whole is also represented by a vertical line of unit length.

As for the comparison between the local results ($a^{(q)}$) and global results (a), relative to occupancy times and numbers, it appears that the locals also indicate very short effective intervals when compared to the total intervals linked, given the small values of the effective ratios $f_{\sigma_t}^{(q)}$ and $f_{\sigma_n}^{(q)}$, histograms at top right in each of the four figures. Although these values are always greater than the corresponding values of f_{σ_t} and f_{σ_n} attached to one or the other of the attractors, local and global values have the same order of magnitude, order common to the four histograms.

Regarding the local and global results ($b^{(q)}$) and (b), the effective cumulative probabilities, relative to occupancy times and numbers, are also very close to 1, histograms at bottom right in each of the four figures. Local values are systematically higher than all locations,

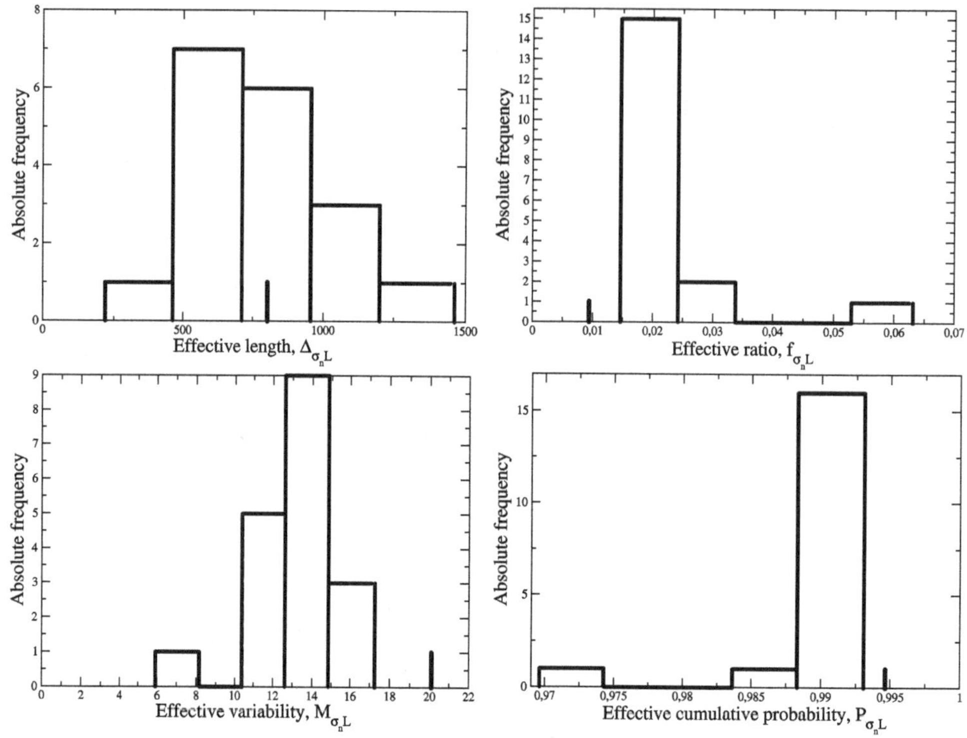

Figure 7. Occupancy numbers: histograms associated with the 18 hypercubes in the phase space in which the Lozi's strange attractor is immersed. Top left: the effective lengths, $\Delta_{\sigma_n}^{(q)}$, beyond the corresponding length, Δ_{σ_n}, linked to the same attractor taken as a whole (black vertical trace with unit length). Each of the values is designated here by $\Delta_{\sigma_n L}$. Top right: the effective ratios, $f_{\sigma_n}^{(q)}$, beyond the corresponding ratio, f_{σ_n}, linked to the same attractor taken as a whole (black vertical trace with unit length). Each of the values is designated here by $f_{\sigma_n L}$. Botton left: the effective variability, $M_{\sigma_n}^{(q)}$, beyond the corresponding variability, M_{σ_n}, linked to the same attractor taken as a whole (black vertical trace with unit length). Each of the values is designated here by $M_{\sigma_n L}$. Botton right: the effective cumulative probability, $P_{\sigma_n}^{(q)}$, beyond the corresponding probability, P_{σ_n}, linked to the same attractor taken as a whole (black vertical trace with unit length). Each of the values is designated here by $P_{\sigma_n L}$.

except for those linked to Hénon attractor occupation times, in which case the local value is among the highest.

From the results $(a^{(q)})$ and $(b^{(q)})$ comes the result $(c^{(q)})$. In turn, the latter agrees with (c) which, finally, comes from (a) and (b). The result $(c^{(q)})$ establishes that only a small fraction of the possible values for occupancy times and occupancy numbers have high probabilities of occurrence, the vast majority of possible values have practically zero probabilities of occurrence.

The results $(d^{(q)})$ and (d), both for occupancy times and for occupancy numbers, are in agreement with each other, in the same order of magnitude order common to the four histograms, at bottom left in each of the four figures. Global values are greater than the local ones in three of the four histograms, the exception concerns the effective variability of occupation times associated with the attractor of Hénon. In particular, the values linked

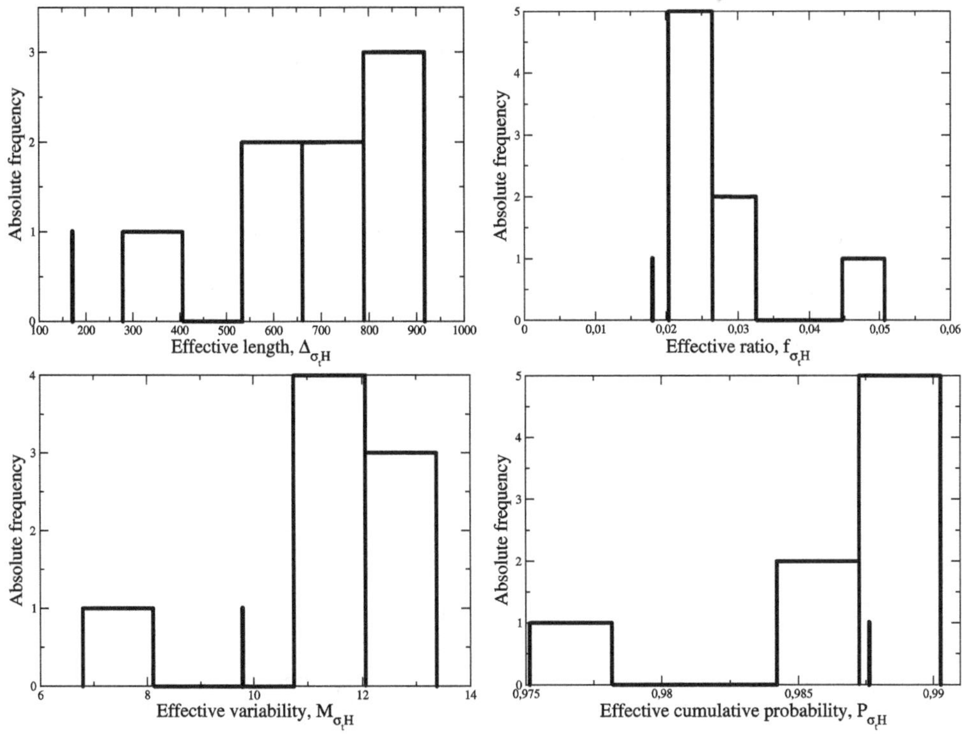

Figure 8. Occupancy times: histograms associated with the eight hypercubes in the phase space in which the Hénon's strange attractor is immersed. Top left: the effective lengths, $\Delta_{\sigma_t}^{(q)}$, beyond the corresponding length, Δ_{σ_t}, linked to the same attractor taken as a whole (black vertical trace with unit length). Each of the values is designated here by $\Delta_{\sigma_t H}$. Top right: the effective ratios, $f_{\sigma_t}^{(q)}$, beyond the corresponding ratio, f_{σ_t}, linked to the same attractor taken as a whole (black vertical trace with unit length). Each of the values is designated here by $f_{\sigma_t H}$. Botton left: the effective variability, $M_{\sigma_t}^{(q)}$, beyond the corresponding variability, M_{σ_t}, linked to the same attractor taken as a whole (black vertical trace with unit length). Each of the values is designated here by $M_{\sigma_t H}$. Botton right: the effective cumulative probability, $P_{\sigma_t}^{(q)}$, beyond the corresponding probability, P_{σ_t}, linked to the same attractor taken as a whole (black vertical trace with unit length). Each of the values is designated here by $P_{\sigma_t H}$.

to the number of occupations consist of measurements of the multifractalities of small portions of the involved attractor, or of the entire object.

Finally, regarding the effective length values, for occupancy times and numbers, despite the greater dispersion of local values, the difference between the largest and smallest value, in each of the four histograms at the top left of each figure, still lie in intervals characterized by the same order of magnitude, order common to the four histograms.

The proposed self-similarity is then verified. It consists in a similarity between the local and global statistical properties, something different from the strict equality between the same properties, which would suppose rigorous equality between each of the local statistical distributions and the corresponding global in each case. As discussed in the article [3], I don't expect strange attractors that are strictly (mono) fractals. Attractors with non-pronounced multifractalities may occur, associated with the lowest existing values of M_{σ_n},

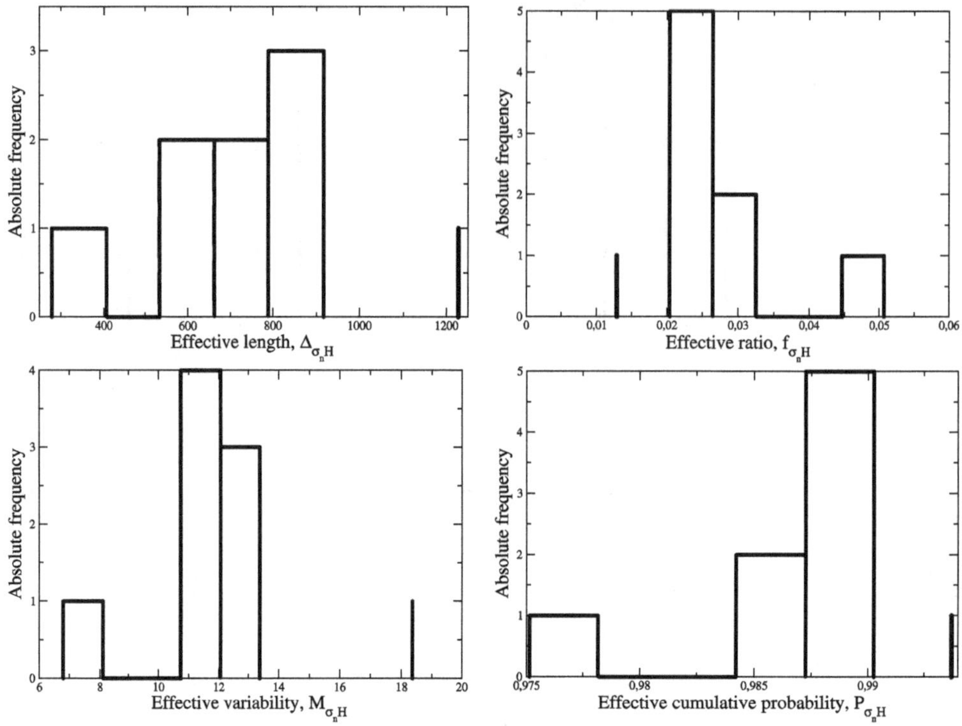

Figure 9. Occupancy numbers: histograms associated with the eight hypercubes in the phase space in which the Hénon's strange attractor is immersed. Top left: the effective lengths, $\Delta_{\sigma_n}^{(q)}$, beyond the corresponding length, Δ_{σ_n}, linked to the same attractor taken as a whole (black vertical trace with unit length). Each of the values is designated here by $\Delta_{\sigma_n H}$. Top right: the effective ratios, $f_{\sigma_n}^{(q)}$, beyond the corresponding ratio, f_{σ_n}, linked to the same attractor taken as a whole (black vertical trace with unit length). Each of the values is designated here by $f_{\sigma_n H}$. Botton left: the effective variability, $M_{\sigma_n}^{(q)}$, beyond the corresponding variability, M_{σ_n}, linked to the same attractor taken as a whole (black vertical trace with unit length). Each of the values is designated here by $M_{\sigma_n H}$. Botton right: the effective cumulative probability, $P_{\sigma_n}^{(q)}$, beyond the corresponding probability, P_{σ_n}, linked to the same attractor taken as a whole (black vertical trace with unit length). Each of the values is designated here by $P_{\sigma_n H}$.

situations in which the similarities between the local and global properties are the greatest possible.

It is true that the small amount of hypercubes considered in each case, much smaller than the total amount of hypercubes used to cover the entire attractor, does not allow a safe statement as to the validity of the results for all of them. On the other hand, it cannot be said that exactly few hypercubes, or a fraction of them, were taken for which such properties are valid. Clarification on what actually happens should come from other works on the subject. One possibility is to carry out a new study based on hypercubes chosen according to criteria that ensure the selection of those associated with the highest occupancy numbers in each case. I'm currently working on code designed to implement such a check.

The confirmation of the self-similarity proposed here, which would imply the confirmation of similarity between properties of the different hypercubes not yet addressed in an objective and direct way, would lead to the addition of a new method for the quantitative determination of the multifractality of strange attractors to the already well-established classical methods, the singularity spectrum and the generalized dimensions. Addition that should bring new insights about these fractal sets derived from distributions given by power laws, these in turn derived from the consideration of the chaotic advection of phase points instead of the classical orbital approach.

5. Conclusion

From the consideration of the chaotic advections of a large number of phase points pertinent to the phase space in which the strange attractor of Lozi or the strange attractor of Hénon is immersed, I verified the emergence of reliable probability distributions of the occupancy times and occupancy numbers, both given by power laws. Such distributions allow definitions of the following specific statistical properties: effective lengths, effective ratios, effective variabilities, and effective cumulative probabilities. These were then compared with the corresponding ones linked to the corresponding attractor taken as a whole, global properties, in order to partially verify the existence of statistical self-similarity regarding the two attractors. The results, although not fully conclusive, are convincing enough to stimulate further work on the subject. If not only the similarities between local and global properties are confirmed in each case, but also the similarities between the local ones, a new method for the quantitative characterization of strange attractors would be established, which would be added to the well-known methods based on the singularity spectrum and generalized dimensions. The codes employed to perform all statistics apply to evaluations of spatial distributions of points, both time-dependent and non-time-dependent distributions. They can be applied to any set of points immersed in one-dimensional or two-dimensional spaces whose positions are known. Therefore they can be applied to strange attractors even if the equations underlying them are not known. Furthermore, they can also be applied to strange attractors constructed from a time series using the well-known Takens Method (Method of Time Delays). They can also be applied to strange attractors from continuous dynamic systems. Finally, it is entirely possible, although not without a lot of work, to create new codes capable of applying statistics to sets of points immersed in spaces with three, four, or more dimensions.

Notes

1. From this point on, the subscript indexes n and t will respectively designate occupancy numbers and occupancy times.
2. From this point forward the superscript index (q) will refer to quantities linked to some hypercube, as opposed to the quantities associated with the corresponding attractor taken as a whole.

Acknowledgments

Thanks to Valdir Barbosa Bezerra, Full Professor at the Federal University of Paraí ba, Brazil, for his continuous encouragement and guidance at crucial times and to René Pierre Lozi, Professor Emeritus at Université Côte d'Azur, France, for his suggestions and encouragement to publish in this journal.

Disclosure statement

No potential conflict of interest was reported by the author.

References

[1] A.B. Chhabra and R.V. Jensen, *Direct determination of the $f(\alpha)$ singularity spectrum*, Phys. Rev. Lett. 62 (1989), pp. 1327–1330.
[2] A.B. Chhabra, C. Meneveau, R.V. Jensen, and K.R. Sreenivasan, *Direct determination of the $f(\alpha)$ singularity spectrum and its application to fully developed turbulence*, Phys. Rev. A 40 (1989), pp. 5284–5294.
[3] A.M. de P. Viveiros, *Non-orbital characterizations of strange attractors: Effective intervals and multifractality measures*, Chaos 31 (2021), Article ID 033139.
[4] P. Grassberger, *Generalized dimensions of strange attractors*, Phys. Lett. A 97 (1983), pp. 227–230.
[5] T.C. Halsey, M.H. Jensen, L.P. Kadanoff, I. Procaccia, and B.I. Shraiman, *Fractal measures and their singularities: The characterization of strange sets*, Phys. Rev. A 33 (1986), pp. 1141–1151.
[6] H.G.E. Hentschel and I. Procaccia, *The infinite number of generalized dimensions of fractals and strange attractors*, Physica D 8 (1983), pp. 435–444.

OPEN ACCESS

Markov partition in the attractor of Lozi maps

Diogo Baptista

ABSTRACT

In this paper, we study iterations of two-dimensional maps, in particular iterations of Lozi maps in the region of the parameter space where it has a strange attractor. Using symbolic dynamics techniques for two-dimensional maps, based on the kneading theory of Milnor and Thurston and also in the symbolic dynamic formalism developed by Sousa Ramos, through the kneading sequence for the Lozi maps, we characterize the region in the parameter space that contains the kneading curves and present a method to define a Markov partition for the Lozi attractors. Consequently, the topological entropy for the Lozi map is computed.

1. Introduction

Consider the following parametrized family of Lozi maps:

$$\mathcal{L}_{ab}(x,y) = (1 - a\,|x| + y,\ b\,x), \quad (x,y) \in \mathbb{R}^2. \tag{1}$$

These piecewise linear plane invertible maps were introduced in 1978 by René Lozi [5] as a simplified version of the parametrized family of Hénon maps, without loosing the capacity to show chaotic behaviour. Since then, Lozi maps has been an important tool to understand complex behaviour of iterated maps on the plane.

Since it is an invertible map, it is possible to define its inverse, $\mathcal{L}_{ab}^{-1}$, given by

$$\mathcal{L}_{ab}^{-1}(x,y) = \left(\frac{y}{b}, x - 1 + a\left|\frac{y}{b}\right|\right).$$

In 1992 [6] Liu et al. showed that for parameters (a, b) satisfying

$$\begin{cases} 0 < b < 1\,; \\ a > b + 1\,; \\ a > 2 - b/2\,, \end{cases} \tag{2}$$

every map $\mathcal{L}_{ab}$ has indeed a strange attractor. Let $\triangle_{\mathcal{L}}$ denote the corresponding region of the parameter space; see Figure 1.

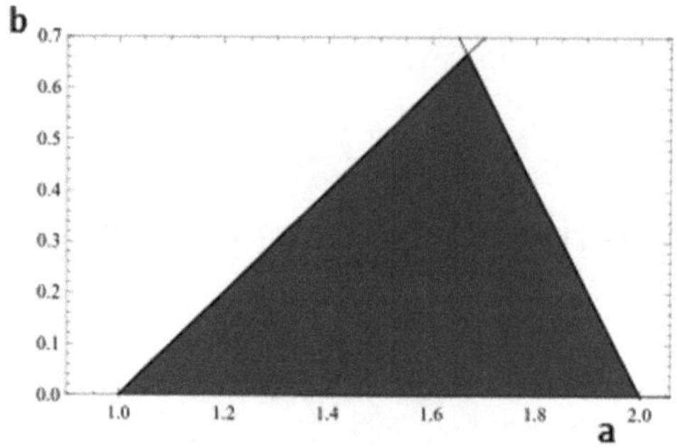

Figure 1. The parameter space region for which $\mathcal{L}_{ab}$ has a strange attractor.

From this point the Lozi maps considered will have parameter values a and b belonging to $\triangle_{\mathcal{L}}$. For simplification purposes, $\mathcal{L}_{ab}$ will be denoted by $\mathcal{L}$.

Earlier, Michał Misiurewicz [7] introduced a region $\nabla_{\mathcal{L}}$ in $\mathbb{R}^2$ satisfying

$$\mathcal{L}(\nabla_{\mathcal{L}}) \subseteq \nabla_{\mathcal{L}}, \tag{3}$$

allowing to present the Lozi map attractor, $\mathcal{A}_{\mathcal{L}}$, from its successive forward iteration,

$$\mathcal{A}_{\mathcal{L}} = \bigcap_{n=0}^{\infty} \mathcal{L}^n(\nabla_{\mathcal{L}}). \tag{4}$$

Considering the fixed point that lies in the first quadrant

$$A = \left(\frac{1}{a+1-b}, \frac{b}{a+1-b}\right) \quad \text{if } b < a+1,$$

and computing its local stability by evaluating the eigenvalues of the jacobian matrix of the map at the fixed point, in the domain of the parameters for $a > |b| + 1$, the fixed point is a saddle point. Let $L^s(A)$ and $L^u(A)$ be the stable and unstable manifolds of A, respectively.

Definition 1.1: The Lozi map fundamental domain $\nabla_{\mathcal{L}}$ is the triangle with vertices **I**, $\mathcal{L}(I)$, and $\mathcal{L}^2(I)$, where

$$I = \left(\frac{2 + a + \sqrt{a^2 + 4b}}{2(1 + a - b)}, 0\right) \tag{5}$$

is the intersection point of the unstable manifold of the fixed point A, L^u_A, with the horizontal axis.

Next, some definitions of Lozi maps symbolic dynamics are presented, following [4].

Definition 1.2: Given $X \in \mathbb{R}$, the itinerary of X is the bi-infinite symbolic sequence

$$\mathsf{it}(X) = \ldots \delta_{-2}\delta_{-1} \cdot \delta_0\delta_1\delta_2 \ldots, \tag{6}$$

where the symbols δ_n, for $n \in \mathbb{Z}$, are defined as

$$\delta_n \equiv \begin{cases} -1, & \mathcal{L}^n(X)_x < 0, \\ \star, & \mathcal{L}^n(X)_x = 0, \\ +1, & \mathcal{L}^n(X)_x > 0, \end{cases} \tag{7}$$

where X_x corresponds to the x component of X.

By the previous definition, $\mathsf{it}(X)$ is a bi-infinite symbolic sequence of $\{-1, *, +1\}^{\mathbb{Z}}$. It is important to distinguish the itinerary for the past and the itinerary for the future of a point X; let $\underline{\delta}(X) = \cdots \delta_{-2}\delta_{-1} \cdot \delta_0\delta_1\delta_2 \cdots$ be an itinerary of X. The symbolic subsequence $\underline{\delta}^u(X) = \cdots \delta_{-2}\delta_{-1}$ is the itinerary for the past of X and the symbolic subsequence $\underline{\delta}^s(X) = \delta_0\delta_1\delta_2 \cdots$ is the itinerary for the future. Thus, $\underline{\delta}(X) = \underline{\delta}^u(X) \cdot \underline{\delta}^s(X)$.

2. Critical set

Let $\mathcal{V}$ be the vertical axis, such that

$$\mathcal{V} = \mathcal{V}^{+1} \cup \mathcal{V}^{-1},$$

where $\mathcal{V}^{+1}$ is the nonnegative part of the vertical axis and $\mathcal{V}^{-1}$ is the negative part of the vertical axis.

According to the construction process suggested by Ishii [4] and Baptista et al. [3], some points belonging to $\mathcal{A}_{\mathcal{L}} \cap \mathcal{V}$play a very important role in the study of Lozi's maps dynamics, especially the point $\mathcal{L}^{-1}(I)$. The orbit associated with the forward itinerary of the point I is the orbit whose itinerary, for the future, corresponds to the maximum symbolic sequence within the set of all admissible symbolic sequences for the future.

Definition 2.1: Given a Lozi map $\mathcal{L}$, its kneading sequence $K(\mathcal{L})$ is the itinerary for the future of I, i.e.

$$K(\mathcal{L}) = \underline{\delta}^s(I).$$

Definition 2.2: Let $(a, b) \in \Delta_{\mathcal{L}}$. Given a Lozi map $\mathcal{L}$, its pruned kneading sequence, $k_n(\mathcal{L})$, is the initial subsequence of symbols of $K(\mathcal{L})$ such that the first symbol $\bigstar$ appears at position n, that is, $\mathcal{L}^{n-1}(I)_x = 0$.

From the inverse map $\mathcal{L}^{-1}$, two maps can be introduced, $\mathcal{L}^{-1}_{+1}$ and $\mathcal{L}^{-1}_{-1}$, as follows:

$$\mathcal{L}^{-1}_{+1}(x, y) = \left(\frac{y}{b}, x - 1 + a\frac{y}{b}\right)$$

and

$$\mathcal{L}^{-1}_{-1}(x, y) = \left(\frac{y}{b}, x - 1 - a\frac{y}{b}\right).$$

Using these two maps and a pruned kneading sequence, $k_n(\mathcal{L})$, we will define a certain type of line segments, in $\nabla_{\mathcal{L}}$, as follows:

Definition 2.3: Let $(a,b) \in \Delta_{\mathcal{L}}$ and $\underline{\delta}^s = k_n(\mathcal{L}) = \delta_0\delta_1 \cdots \delta_{n-2}*$ for $n > 3$.
The line segments $\mathcal{S}^{\delta_i \cdots \delta_{n-2}*}$, with $2 \le i \le n-2$, in $\nabla_{\mathcal{L}}$, are defined by

$$\mathcal{S}^{\delta_{n-2}*} = \mathcal{L}^{-1}_{\delta_{n-2}}(\nabla_{\mathcal{L}} \cap \mathcal{V}^{\delta_{n-2}}) \cap \nabla_{\mathcal{L}};$$

$$\mathcal{S}^{\delta_{n-3}\delta_{n-2}*} = \mathcal{L}^{-1}_{\delta_{n-3}}(\mathcal{S}^{\delta_{n-2}*}) \cap \nabla_{\mathcal{L}};$$

$$\cdots$$

$$\mathcal{S}^{\delta_2 \cdots \delta_{n-2}*} = \mathcal{L}^{-1}_{\delta_2}(\mathcal{S}^{\delta_3 \cdots \delta_{n-2}*}) \cap \nabla_{\mathcal{L}}.$$

After the definition of the line segments $\mathcal{S}^{\delta_i \cdots \delta_{n-2}*}$ with $2 \le i \le n-2$, for $n > 3$, its possible to define the critical set for $\mathcal{L}$ when $(a,b) \in \Delta_{\mathcal{L}}$.

Definition 2.4: Let $(a,b) \in \Delta_{\mathcal{L}}$ and $\underline{\delta}^s = k_n(\mathcal{L})$. The critical set $\mathcal{C}_{\underline{\delta}^s}(\mathcal{L})$ is defined as the set containing the line segments $\mathcal{S}^{\delta_i \cdots \delta_{n-2}*}$, with $2 \le i \le n-2$, the vertical line segment $\mathcal{S}^* = \mathcal{V} \cap \nabla_{\mathcal{L}}$ and the points $\mathcal{L}(I)$ and I.

The following proposition presents the slope of each line segment $\mathcal{S}^{\delta_i \cdots \delta_{n-2}*}$, with $2 \le i \le n-2$ (for $n > 3$).

Proposition 2.5: *Let $(a,b) \in \Delta_{\mathcal{L}}$ and $\underline{\delta}^s = k_n(\mathcal{L}) = \delta_0\delta_1 \cdots \delta_{n-2}*$. The line segments $\mathcal{S}^{\delta_i \cdots \delta_{n-2}*}$, with $0 \le i \le n-2$, are given by*

$$\mathcal{S}^{\delta_i \cdots \delta_{n-2}*} = \nabla_{\mathcal{L}} \cap \left\{(x,y) : y = m_i(x - x_i) + y_i\right\},$$

where $(x_i, y_i) = \mathcal{L}^i(I)$ and

$$m_i = \begin{cases} a\delta_{n-2}, & i = n-2, \\ a\delta_i + \dfrac{b}{m_{i+1}}, & 0 \le i < n-2. \end{cases}$$

Proof: Let $\underline{\delta}^s = k_n(\mathcal{L}) = \delta_0\delta_1 \cdots \delta_{n-2}*$ be the pruned kneading sequence of $\mathcal{L}$ with dimension n.

Consider the jacobian matrix of $\mathcal{L}^{-1}$ at the point $X \in \mathbb{R}$, defined as follows:

$$D\mathcal{L}^{-1}_{+1} = \begin{bmatrix} 0 & \dfrac{1}{b} \\ 1 & \dfrac{a}{b} \end{bmatrix} \quad \text{if } X_y > 0$$

and

$$D\mathcal{L}^{-1}_{-1} = \begin{bmatrix} 0 & \dfrac{1}{b} \\ 1 & \dfrac{-a}{b} \end{bmatrix} \quad \text{if } X_y < 0,$$

where X_y corresponds to the y component of X.

Let $\vec{v} = (0,1)$ be the vector of the line segment $\mathcal{V}^{\delta_{n-2}}$. The vector of a line segment $\mathcal{L}^{-1}_{\delta_{n-2}}(\mathcal{V}^{\delta_{n-2}}) \cap \nabla_{\mathcal{L}}$ is given by

$$\vec{v}^{\delta_{n-2}} = \vec{v}\left(D\mathcal{L}^{-1}_{\delta_{n-2}}\right)^T = \begin{bmatrix} 0 & 1 \end{bmatrix} \begin{bmatrix} 0 & 1 \\ \dfrac{1}{b} & \dfrac{a\delta_{n-2}}{b} \end{bmatrix}$$

$$= \begin{bmatrix} \frac{1}{b} & \frac{a}{b}\delta_{n-2} \end{bmatrix}.$$

Depending on the value of δ_{n-2}, the point $\mathcal{L}^{n-1}(I)$ will belong to $\mathcal{V}^{+1}$ or $\mathcal{V}^{-1}$. By the definition of $\mathcal{L}_{\delta_{n-2}}^{-1}$, $\mathcal{S}^{\delta_{n-2}*} = \mathcal{L}_{\delta_{n-2}}^{-1}(\mathcal{V}^{\delta_{n-2}}) \cap \nabla_{\mathcal{L}}$ will be a line segment that passes through the point $\mathcal{L}^{n-2}(I)$, with vector $\vec{v}^{\delta_{n-2}} = (\frac{1}{b}, \frac{a}{b}\delta_{n-2})$ and slope $m_{n-2} = a\delta_{n-2}$.

Let $\vec{v}^{\delta_{n-2}} = (v_x, v_y)$. As

$$\begin{bmatrix} v_x & v_y \end{bmatrix} \begin{bmatrix} 0 & 1 \\ \frac{1}{b} & \frac{a\delta_{n-3}}{b} \end{bmatrix} = \begin{bmatrix} \frac{1}{b}v_y & v_x + \frac{a}{b}v_y\delta_{n-3} \end{bmatrix},$$

the slope of the line segment $\mathcal{S}^{\delta_{n-3}\delta_{n-2}*} = \mathcal{L}_{\delta_{n-3}}^{-1}(\mathcal{S}^{\delta_{n-2}*}) \cap \nabla_{\mathcal{L}}$ will be

$$\begin{aligned} m_{n-3} &= b\frac{v_x}{v_y} + a\delta_{n-3} \\ &= a\delta_{n-3} + \frac{b}{m_{n-2}}. \end{aligned}$$

Therefore, for $2 \le i \le n-2$, the line segment $\mathcal{S}^{\delta_i \cdots \delta_{n-2}*}$ that passes through the point $(x_i, y_i) = \mathcal{L}^i(I)$ has the vector given by

$$\vec{v}^{\delta_i \cdots \delta_{n-2}} = \vec{v}^{\delta_{i+1} \cdots \delta_{n-2}} \left(D\mathcal{L}_{\delta_i}^{-1} \right)^T$$

and slope given by

$$m_i = a\delta_i + \frac{b}{m_{i+1}}.$$

■

Proposition 2.6: *Let $(a, b) \in \Delta_{\mathcal{L}}$, $\underline{\delta}^s = k_n(\mathcal{L})$. The intersection of the line segment $\mathcal{S}^{\delta_i \cdots \delta_{n-2}*}$ with the x-axis is the point $(\xi_i, 0)$ where*

$$\xi_i = \begin{cases} 0 & \text{if } i = n-1. \\ \frac{1}{m_i}(1 - \xi_{i+1}) & \text{if } 2 \le i \le n-2. \end{cases}$$

Proof: The line segment $\mathcal{S}^*$ (when $i = n-1$) is the vertical line, and its intersection with the *x-axis* occurs at the point $(\xi_{n-1}, 0) = (0, 0)$. The pre-image of $(\xi_{n-1}, 0)$ is $\mathcal{L}_{\delta_{n-2}}^{-1}(0, 0) = (-1, 0)$. Let m_i be the slope of the line segments $\mathcal{S}^{\delta_i \cdots \delta_{n-2}*}$ with $2 \le i \le n-2$, for $n > 3$. Therefore, the line segment that passes through the points $(0, -1)$ and $\mathcal{L}^{n-2}(I)$ has slope m_{n-2} and can be given by $y = m_{n-2}x - 1$. Since this line segment contains the segment $\mathcal{S}^{\delta_{n-2}*}$, the intersection of the line segment $\mathcal{S}^{\delta_{n-2}*}$ with the *x-axis* occurs at the point $(\xi_{n-2}, 0) = (\frac{1}{m_{n-2}}, 0)$.

Continuing this process, the pre-image of the point $(\xi_{n-2}, 0)$ is the point $(0, \xi_{n-2} - 1)$. Therefore, the line segment $\mathcal{S}^{\delta_{n-3}\delta_{n-2}*}$ is contained in a line segment that passes through the point $(0, \xi_{n-2} - 1)$ with slope m_{n-3} and intersects the *x-axis* at the point $(\xi_{n-3}, 0) = (\frac{1}{m_{n-3}}(1 - \xi_{n-2}), 0)$. In general, the pre-image of a point $(\xi, 0)$ is the point $(0, \xi - 1)$ and consequently the line segment $\mathcal{S}^{\delta_i \cdots \delta_{n-2}*}$ with $2 \leq i \leq n-2$, for $n > 3$, intersects the *x-axis* at the point $(\xi_i, 0)$, where

$$\xi_i = \frac{1}{m_i}(1 - \xi_{i+1}) . \qquad \blacksquare$$

Considering Propositions 2.5 and 2.6, the next result follows.

Proposition 2.7: *Let $(a, b) \in \Delta_{\mathcal{L}}$ and $\underline{\delta}^s = k_n(\mathcal{L})$. The line segments $\mathcal{S}^{\delta_i \cdots \delta_{n-2}*}$, with $2 \leq i \leq n-2$, for $n > 3$, and the vertical line segment $\mathcal{S}^*$ do not intersect.*

2.1. Order relation in the set $\{-1, *, +1\}^{\mathbb{N}}$

From [4], the attractor points with the same fixed itinerary for the past $\underline{\varepsilon}^u$ lie in a segment L^u, called the unstable leaf, completely characterized by the parameter values (a, b) and $\underline{\varepsilon}^u$. Moreover, attractor points with the same fixed itinerary for the future $\underline{\varepsilon}^s$ lie in a segment L^s, called the stable leaf, also completely characterized by (a, b) and $\underline{\varepsilon}^s$. If a point $X \in \mathcal{A}_{\mathcal{L}}$ has a fixed itinerary $\underline{\varepsilon}$, X is the intersection of the line segments previously defined, denoted by $\mathsf{L}^u(X)$ and $\mathsf{L}^s(X)$, respectively.

In order to rewrite the order relation established by Ishii [4] for the set $\{-1, *, +1\}^{\mathbb{N}}$, consider the following definition.

Definition 2.8: Let $\underline{\varepsilon}^s$ and $\underline{\delta}^s$ be two symbolic subsequences for the future.
$\underline{\varepsilon}^s <_s \underline{\delta}^s$ if one of the following conditions is satisfied:

(1) $\varepsilon_0 < \delta_0$;
(2) $\varepsilon_0 \cdots \varepsilon_{i-1} = \delta_0 \cdots \delta_{i-1}$ and $\varepsilon_i \neq \delta_i$,:

(i) the number of $+1's$ in $\cdot\varepsilon_0 \cdots \varepsilon_{i-1}$ is even and $\varepsilon_i < \delta_i$;
(ii) the number of $+1's$ in $\cdot\varepsilon_0 \cdots \varepsilon_{i-1}$ is odd and $\varepsilon_i > \delta_i$;

where the order on the symbols is $-1 < * < +1$.
If $\varepsilon_i = \delta_i$ for all $i \geq 0$, $\underline{\varepsilon}^s = \underline{\delta}^s$.

From the previous definition it can be establish that, for any point $X \in \nabla_{\mathcal{L}}$,

$$\underline{\delta}^s(\mathcal{L}(I)) <_s \underline{\delta}^s(X) <_s \underline{\delta}^s(I) . \tag{8}$$

Considering the set $\{\sigma^i(\underline{\delta}^s)\}_{i=0}^{n-1}$, where σ is the shift-operator, the set $\{\xi_i\}_{i=2}^{n-2} \cup \{I_x, \mathcal{L}(I)_x\}$ and denoting by p as a permutation in the set $\{2, \ldots, n-1\}$ such that

$$\sigma(\underline{\delta}^s) <_s \sigma^{p(2)}(\underline{\delta}^s) <_s \cdots <_s \sigma^{p(n-1)}(\underline{\delta}^s) <_s \underline{\delta}^s \tag{9}$$

it follows that

$$\mathcal{L}(I)_x \langle \mathcal{S}^{\sigma^{p(2)}(\underline{\delta}^s)} \langle \cdots \langle \mathcal{S}^{\sigma^{p(n-1)}(\underline{\delta}^s)} \langle I_x, \tag{10}$$

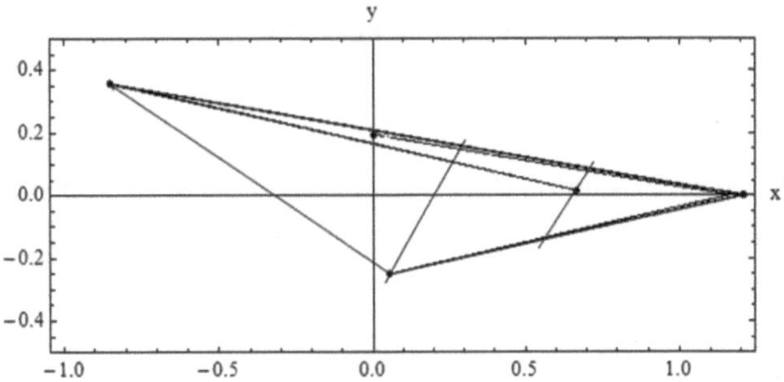

Figure 2. The critical set and the Lozi map attractor for the pruned kneading sequence $\delta^s = +1 - 1 + 1 + 1*$.

and where $A \langle B$ means that A is on the left side of B, and consequently

$$\mathcal{L}(I)_x < \xi_{p(2)} < \cdots < \xi_{p(n-1)} < I_x. \tag{11}$$

Figure 2 illustrates the critical set, the attractor and the invariant triangle $\nabla_{\mathcal{L}}$, considering the pruned kneading sequence $\underline{\delta}^s = +1 - 1 + 1 + 1*$.

2.2. *Markov partition on $\mathcal{A}_{\mathcal{L}}$*

Using the elements of the critical set $\mathcal{C}_{\underline{\delta}^s}(\mathcal{L})$, where $\underline{\delta}^s = k_n(\mathcal{L})$, it is possible to construct a partition, with $n-1$ regions, for the set $\nabla_{\mathcal{L}}$ and therefore to $\mathcal{A}_{\mathcal{L}}$.

Considering the Definition 2.8 and the order relations given in (9), (10) and (11) we define the $n-1$ regions on the triangle $\nabla_{\mathcal{L}}$ as follows:

let p be a permutation in the set $\{2, \ldots, n-1\}$ such that

$$\mathcal{L}(I)_x \langle \mathcal{S}^{\sigma^{p(2)}(\underline{\delta}^s)} \langle \cdots \langle \mathcal{S}^{\sigma^{p(n-1)}(\underline{\delta}^s)} \langle I_x.$$

The $n-1$ regions on the triangle $\nabla_{\mathcal{L}}$ are defined as follows:

$$R_1 = \left\{(x, y) \in \nabla_{\mathcal{L}} : (x, y) \ \langle \mathcal{S}^{\sigma^{p(2)}(\underline{\delta}^s)}\right\}$$

$$\vdots$$

$$R_k = \left\{(x, y) \in \nabla_{\mathcal{L}} : \mathcal{S}^{\sigma^{p(k)}(\underline{\delta}^s)} \ \underline{\langle} \ (x, y) \ \langle \mathcal{S}^{\sigma^{p(k+1)}(\underline{\delta}^s)}\right\}, \ k = 3, \ldots, n-2$$

$$\vdots$$

$$R_{n-1} = \left\{(x, y) \in \nabla_{\mathcal{L}} : \mathcal{S}^{\sigma^{p(n-1)}(\underline{\delta}^s)} \ \underline{\langle} \ (x, y)\right\},$$

where $A \ \underline{\langle} \ (x, y)$ means that (x, y) is on the right side of A or (x, y) belongs to A and $(x, y) \langle A$ means that (x, y) is on the left side of A.

Considering the previous construction the next result follows:

Theorem 2.9: *The set $\mathcal{P} = \{R_1, \ldots, R_{n-1}\}$ is a topological partition of $\nabla_{\mathcal{L}}$.*

Let X belong to the interior of a set R_i. Suppose that

$$\mathcal{L}(R_i) \subset R_j,$$

where R_j is single region or a union of two or more regions of the topological partition $\mathcal{P}$. Assume, without loss of generality, that it is a single region. Thus,

$$\mathcal{L}(X) \in R_j.$$

Moreover, by Ishii [4], the unstable leaf and the stable leaf of X lie in a segment denoted by $\mathsf{L}^u(X)$ and $\mathsf{L}^s(X)$, respectively, and X is the intersection point of the segments previously defined. Therefore, and due to the Lozi map linearity, we can state that

$$\mathcal{L}\left(L^u(X) \cap R_i\right) \supset L^u(\mathcal{L}(X)) \cap R_j$$

and

$$\mathcal{L}\left(L^s(X) \cap R_i\right) \subset L^s(\mathcal{L}(X)) \cap R_j.$$

Thus, by Bowen [2], it follows:

Theorem 2.10: *The topological partition $\{R_1, \ldots, R_{n-1}\}$ provides a Markov partition of $\nabla_{\mathcal{L}}$ associated to $\underline{\delta}^s = k_n(\mathcal{L})$.*

3. Transition matrix and topological entropy

Definition 3.1: Let $\{R_1, \ldots, R_{n-1}\}$ be a Markov partition of $\nabla_{\mathcal{L}}$ associated to $\underline{\delta}^s = k_n(\mathcal{L})$. The transition matrix over $\nabla_{\mathcal{L}}$ is a square matrix $A_{\underline{\delta}^s}$, of order $n-1$, such that

$$A_{\underline{\delta}^s} = \left(a_{ij}\right) = \begin{cases} 1 & \text{if } \mathcal{L}(R_i) \cap R_j \neq \emptyset \\ 0 & \text{if } \mathcal{L}(R_i) \cap R_j = \emptyset \end{cases}.$$

Following [8] and [1], the topological entropy of $\mathcal{L}$ can be calculated using the corresponding transition matrix. This result can be stated as follows:

Proposition 3.2: *Let $\underline{\delta}^s = k_n(\mathcal{L}) = \delta_0\delta_1 \cdots \delta_{n-2}*$ and $(a, b) \in \Delta_{\mathcal{L}}$. Let $A_{\underline{\delta}^s}$ be the transition matrix associated to $\underline{\delta}^s$. The topological entropy of $\mathcal{L}$ is given by*

$$h_{top}(\mathcal{L}) = \log\left(\lambda_{max}\left(A_{\underline{\delta}^s}\right)\right),$$

where $\lambda_{max}(A_{\underline{\delta}^s})$ is the spectral radius of $A_{\underline{\delta}^s}$.

Example 3.3: Let consider $(a, b) \approx (1.518, 0.185728)$ and the pruned kneading sequence associated

$$\underline{\delta}^s = +1 - 1 + 1 + 1*$$

The itinerary given by the pruned kneading sequence $\underline{\delta}$ for $(a, b) \approx (1.518, 0.185728)$ is given in Figure 3.

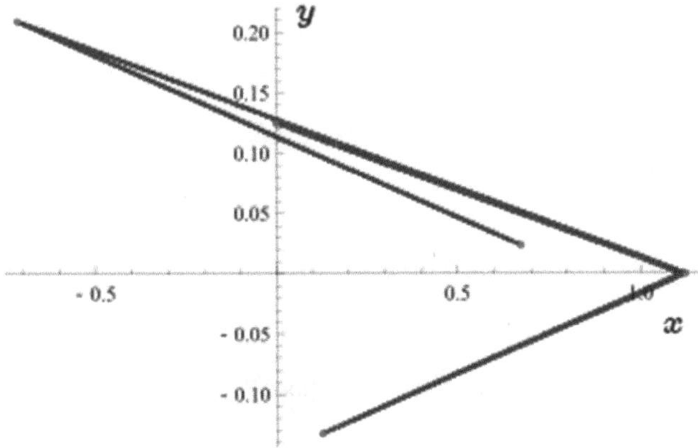

Figure 3. Itinerary given by the pruned kneading sequence and Lozi attractor for $(a, b) \approx (1.518, 0.185728)$.

Applying the shift map to the pruned kneading sequence $\underline{\delta}^s$, it follows:

$$\begin{aligned} \underline{\delta}^s &= +1 - 1 + 1 + 1* \\ \sigma\left(\underline{\delta}^s\right) &= -1 + 1 + 1* \\ \sigma^2\left(\underline{\delta}^s\right) &= +1 + 1* \\ \sigma^3\left(\underline{\delta}^s\right) &= +1* \\ \sigma^4\left(\underline{\delta}^s\right) &= * \end{aligned}$$

By Definition 2.8, the following order relation is valid

$$\sigma\left(\underline{\delta}^s\right) <_s \sigma^4\left(\underline{\delta}^s\right) <_s \sigma^2\left(\underline{\delta}^s\right) <_s \sigma^3\left(\underline{\delta}^s\right) <_s \underline{\delta}^s$$

and consequently

$$\mathcal{L}(I)_x \langle \mathcal{S}^* \langle \mathcal{S}^{+1+1*} \langle \mathcal{S}^{+1*} \langle I_x.$$

The partition of $\nabla_{\mathcal{L}}$ is given through the regions (Figure 4)

$$\begin{aligned} R_1 &= \left\{(x, y) \in \nabla_{\mathcal{L}} : (x, y) \langle \mathcal{S}^*\right\} \\ R_2 &= \left\{(x, y) \in \nabla_{\mathcal{L}} : \mathcal{S}^* \underline{\langle} (x, y) \langle \mathcal{S}^{+1+1*}\right\} \\ R_3 &= \left\{(x, y) \in \nabla_{\mathcal{L}} : \mathcal{S}^{+1+1*} \underline{\langle} (x, y) \langle \mathcal{S}^{+1*}\right\} \\ R_4 &= \left\{(x, y) \in \nabla_{\mathcal{L}} : \mathcal{S}^{+1*} \underline{\langle} (x, y)\right\} \end{aligned}$$

which can be associated to the following transition matrix $A_{\underline{\delta}^s}$ given by

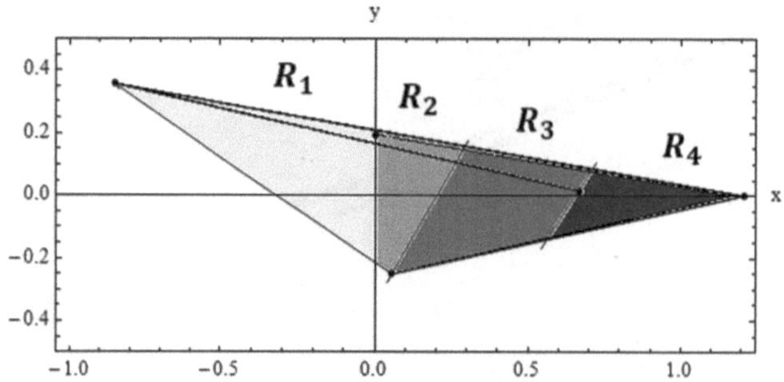

Figure 4. Itinerary of point I, the critical set and the regions $R_j, j = 1, 2, 3, 4$.

$$A_{\underline{\delta}^s} = \begin{bmatrix} 0 & 0 & 1 & 1 \\ 0 & 0 & 0 & 1 \\ 0 & 1 & 1 & 0 \\ 1 & 0 & 0 & 0 \end{bmatrix}.$$

The transition matrix $A_{\underline{\delta}^s}$ has maximum eigenvalue $\lambda_{\max} \approx 1.51288$ and the topological entropy is approximately

$$h_{top} \approx \log(1.51288) \approx 0.41402.$$

Example 3.4: Consider $(a, b) \approx (1.887, 0.0957356)$ and the pruned kneading sequence associated

$$\underline{\delta}^s = +1 - 1 - 1 - 1 - 1 + 1 - 1\, *.$$

Applying the shift map to $\underline{\delta}^s$

$$\begin{aligned} \underline{\delta}^s &= +1 - 1 - 1 - 1 - 1 + 1 - 1* \\ \sigma(\underline{\delta}^s) &= -1 - 1 - 1 - 1 + 1 - 1* \\ \sigma^2(\underline{\delta}^s) &= -1 - 1 - 1 + 1 - 1* \\ \sigma^3(\underline{\delta}^s) &= -1 - 1 + 1 - 1* \\ \sigma^4(\underline{\delta}^s) &= -1 + 1 - 1* \\ \sigma^5(\underline{\delta}^s) &= +1 - 1* \\ \sigma^6(\underline{\delta}^s) &= -1* \\ \sigma^7(\underline{\delta}^s) &= * \end{aligned}$$

that verifies the following order relationship:

$$\sigma(\underline{\delta}^s) <_s \sigma^2(\underline{\delta}^s) <_s \sigma^3(\underline{\delta}^s) <_s \sigma^6(\underline{\delta}^s) <_s \sigma^4(\underline{\delta}^s) <_s \sigma^7(\underline{\delta}^s) <_s \sigma^5(\underline{\delta}^s) <_s \underline{\delta}^s$$

and

$$\mathcal{L}(I)_x \,\langle S^{\sigma^2(\underline{\delta}^s)} \langle S^{\sigma^3(\underline{\delta}^s)} \langle S^{\sigma^6(\underline{\delta}^s)} \langle S^{\sigma^4(\underline{\delta}^s)} \langle S^{\sigma^7(\underline{\delta}^s)} \langle S^{\sigma^5(\underline{\delta}^s)} \langle I_x.$$

The partition of $\nabla_{\mathcal{L}}$ is given through the regions (Figure 5)

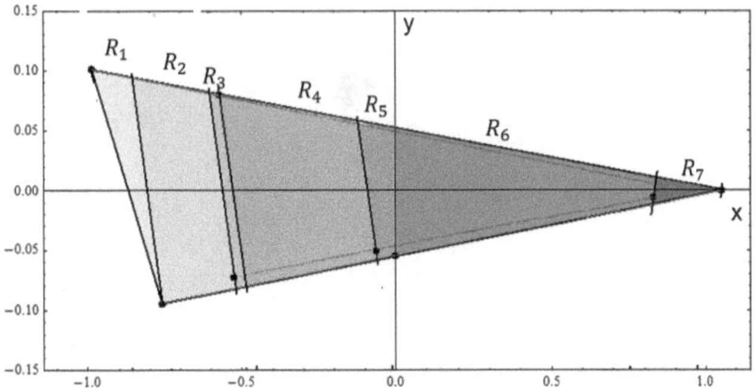

Figure 5. Itinerary and the partition associated to the pruned kneading sequence $\delta^s = +1-1-1-1-1+1-1*$ when $(a,b) \approx (1.887, 0.0957356)$.

$$R_1 = \left\{(x,y) \in \nabla_{\mathcal{L}} : (x,y) \langle \mathcal{S}^{\sigma^2(\underline{\delta}^s)}\right\},$$
$$R_2 = \left\{(x,y) \in \nabla_{\mathcal{L}} : \mathcal{S}^{\sigma^2(\underline{\delta}^s)} \underline{\langle} \ (x,y) \langle \mathcal{S}^{\sigma^3(\underline{\delta}^s)}\right\},$$
$$R_3 = \left\{(x,y) \in \nabla_{\mathcal{L}} : \mathcal{S}^{\sigma^3(\underline{\delta}^s)} \underline{\langle} \ (x,y) \langle \mathcal{S}^{\sigma^6(\underline{\delta}^s)}\right\},$$
$$R_4 = \left\{(x,y) \in \nabla_{\mathcal{L}} : \mathcal{S}^{\sigma^6(\underline{\delta}^s)} \underline{\langle} \ (x,y) \langle \mathcal{S}^{\sigma^4(\underline{\delta}^s)}\right\},$$
$$R_5 = \left\{(x,y) \in \nabla_{\mathcal{L}} : \mathcal{S}^{\sigma^4(\underline{\delta}^s)} \underline{\langle} \ (x,y) \langle \mathcal{S}^{\sigma^7(\underline{\delta}^s)}\right\},$$
$$R_6 = \left\{(x,y) \in \nabla_{\mathcal{L}} : \mathcal{S}^{\sigma^7(\underline{\delta}^s)} \underline{\langle} \ (x,y) \langle \mathcal{S}^{\sigma^5(\underline{\delta}^s)}\right\},$$
$$R_7 = \left\{(x,y) \in \nabla_{\mathcal{L}} : \mathcal{S}^{\sigma^5(\underline{\delta})} \underline{\langle} \ (x,y)\right\},$$

which can be associated to the following transition matrix $A_{\underline{\delta}^s}$:

$$A_{\underline{\delta}^s} = \begin{bmatrix} 0 & 1 & 0 & 0 & 0 & 0 & 0 \\ 0 & 0 & 1 & 1 & 0 & 0 & 0 \\ 0 & 0 & 0 & 0 & 1 & 0 & 0 \\ 0 & 0 & 0 & 0 & 0 & 1 & 0 \\ 0 & 0 & 0 & 0 & 0 & 0 & 1 \\ 0 & 0 & 0 & 1 & 1 & 1 & 1 \\ 1 & 1 & 1 & 0 & 0 & 0 & 0 \end{bmatrix}.$$

The transition matrix $A_{\underline{\delta}^s}$ has a maximum eigenvalue $\lambda_{\max} \approx 1.9363$ and the topological entropy is approximately

$$h_{top} = \log(1.9363) \approx 0.66078.$$

3.1. Isentropics curves for Lozi maps

From [3], a pruned kneading sequence corresponds to an algebraic condition on the parameters and, therefore, to a curve on the parameter plane. For example, a simple computation shows that the kneading curve for the pruned kneading sequence $\underline{\delta}^s = +1-1\star$

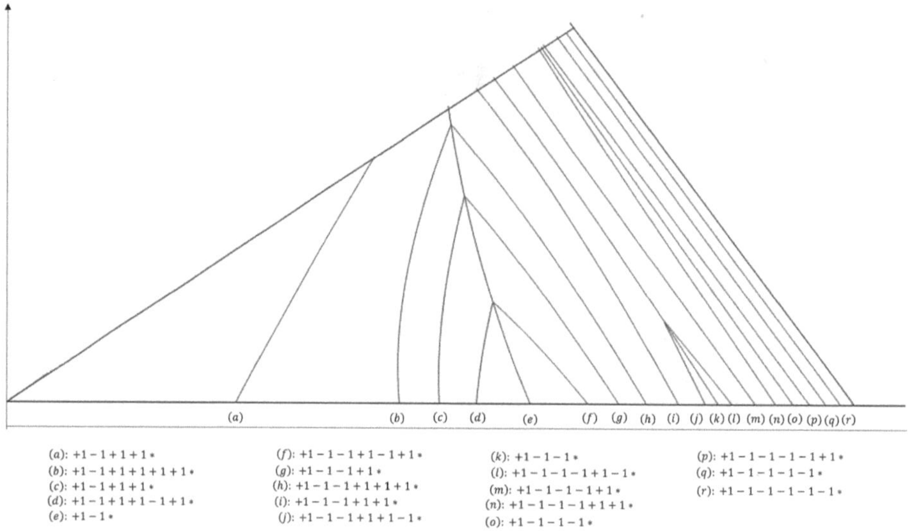

Figure 6. Isentropic curves associated to a pruned kneading sequence.

is described by the following condition:

$$2 + 4a - a^3 - ab - a^2\sqrt{a^2 + 4b} + b\sqrt{a^2 + 4b} = 0.$$

Definition 3.5: Given a pruned kneading sequence $\underline{\delta}^s$ of dimension n, its kneading curve $\Gamma(\underline{\delta}^s)$ is the parameter space curve corresponding to parameter values such that $k_n(\mathcal{L}) = \underline{\delta}^s$.

Following the work [3], and considering Proposition 3.2, the topological entropy of $\mathcal{L}$ along the curve $\Gamma(\underline{\delta}^s)$ is equal to $\log(a_{\underline{\delta}^s})$, where $(a_{\underline{\delta}^s}, 0)$ is the point of the horizontal axis given by the limit, as $b \to 0$, of $\Gamma(\underline{\delta}^s)$ and therefore

$$0 \leq h_{top}(\mathcal{L}) \leq \log(2).$$

Figure 6 illustrates the kneading curves (isentropic curves) for pruned kneading sequence up to dimension 7.

Acknowledgments

The author expresses his gratitude to Ricardo Severino and Sandra Vinagre for all their support and encouragement in obtaining these results.

Disclosure statement

No potential conflict of interest was reported by the author(s).

Funding

This work was partially supported by projects UIBD/00308/2020 and UIDB/04-674/2020, by FCT-Portuguese Foundation for Science and Technology

ORCID

Diogo Baptista http://orcid.org/0000-0001-5997-5302

References

[1] V.M. Alekseev and M.V. Yakobson, Symbolic dynamics and hyperbolic dynamic systems, Phys. Rep.75 (1981), pp. 290–325.
[2] Rufus Bowen, Markov partitions for axiom A diffeomorphisms, Amer. J. Math. 92 (1970), pp. 725–747.
[3] Diogo Baptista, Ricardo Severino, and Sandra Vinagre, Kneading curves for Lozi maps, Grazer Math. Ber. 354 (2009), pp. 6–14.
[4] Y. Ishii, Towards a kneading theory for Lozi mappings I: A solution of pruning front conjecture and the first tangency problem, Nonlinearity 10 (1997), pp. 731–747.
[5] R. Lozi, Un attracteur étrange du type attracteur de Hénon, J. Physique (Paris) 39 (1978), pp. 9–10.
[6] Z. Liu, H. Xie, Z. Zhu, and Q. Lu, The strange attractor of the Lozi mapping, Int. J. Bifur. Chaos Appl. Sci. Engrg. 2 (1992), pp. 831–839.
[7] M. Misiurewicz, *Strange attractors for the Lozi mappings*, in *Nonlinear Dynamics*, R.G. Helleman, ed., The New York Academy of Sciences, New York, 1980.
[8] Peter Walters, *An introduction to ergodic theory*, in *Graduate Texts in Mathematics*, Vol. 79, Springer-Verlag, New York, 1982.

Lozi map embedded into the 2D border collision normal form

Irina Sushko, Viktor Avrutin and Laura Gardini

ABSTRACT
A 2D piecewise linear continuous two-parameter map known as the Lozi map is a special case of the 2D border collision normal form depending on four parameters. In the present paper, we investigate how the bifurcation structure of the Lozi map is incorporated into the bifurcation structure of the 2D border collision normal form using an analytical representation of the boundaries of the largest periodicity regions related to the cycles with rotation number $1/n$, $n \geqslant 3$. At the centre bifurcation boundary of the stability domain of the fixed point both maps are conservative which leads to a quite intricate bifurcation structure near this boundary.

1. Introduction

Various bifurcation phenomena in non-smooth dynamical systems are nowadays actively studied from both applied and theoretical points of view (see, e.g. [8,10,12,14,23,28,38]). Border collision bifurcations, which are characteristic for piecewise smooth maps, occur when an invariant set of such a map collides with a border (switching manifold) between the partitions in the state space associated with different smooth functions. In continuous maps, in particular, in one-dimensional (1D) and two-dimensional (2D) maps, border collision bifurcations can be investigated by means of related normal forms[1] [29,30]. For 1D maps, this normal form is the skew tent map – a 1D piecewise linear continuous map defined by two linear functions – whose dynamics depending on the slopes of the linear branches has been completely described (see, e.g. [3]). For 2D maps, the border collision normal form is a 2D continuous map F defined by two linear maps [7,29,32,36]. This map depends on four parameters (traces $\tau_{\mathcal{L}}$, $\tau_{\mathcal{R}}$, and determinants $\delta_{\mathcal{L}}$, $\delta_{\mathcal{R}}$ of the Jacobian matrices of two linear maps defined in two partitions separated by the axes $x = 0$), and possesses a quite interesting and complicated dynamics which is not yet completely studied. In fact, besides border collision bifurcations of various kinds, in map F one can observe robust chaotic attractors [9], dangerous bifurcations [5,17,18,20], multiple-attractor bifurcations [4,15,31], degenerate bifurcations [2,37] and other non-standard phenomena, leading to intricate bifurcation structures in the parameter space.

A special attention has been paid to the bifurcation structure in the parameter space of map F near the boundary of the stability domain of its fixed point, associated with a centre bifurcation [19,32,33,36,39]. Recall that a centre bifurcation is a degenerate Neimark-Sacker bifurcation, occurring if a pair of complex-conjugate eigenvalues of a fixed point or cycle of a piecewise smooth map crosses the unit circle in case when the related non-degeneracy conditions are violated [36]. In particular, in [35] we investigated the bifurcation structure near the centre bifurcation boundary in the parameter plane of the well-known Lozi map [25], denoted below as map G which is a special case of the border collision normal form F with $\delta_{\mathcal{L}} = \delta_{\mathcal{R}}$ and $\tau_{\mathcal{L}} = -\tau_{\mathcal{R}}$. This case is interesting because at the centre bifurcation boundary map G is conservative [1,11,24], and near this boundary the fixed point can coexist with many attractors, including attracting cycles of various periods as well as chaotic attractors. Recall that the Lozi map is a 2D piecewise linear map associated with the Hénon map [21], and among 2D maps it is one of the most studied. Note, however, that the researchers were mainly interested in a part of its parameter plane related to two saddle fixed points and chaotic dynamics [6,13,16,22,26,27].

To some extent, the present paper is a follow-up to our previous work [35]. Our aim is to show how the bifurcation structure near the centre bifurcation boundary of the fixed point of the Lozi map G, as described in [35], is incorporated into the bifurcation structure of the 2D border collision normal form F. To this end, in Section 2, after introducing map F and map G, we present some examples of bifurcation structures in these maps, associated with centre bifurcations of their fixed points. Then, in Section 3, for map F in case $\delta_{\mathcal{L}} = \delta_{\mathcal{R}}$, we obtain an analytical representation of the main periodicity regions related to attracting n-cycles with rotation number[2] $1/n$ (similar results for map F depending on all four parameters are presented in the Appendix). In Section 4, we analyse how the bifurcation structure of the Lozi map G is embedded into the bifurcation structure of map F. Section 5 concludes.

2. Bifurcation structure of the parameter space near the centre bifurcation

The 2D border collision normal form $F : \mathbb{R}^2 \to \mathbb{R}^2$ is defined by two linear maps, $F_{\mathcal{L}}$ and $F_{\mathcal{R}}$, in two half-planes, $D_{\mathcal{L}} = \{(x,y) \mid x \leqslant 0\}$ and $D_{\mathcal{R}} = \{(x,y) \mid x \geqslant 0\}$, as follows:

$$F : \begin{pmatrix} x \\ y \end{pmatrix} \mapsto \begin{cases} F_{\mathcal{L}}(x,y) & \text{if } (x,y) \in D_{\mathcal{L}} \\ F_{\mathcal{R}}(x,y) & \text{if } (x,y) \in D_{\mathcal{R}} \end{cases} \tag{1a}$$

with

$$F_{\mathcal{L}} : \begin{pmatrix} x \\ y \end{pmatrix} \mapsto \underbrace{\begin{pmatrix} \tau_{\mathcal{L}} & 1 \\ -\delta_{\mathcal{L}} & 0 \end{pmatrix}}_{\mathcal{A}_{\mathcal{L}}} \begin{pmatrix} x \\ y \end{pmatrix} + \begin{pmatrix} \mu \\ 0 \end{pmatrix} \tag{1b}$$

$$F_{\mathcal{R}} : \begin{pmatrix} x \\ y \end{pmatrix} \mapsto \underbrace{\begin{pmatrix} \tau_{\mathcal{R}} & 1 \\ -\delta_{\mathcal{R}} & 0 \end{pmatrix}}_{\mathcal{A}_{\mathcal{R}}} \begin{pmatrix} x \\ y \end{pmatrix} + \begin{pmatrix} \mu \\ 0 \end{pmatrix} \tag{1c}$$

Here, μ, $\tau_{\mathcal{L}}$, $\tau_{\mathcal{R}}$, $\delta_{\mathcal{L}}$, $\delta_{\mathcal{R}}$ are real parameters with $\tau_{\mathcal{L}}$, $\tau_{\mathcal{R}}$ being the traces and $\delta_{\mathcal{L}}$, $\delta_{\mathcal{R}}$ the determinants of the matrices $\mathcal{A}_{\mathcal{L}}$ and $\mathcal{A}_{\mathcal{R}}$, respectively.

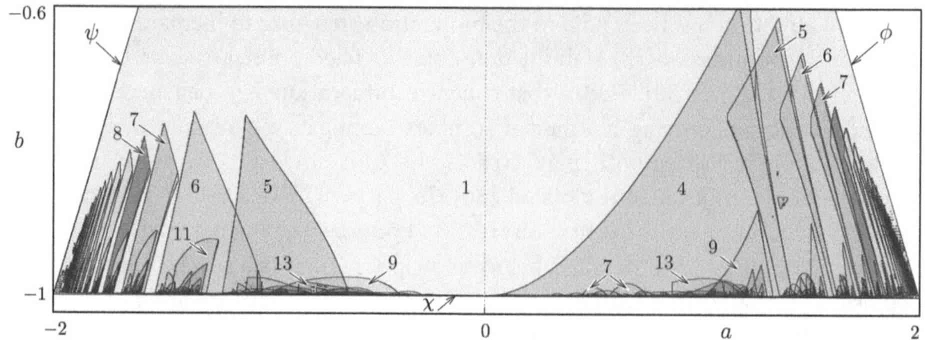

Figure 1. Lozi map. Bifurcations structure (obtained numerically) of the (a, b)-parameter plane near the centre bifurcation line $b = -1$. Some of the periods are marked; the main n-periodicity regions overlapping with the stability domain of the fixed point are shown in different colours for $n = 4, \ldots, 8$; other kinds of coexistence are marked by different tonalities of orange; the boundaries of the stability region of the fixed point $\mathcal{O}_{\mathcal{R}}$ given by the lines $\phi : b = 1 - a$ (degenerate transcritical bifurcation), $\psi : b = 1 + a$ (degenerate flip bifurcation) and $\chi : b = -1$ (center bifurcation) are also indicated.

Since μ is a scaling factor, considering the case $\mu > 0$ we can fix $\mu = 1$. In a special case of map F with $\delta_{\mathcal{L}} = \delta_{\mathcal{R}}, \tau_{\mathcal{L}} = -\tau_{\mathcal{R}}$, introducing the notations $\tau_{\mathcal{L}} = -\tau_{\mathcal{R}} = a, \delta_{\mathcal{L}} = \delta_{\mathcal{R}} = -b$, we obtain the well-known Lozi map:

$$G : \begin{pmatrix} x \\ y \end{pmatrix} \mapsto \begin{cases} G_{\mathcal{L}}(x, y) & \text{if } (x, y) \in D_{\mathcal{L}} \\ G_{\mathcal{R}}(x, y) & \text{if } (x, y) \in D_{\mathcal{R}} \end{cases} \tag{2a}$$

where

$$G_{\mathcal{L}} : \begin{pmatrix} x \\ y \end{pmatrix} \mapsto \begin{pmatrix} a & 1 \\ b & 0 \end{pmatrix} \begin{pmatrix} x \\ y \end{pmatrix} + \begin{pmatrix} 1 \\ 0 \end{pmatrix} \tag{2b}$$

$$G_{\mathcal{R}} : \begin{pmatrix} x \\ y \end{pmatrix} \mapsto \begin{pmatrix} -a & 1 \\ b & 0 \end{pmatrix} \begin{pmatrix} x \\ y \end{pmatrix} + \begin{pmatrix} 1 \\ 0 \end{pmatrix} \tag{2c}$$

In [35], we investigated the bifurcation structure of the (a, b)-parameter plane of the Lozi map near the centre bifurcation line $b = -1$, at which the map is conservative. This structure is illustrated in Figure 1. Its characteristic feature is that for the parameter region $\{(a, b) \mid b < 1 - a, b < 1 + a, b > -1\}$, the attracting fixed point

$$\mathcal{O}_{\mathcal{R}} = \left(\frac{1}{1 + a - b}, \frac{b}{1 + a - b} \right) \tag{3}$$

can coexist with one or several attracting cycles of various periods, as well as with chaotic attractors. Moreover, the number of coexisting attractors increases when the parameter point approaches the centre bifurcation line. A few examples of the coexisting attractors together with their basins are shown in Figure 2, where in (a) an attracting fixed point coexists with attracting 6-, 11-, 16-cycles, while in (b) the parameters are slightly moved in such a way that the 16-cycle undergoes a degenerate flip bifurcation leading to a 32-band chaotic attractor which then, after a homoclinic bifurcation, transforms into a 16-band chaotic attractor (the inset in Figure 2(b) shows one band of this attractor magnified).

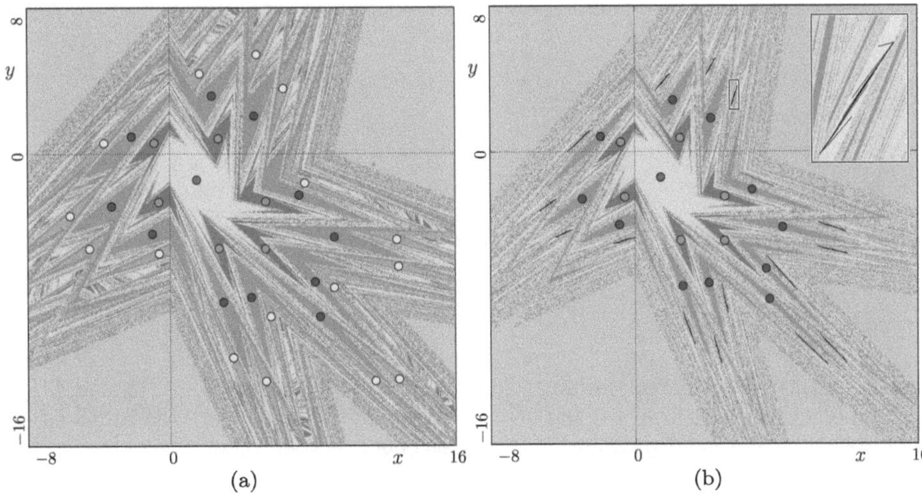

Figure 2. Lozi map. Basins of attracting fixed point and 6-, 11-, 16-cycles in (a), 6-, 11-cycles and 16-band chaotic attractor in (b). Here $a = -1.3$, $b = -0.97$ in (a), $a = -1.287$, $b = -0.97$ in (b). Gray regions correspond to diverging trajectories.

A complete description of the bifurcation structure of the Lozi map is a challenging task, and as a first step in this direction, in [35] we obtain analytical representation of the boundaries of the so-called main periodicity regions related to attracting cycles with rotation numbers $1/n$, $n \geq 4$, and symbolic sequences $\mathcal{L}^2\mathcal{R}^{n-2}$ and $\mathcal{R}\mathcal{L}^{n-1}$ (see Figure 3). These boundaries correspond to border collision, degenerated transcritical, degenerate flip and centre bifurcations (for a description of the degenerate bifurcations see [37]). As mentioned in the Introduction, our aim in the present paper is to discuss how the bifurcation structure of the Lozi map is embedded into the bifurcation structure of the border collision normal form F. Since the map G satisfies $\det G_{\mathcal{L}} = \det G_{\mathcal{R}}$, it is reasonable to consider a special case of map F with $\det F_{\mathcal{L}} = \det F_{\mathcal{R}} := \delta$. An additional advantage is that in this way we have a 3D parameter space $(\tau_{\mathcal{L}}, \tau_{\mathcal{R}}, \delta)$ which is more easy to visualize. An example is shown in Figure 4, with a horizontal cross-section corresponding to $\delta = 0.8$ and a vertical cross-section related to $\tau_{\mathcal{L}} = -\tau_{\mathcal{R}}$, i.e. to the parameter plane of the Lozi map. In the next section, we give analytical representations of the boundaries of the main periodicity regions of map F in case $\delta_{\mathcal{L}} = \delta_{\mathcal{R}} = \delta$ and explain how the symmetry of the bifurcation surfaces in the $(\tau_{\mathcal{L}}, \tau_{\mathcal{R}}, \delta)$-parameter space with respect to the plane $\tau_{\mathcal{L}} = \tau_{\mathcal{R}}$ can be used to simplify calculations. Note that for $\delta_{\mathcal{L}} = \delta_{\mathcal{R}}$, $\tau_{\mathcal{L}} = \tau_{\mathcal{R}}$ map F is linear.

3. Main periodicity regions

Let parameters of map F with $\delta_{\mathcal{L}} = \delta_{\mathcal{R}} = \delta$ belong to the following domain:

$$S = \{(\tau_{\mathcal{L}}, \tau_{\mathcal{R}}, \delta) \mid |\tau_{\mathcal{R}}| < 1 + \delta,\ |\tau_{\mathcal{L}}| < 1 + \delta,\ 0 < \delta < 1\} \tag{4}$$

For $(\tau_{\mathcal{L}}, \tau_{\mathcal{R}}, \delta) \in S$, map F has an attracting fixed point

$$\mathcal{O}_{\mathcal{R}} = \left(\frac{1}{1 - \tau_{\mathcal{R}} + \delta}, \frac{-\delta}{1 - \tau_{\mathcal{R}} + \delta}\right) \tag{5}$$

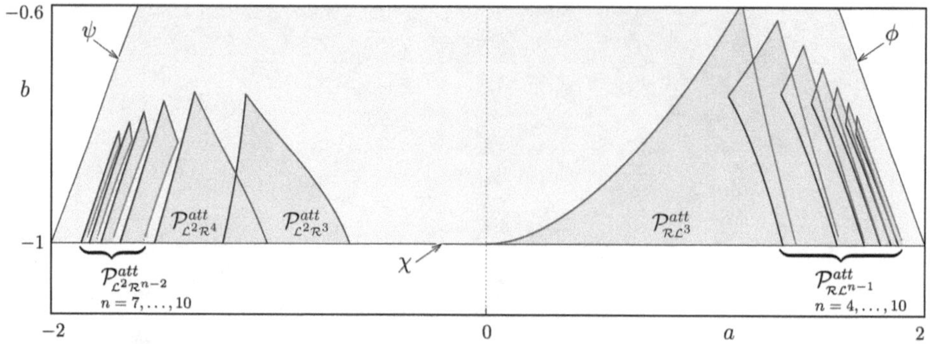

Figure 3. Lozi map. Stability regions (obtained analytically) $\mathcal{P}^{att}_{\mathcal{L}^2\mathcal{R}^{n-2}}$, $n = 5, \ldots, 10$, and $\mathcal{P}^{att}_{\mathcal{R}\mathcal{L}^{n-1}}$, $n = 4, \ldots, 10$, of the cycles $\mathcal{O}_{\mathcal{L}^2\mathcal{R}^{n-2}}$ and $\mathcal{O}_{\mathcal{R}\mathcal{L}^{n-1}}$ with rotation numbers $1/n$. The border collision, degenerate transcritical and degenerate flip bifurcation boundaries are shown in red, blue and green, respectively.

which can coexist with other attractors. At the boundary of S defined by $\delta = 1$ and related to the centre bifurcation of $\mathcal{O}_{\mathcal{R}}$, map F is conservative. As illustrated in Figure 4, near this boundary the bifurcation structure is quite complicated, being characterized by many overlapping regions associated with attracting cycles. In this section, we consider two families of the largest periodicity regions associated with attracting cycles $\mathcal{O}_{\mathcal{L}^2\mathcal{R}^{n-2}}$ (with their complementary saddle cycles $\mathcal{O}_{\mathcal{L}\mathcal{R}^{n-1}}$) and $\mathcal{O}_{\mathcal{R}\mathcal{L}^{n-1}}$ (with complementary saddle cycles $\mathcal{O}_{\mathcal{R}^2\mathcal{L}^{n-2}}$). Since the pair of 3-cycles $\mathcal{O}_{\mathcal{L}^2\mathcal{R}^n}$ and $\mathcal{O}_{\mathcal{R}^2\mathcal{L}}$, as well as the 4-cycle $\mathcal{O}_{\mathcal{L}^2\mathcal{R}^2}$, belong to both families, they are treated separately, while below we consider the generic case $n \geqslant 5$.

3.1. *First family of cycles:* $\mathcal{O}_{\mathcal{L}\mathcal{R}^{n-1}}$ *and* $\mathcal{O}_{\mathcal{L}^2\mathcal{R}^{n-2}}$

We first obtain the existence regions for the cycles $\mathcal{O}_{\mathcal{L}\mathcal{R}^{n-1}}$ and $\mathcal{O}_{\mathcal{L}^2\mathcal{R}^{n-2}}$, $n \geqslant 5$, as well as their stability conditions. Below, these results are presented in short, given that they are obtained following a similar way of reasoning as it is done for the Lozi map in [35] (see also [36]).

Let $(x'_0, y'_0) \in D_{\mathcal{L}}$, i.e. $x'_0 \leqslant 0$, and $(x'_i, y'_i) \in D_{\mathcal{R}}$, i.e. $x'_i \geqslant 0$, $i = 1, \ldots, n-1$, be the points of a cycle $\mathcal{O}_{\mathcal{L}\mathcal{R}^{n-1}}$ of map F. Under parameter variation, these inequalities may be violated either via a degenerate transcritical or via a border collision bifurcation. In the latter case, one or several periodic points collide with the border $\{(x, y) \mid x = 0\}$ between the partitions $D_{\mathcal{L}}$ and $D_{\mathcal{R}}$, that may lead either to disappearance of the cycle or to a change of its symbolic sequence. Since we consider a cycle with rotation number $1/n$, the points which may collide with the border under parameter variation are (x'_0, y'_0), (x'_1, y'_1) and (x'_{n-1}, y'_{n-1}). Here x'_0, x'_1 and x'_{n-1} can be obtained from $(x'_0, y'_0) = F^{n-1}_{\mathcal{R}} \circ F_{\mathcal{L}}(x'_0, y'_0)$, $(x'_1, y'_1) = F_{\mathcal{L}} \circ F^{n-1}_{\mathcal{R}}(x'_1, y'_1)$ and $(x'_{n-1}, y'_{n-1}) = F^{n-2}_{\mathcal{R}} \circ F_{\mathcal{L}} \circ F_{\mathcal{R}}(x'_{n-1}, y'_{n-1})$, respectively. In terms of the parameters, the conditions $x'_0 \leqslant 0$, $x'_1 \geqslant 0$ and $x'_{n-1} \geqslant 0$ can be expressed as follows:

$$x'_0 = \frac{\Pi_{\mathcal{R}^n}(1)}{\Pi_{\mathcal{L}\mathcal{R}^{n-1}}(1)\Pi_{\mathcal{R}}(1)} \leqslant 0 \tag{6}$$

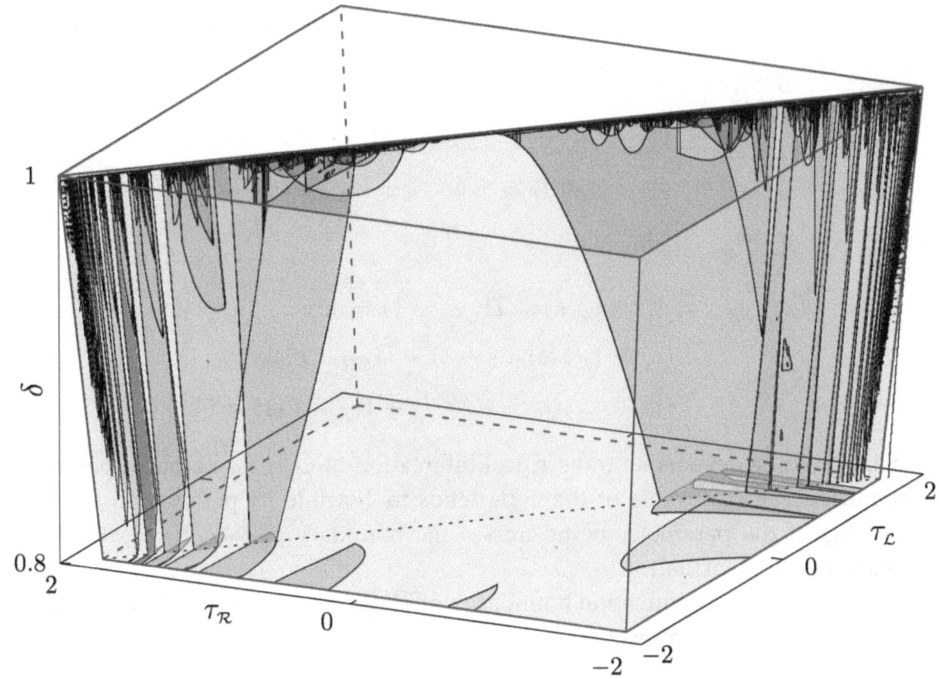

Figure 4. Map F for $\delta_{\mathcal{L}} = \delta_{\mathcal{R}} = \delta$. Two cross-sections of the $(\tau_{\mathcal{L}}, \tau_{\mathcal{R}}, \delta)$-parameter space near the centre bifurcation plane defined by $\delta = 1$: horizontal cross-section corresponds to $\delta = 0.8$ and vertical cross-section is related to $\tau_{\mathcal{L}} = -\tau_{\mathcal{R}}$, i.e. to the Lozi map.

$$x_1' = \frac{(\tau_{\mathcal{L}} - \tau_{\mathcal{R}})\,(\delta r_{n-2} + 1) + \Pi_{\mathcal{L}\mathcal{R}^{n-1}}(1)}{\Pi_{\mathcal{L}\mathcal{R}^{n-1}}(1)\Pi_{\mathcal{R}}(1)} \geqslant 0 \tag{7}$$

and

$$x_{n-1}' = \frac{(\tau_{\mathcal{L}} - \tau_{\mathcal{R}})(r_{n-2} + \delta^{n-1}) + \Pi_{\mathcal{L}\mathcal{R}^{n-1}}(1)}{\Pi_{\mathcal{L}\mathcal{R}^{n-1}}(1)\Pi_{\mathcal{R}}(1)} \geqslant 0 \tag{8}$$

where r_k is obtained from

$$r_k = \tau_{\mathcal{R}} r_{k-1} - \delta r_{k-2} \quad \text{with} \quad r_0 = 1,\ r_1 = \tau_{\mathcal{R}},\ k = 2, \ldots \tag{9}$$

and

$$\begin{aligned}
\Pi_{\mathcal{R}}(\lambda) &= \lambda^2 - \tau_{\mathcal{R}}\lambda + \delta \\
\Pi_{\mathcal{R}^n}(\lambda) &= \lambda^2 - (r_n - \delta r_{n-2})\lambda + \delta^n \\
\Pi_{\mathcal{L}\mathcal{R}^{n-1}}(\lambda) &= \lambda^2 - (\tau_{\mathcal{L}} r_{n-1} - 2\delta r_{n-2})\lambda + \delta^n
\end{aligned}$$

are the characteristic polynomials of the matrices $\mathcal{A}_{\mathcal{R}}$, $\mathcal{A}_{\mathcal{R}}^n$ and $\mathcal{A}_{\mathcal{L}}\mathcal{A}_{\mathcal{R}}^{n-1}$, respectively.

In the considered case $(\tau_{\mathcal{L}}, \tau_{\mathcal{R}}, \delta) \in S$, we have that $\Pi_{\mathcal{R}}(1) = 1 - \tau_{\mathcal{R}} + \delta > 0$ and also $\Pi_{\mathcal{R}^n}(1) > 0$. Thus, to satisfy (6), it must hold $\Pi_{\mathcal{L}\mathcal{R}^{n-1}}(1) < 0$, and to satisfy (7) and (8) the corresponding numerators in (7) and (8) must be negative or equal to 0. Therefore, the existence region $\mathcal{P}^{ex}_{\mathcal{LR}^{n-1}}$ of the cycle $\mathcal{O}_{\mathcal{LR}^{n-1}}$ can be defined as follows:

$$\begin{aligned}\mathcal{P}^{ex}_{\mathcal{LR}^{n-1}} = \{(\tau_{\mathcal{L}}, \tau_{\mathcal{R}}, \delta) \mid \Pi_{\mathcal{LR}^{n-1}}(1) < 0,\\ (\tau_{\mathcal{L}} - \tau_{\mathcal{R}})(\delta r_{n-2} + 1) + \Pi_{\mathcal{LR}^{n-1}}(1) \leqslant 0,\\ (\tau_{\mathcal{L}} - \tau_{\mathcal{R}})(r_{n-2} + \delta^{n-1}) + \Pi_{\mathcal{LR}^{n-1}}(1) \leqslant 0\}\end{aligned} \tag{10}$$

The boundary of $\mathcal{P}^{ex}_{\mathcal{LR}^{n-1}}$ defined as

$$\begin{aligned}\tau_{\mathcal{LR}^{n-1}} = \{(\tau_{\mathcal{L}}, \tau_{\mathcal{R}}, \delta) \mid \Pi_{\mathcal{LR}^{n-1}}(1) = 0,\\ (\tau_{\mathcal{L}} - \tau_{\mathcal{R}})(\delta r_{n-2} + 1) + \Pi_{\mathcal{LR}^{n-1}}(1) < 0,\\ (\tau_{\mathcal{L}} - \tau_{\mathcal{R}})(r_{n-2} + \delta^{n-1}) + \Pi_{\mathcal{LR}^{n-1}}(1) < 0\}\end{aligned} \tag{11}$$

is associated with a degenerate transcritical bifurcation of $\mathcal{O}_{\mathcal{LR}^{n-1}}$ because approaching this boundary one eigenvalue of the cycle tends to 1, while its points tend to infinity (see (6)–(8)). If the parameter point crosses the boundary $\tau_{\mathcal{LR}^{n-1}}$, the cycle $\mathcal{O}_{\mathcal{LR}^{n-1}}$ disappears (becomes virtual).

Two border collision bifurcation boundaries of $\mathcal{P}^{ex}_{\mathcal{LR}^{n-1}}$ associated with the conditions $x'_1 = 0$ and $x'_{n-1} = 0$ are, respectively,

$$\xi^1_{\mathcal{LR}^{n-1}} = \left\{(\tau_{\mathcal{L}}, \tau_{\mathcal{R}}, \delta) \in \mathcal{P}^{ex}_{\mathcal{LR}^{n-1}} \mid (\tau_{\mathcal{L}} - \tau_{\mathcal{R}})(\delta r_{n-2} + 1) + \Pi_{\mathcal{LR}^{n-1}}(1) = 0\right\} \tag{12a}$$

$$\xi^{n-1}_{\mathcal{LR}^{n-1}} = \left\{(\tau_{\mathcal{L}}, \tau_{\mathcal{R}}, \delta) \in \mathcal{P}^{ex}_{\mathcal{LR}^{n-1}} \mid (\tau_{\mathcal{L}} - \tau_{\mathcal{R}})(r_{n-2} + \delta^{n-1}) + \Pi_{\mathcal{LR}^{n-1}}(1) = 0\right\} \tag{12b}$$

where the upper indexes 1 and $n-1$ correspond to the lower indexes of the colliding periodic points, x'_1 and x'_{n-1}. Note, however, that for the considered parameter domain S, in particular, for $\delta < 1$, among the two border collision bifurcation boundaries only $\xi^{n-1}_{\mathcal{LR}^{n-1}}$ serves as a boundary of $\mathcal{P}^{ex}_{\mathcal{LR}^{n-1}}$. In fact, it can be shown that in such a case the condition $x'_{n-1} \geqslant 0$ implies $x'_1 \geqslant 0$. In contrast, for $\delta > 1$ it is $\xi^1_{\mathcal{LR}^1}$ that represents a boundary of $\mathcal{P}^{ex}_{\mathcal{LR}^{n-1}}$. Both boundaries are involved if map F with $\delta_{\mathcal{L}} \neq \delta_{\mathcal{R}}$ is considered (see, e.g. [36]).

Given that in the complete existence region $\mathcal{P}^{ex}_{\mathcal{LR}^{n-1}}$ of the cycle $\mathcal{O}_{\mathcal{LR}^{n-1}}$ it holds $\Pi_{\mathcal{LR}^{n-1}}(1) < 0$, one eigenvalue of this cycle is larger than 1, and since $0 < \delta < 1$, its second eigenvalue is positive and less than 1, that is, the cycle $\mathcal{O}_{\mathcal{LR}^{n-1}}$ is a saddle. In contrast, the complementary cycle $\mathcal{O}_{\mathcal{L}^2\mathcal{R}^{n-2}}$ can be attracting.

In fact, let $(x_0, y_0) \in D_{\mathcal{L}}$ (i.e. $x_0 \leqslant 0$), $(x_1, y_1) \in D_{\mathcal{L}}$ ($x_1 \leqslant 0$) and $(x_i, y_i) \in D_{\mathcal{R}}$ ($x_i \geqslant 0$), $i = 2, \dots, n-1$, be the points of a cycle $\mathcal{O}_{\mathcal{L}^2\mathcal{R}^{n-2}}$. Obtaining x_0 and x_1 from $(x_0, y_0) = F^{n-2}_{\mathcal{R}} \circ F^2_{\mathcal{L}}(x_0, y_0)$ and $(x_1, y_1) = F_{\mathcal{L}} \circ F^{n-2}_{\mathcal{R}} \circ F_{\mathcal{L}}(x_1, y_1)$, respectively, and expressing the conditions $x_0 \leqslant 0, x_1 \leqslant 0$ in terms of the parameters, we get the existence region $\mathcal{P}^{ex}_{\mathcal{L}^2\mathcal{R}^{n-2}}$ of the cycle $\mathcal{O}_{\mathcal{L}^2\mathcal{R}^{n-2}}$:

$$\begin{aligned}\mathcal{P}^{ex}_{\mathcal{L}^2\mathcal{R}^{n-2}} = \{(\tau_{\mathcal{L}}, \tau_{\mathcal{R}}, \delta) \mid (\tau_{\mathcal{L}} - \tau_{\mathcal{R}})(\delta r_{n-2} + 1) + \Pi_{\mathcal{LR}^{n-1}}(1) \leqslant 0,\\ (\tau_{\mathcal{L}} - \tau_{\mathcal{R}})(r_{n-2} + \delta^{n-1}) + \Pi_{\mathcal{LR}^{n-1}}(1) \leqslant 0,\\ \Pi_{\mathcal{L}^2\mathcal{R}^{n-2}}(1) > 0\}\end{aligned} \tag{13}$$

where

$$\Pi_{\mathcal{L}^2\mathcal{R}^{n-2}}(\lambda) = \lambda^2 - ((2\tau_{\mathcal{L}} - \tau_{\mathcal{R}})r_{n-1} - ((\tau_{\mathcal{L}} - \tau_{\mathcal{R}})^2 - 2\delta)r_{n-2})\lambda + \delta^n$$

is the characteristic polynomial of the matrix $\mathcal{A}_{\mathcal{L}}^2\mathcal{A}_{\mathcal{R}}^{n-2}$. The boundary

$$\begin{aligned}\tau_{\mathcal{L}^2\mathcal{R}^{n-2}} = \{(\tau_{\mathcal{L}}, \tau_{\mathcal{R}}, \delta) \mid \Pi_{\mathcal{L}^2\mathcal{R}^{n-2}}(1) = 0,\\ (\tau_{\mathcal{L}} - \tau_{\mathcal{R}})(\delta r_{n-2} + 1) + \Pi_{\mathcal{L}\mathcal{R}^{n-1}}(1) < 0,\\ (\tau_{\mathcal{L}} - \tau_{\mathcal{R}})(r_{n-2} + \delta^{n-1}) + \Pi_{\mathcal{L}\mathcal{R}^{n-1}}(1) < 0\}\end{aligned} \tag{14}$$

is related to a degenerate transcritical bifurcation of $\mathcal{O}_{\mathcal{L}^2\mathcal{R}^{n-2}}$, while the border collision bifurcation boundaries of $\mathcal{P}^{ex}_{\mathcal{L}^2\mathcal{R}^{n-2}}$ are the same as those of the region $\mathcal{P}^{ex}_{\mathcal{L}\mathcal{R}^{n-1}}$ (see (12a)). In fact, at this boundary, a fold border collision bifurcation occurs giving rise to a pair of complementary cycles, $\mathcal{O}_{\mathcal{L}\mathcal{R}^{n-1}}$ and $\mathcal{O}_{\mathcal{L}^2\mathcal{R}^{n-2}}$. At the first border collision it holds that $x_0 = x'_{n-1} = 0$ leading to the boundary $\xi^0_{\mathcal{L}^2\mathcal{R}^{n-2}} \equiv \xi^{n-1}_{\mathcal{L}\mathcal{R}^{n-1}}$ (relevant in the considered case $0 < \delta < 1$), while at the second border collision it holds that $x_1 = x'_1 = 0$, leading to the boundary $\xi^1_{\mathcal{L}^2\mathcal{R}^{n-2}} \equiv \xi^1_{\mathcal{L}\mathcal{R}^{n-1}}$ (relevant in case $\delta > 1$).

As for the stability of the cycle $\mathcal{O}_{\mathcal{L}^2\mathcal{R}^{n-2}}$, it can be attracting or saddle. Its stability region can be determined as follows:

$$\mathcal{P}^{att}_{\mathcal{L}^2\mathcal{R}^{n-2}} = \left\{(\tau_{\mathcal{L}}, \tau_{\mathcal{R}}, \delta) \in \mathcal{P}^{ex}_{\mathcal{L}^2\mathcal{R}^{n-2}} \mid \Pi_{\mathcal{L}^2\mathcal{R}^{n-2}}(-1) > 0,\ \delta < 1\right\} \tag{15}$$

where the boundary

$$\chi_{\mathcal{L}^2\mathcal{R}^{n-2}} = \left\{(\tau_{\mathcal{L}}, \tau_{\mathcal{R}}, \delta) \in \mathcal{P}^{ex}_{\mathcal{L}^2\mathcal{R}^{n-2}} \mid \Pi_{\mathcal{L}^2\mathcal{R}^{n-2}}(-1) > 0,\ \delta = 1\right\} \tag{16}$$

is related to a centre bifurcation of $\mathcal{O}_{\mathcal{L}^2\mathcal{R}^{n-2}}$, and the boundary

$$\vartheta_{\mathcal{L}^2\mathcal{R}^{n-2}} = \left\{(\tau_{\mathcal{L}}, \tau_{\mathcal{R}}, \delta) \in \mathcal{P}^{ex}_{\mathcal{L}^2\mathcal{R}^{n-2}} \mid \Pi_{\mathcal{L}^2\mathcal{R}^{n-2}}(-1) = 0,\ \delta < 1\right\} \tag{17}$$

corresponds to a degenerate flip bifurcation of $\mathcal{O}_{\mathcal{L}^2\mathcal{R}^{n-2}}$. After this bifurcation, the cycle $\mathcal{O}_{\mathcal{L}^2\mathcal{R}^{n-2}}$ becomes a saddle and a $2n$-band chaotic attractor appears.

To summarize, we can state the following

Proposition 3.1: *Consider map F given in (1a). Let $\delta_{\mathcal{L}} = \delta_{\mathcal{R}} = \delta$ and $(\tau_{\mathcal{L}}, \tau_{\mathcal{R}}, \delta) \in S$ where S is given in (4). Map F has a cycle $\mathcal{O}_{\mathcal{L}\mathcal{R}^{n-1}}$, $n \geqslant 5$, for $(\tau_{\mathcal{L}}, \tau_{\mathcal{R}}, \delta) \in \mathcal{P}^{ex}_{\mathcal{L}\mathcal{R}^{n-1}}$, where the region $\mathcal{P}^{ex}_{\mathcal{L}\mathcal{R}^{n-1}}$ is defined in (10), and a complementary cycle $\mathcal{O}_{\mathcal{L}^2\mathcal{R}^{n-2}}$, $n \geqslant 5$, for $(\tau_{\mathcal{L}}, \tau_{\mathcal{R}}, \delta) \in \mathcal{P}^{ex}_{\mathcal{L}^2\mathcal{R}^{n-2}}$, where the region $\mathcal{P}^{ex}_{\mathcal{L}^2\mathcal{R}^{n-2}}$ is defined in (13). These regions have the same border collision bifurcation boundary $\xi^0_{\mathcal{L}^2\mathcal{R}^{n-2}} \equiv \xi^{n-1}_{\mathcal{L}\mathcal{R}^{n-1}}$ given in (12b), and different transcritical bifurcation boundaries: $\tau_{\mathcal{L}\mathcal{R}^{n-1}}$ given in (11) for the region $\mathcal{P}^{ex}_{\mathcal{L}\mathcal{R}^{n-1}}$, and $\tau_{\mathcal{L}^2\mathcal{R}^{n-2}}$ given in (14) for the region $\mathcal{P}^{ex}_{\mathcal{L}^2\mathcal{R}^{n-2}}$. The cycle $\mathcal{O}_{\mathcal{L}\mathcal{R}^{n-1}}$ is a saddle in its complete existence region, while $\mathcal{O}_{\mathcal{L}^2\mathcal{R}^{n-2}}$ is attracting in the region $\mathcal{P}^{att}_{\mathcal{L}^2\mathcal{R}^{n-2}}$ given in (15). This region has the degenerate flip bifurcation boundary $\vartheta_{\mathcal{L}^2\mathcal{R}^{n-2}}$ given in (17).*

In Appendix we present the existence and stability conditions for the cycles $\mathcal{O}_{\mathcal{L}\mathcal{R}^{n-1}}$ and $\mathcal{O}_{\mathcal{L}^2\mathcal{R}^{n-2}}$ of map F in a generic case $\delta_{\mathcal{R}} \neq \delta_{\mathcal{L}}$.

3.2. Second family of cycles: $\mathcal{O}_{\mathcal{R}\mathcal{L}^{n-1}}$ and $\mathcal{O}_{\mathcal{R}^2\mathcal{L}^{n-2}}$

The existence regions of the cycles $\mathcal{O}_{\mathcal{R}\mathcal{L}^{n-1}}$ and $\mathcal{O}_{\mathcal{R}^2\mathcal{L}^{n-2}}$ of map F, as well as their stability conditions can be obtained using the symmetry of the bifurcation boundaries with respect to the line $\tau_{\mathcal{L}} = \tau_{\mathcal{R}}$.

Consider, for example, a cycle $\mathcal{O}_{\mathcal{R}\mathcal{L}^{n-1}}$, with the points $(\bar{x}'_0, \bar{y}'_0) \in D_{\mathcal{R}}$, $(\bar{x}'_i, \bar{y}'_i) \in D_{\mathcal{L}}$, $i = 1, \ldots, n-1$. Obviously, all the equations for the bifurcation boundaries obtained in the previous section for the cycles $\mathcal{O}_{\mathcal{L}\mathcal{R}^{n-1}}$ are valid also for the cycles $\mathcal{O}_{\mathcal{R}\mathcal{L}^{n-1}}$ up to an interchange $\mathcal{L} \leftrightarrow \mathcal{R}$ and, in particular, $\tau_{\mathcal{R}} \leftrightarrow \tau_{\mathcal{L}}$. However, the existence conditions are different: For the cycle $\mathcal{O}_{\mathcal{R}\mathcal{L}^{n-1}}$ it must hold $\bar{x}'_0 \geqslant 0$, $\bar{x}'_1 \leqslant 0$ and $\bar{x}'_{n-1} \leqslant 0$, i.e.

$$\bar{x}'_0 = \frac{\Pi_{\mathcal{L}^n}(1)}{\Pi_{\mathcal{R}\mathcal{L}^{n-1}}(1)\Pi_{\mathcal{L}}(1)} \geqslant 0 \tag{18}$$

$$\bar{x}'_1 = \frac{(\tau_{\mathcal{R}} - \tau_{\mathcal{L}})(\delta\ell_{n-2} + 1) + \Pi_{\mathcal{R}\mathcal{L}^{n-1}}(1)}{\Pi_{\mathcal{R}\mathcal{L}^{n-1}}(1)\Pi_{\mathcal{L}}(1)} \leqslant 0 \tag{19}$$

$$\bar{x}'_{n-1} = \frac{(\tau_{\mathcal{R}} - \tau_{\mathcal{L}})(\ell_{n-2} + \delta^{n-1}) + \Pi_{\mathcal{R}\mathcal{L}^{n-1}}(1)}{\Pi_{\mathcal{R}\mathcal{L}^{n-1}}(1)\Pi_{\mathcal{L}}(1)} \leqslant 0 \tag{20}$$

where ℓ_k is obtained from

$$\ell_k = \tau_{\mathcal{L}}\ell_{k-1} - \delta\ell_{k-2} \quad \text{with} \quad \ell_0 = 1,\ \ell_1 = \tau_{\mathcal{L}},\ k = 2, \ldots$$

and

$$\begin{aligned}
\Pi_{\mathcal{L}}(\lambda) &= \lambda^2 - \tau_{\mathcal{L}}\lambda + \delta \\
\Pi_{\mathcal{L}^n}(\lambda) &= \lambda^2 - (\ell_n - \delta\ell_{n-2})\lambda + \delta^n \\
\Pi_{\mathcal{R}\mathcal{L}^{n-1}}(\lambda) &= \lambda^2 - (\tau_{\mathcal{R}}\ell_{n-1} - 2\delta\ell_{n-2})\lambda + \delta^n
\end{aligned}$$

Since for $(\tau_{\mathcal{L}}, \tau_{\mathcal{R}}, \delta) \in S$ it holds that $\Pi_{\mathcal{L}}(1) = 1 - \tau_{\mathcal{L}} + \delta > 0$ and $\Pi_{\mathcal{L}^n}(1) > 0$, to satisfy (18) it must hold $\Pi_{\mathcal{R}\mathcal{L}^{n-1}}(1) > 0$. Therefore, differently from the cycle $\mathcal{O}_{\mathcal{L}\mathcal{R}^{n-1}}$ which is always a saddle, cycle $\mathcal{O}_{\mathcal{R}\mathcal{L}^{n-1}}$ can be attracting. Its existence region can be defined as follows:

$$\begin{aligned}
\mathcal{P}^{ex}_{\mathcal{R}\mathcal{L}^{n-1}} = \{(\tau_{\mathcal{L}}, \tau_{\mathcal{R}}, \delta) \mid\ & (\tau_{\mathcal{R}} - \tau_{\mathcal{L}})(\delta\ell_{n-2} + 1) + \Pi_{\mathcal{R}\mathcal{L}^{n-1}}(1) \leqslant 0, \\
& (\tau_{\mathcal{R}} - \tau_{\mathcal{L}})(\ell_{n-2} + \delta^{n-1}) + \Pi_{\mathcal{R}\mathcal{L}^{n-1}}(1) \leqslant 0, \\
& \Pi_{\mathcal{R}\mathcal{L}^{n-1}}(1) > 0\}
\end{aligned} \tag{21}$$

The boundary of $\mathcal{P}^{ex}_{\mathcal{R}\mathcal{L}^{n-1}}$ associated with a degenerate transcritical bifurcation is defined as

$$\begin{aligned}
\tau_{\mathcal{R}\mathcal{L}^{n-1}} = \{(\tau_{\mathcal{L}}, \tau_{\mathcal{R}}, \delta) \mid\ & \Pi_{\mathcal{R}\mathcal{L}^{n-1}}(1) = 0, \\
& (\tau_{\mathcal{R}} - \tau_{\mathcal{L}})(\delta\ell_{n-2} + 1) + \Pi_{\mathcal{R}\mathcal{L}^{n-1}}(1) < 0, \\
& (\tau_{\mathcal{R}} - \tau_{\mathcal{L}})(\ell_{n-2} + \delta^{n-1}) + \Pi_{\mathcal{R}\mathcal{L}^{n-1}}(1) < 0\}
\end{aligned} \tag{22}$$

and the border collision bifurcation boundaries are

$$\xi^1_{\mathcal{R}\mathcal{L}^{n-1}} = \left\{(\tau_{\mathcal{L}}, \tau_{\mathcal{R}}, \delta) \in \mathcal{P}^{ex}_{\mathcal{R}\mathcal{L}^{n-1}} \mid (\tau_{\mathcal{R}} - \tau_{\mathcal{L}})(\delta\ell_{n-2} + 1) + \Pi_{\mathcal{R}\mathcal{L}^{n-1}}(1) = 0\right\} \tag{23a}$$

$$\xi^{n-1}_{\mathcal{R}\mathcal{L}^{n-1}} = \left\{(\tau_{\mathcal{L}}, \tau_{\mathcal{R}}, \delta) \in \mathcal{P}^{ex}_{\mathcal{R}\mathcal{L}^{n-1}} \mid (\tau_{\mathcal{R}} - \tau_{\mathcal{L}})(\ell_{n-2} + \delta^{n-1}) + \Pi_{\mathcal{R}\mathcal{L}^{n-1}}(1) = 0\right\} \tag{23b}$$

The stability region of $\mathcal{O}_{\mathcal{R}\mathcal{L}^{n-1}}$ can be defined as

$$\mathcal{P}^{att}_{\mathcal{R}\mathcal{L}^{n-1}} = \left\{(\tau_{\mathcal{L}}, \tau_{\mathcal{R}}, \delta) \in \mathcal{P}^{ex}_{\mathcal{R}\mathcal{L}^{n-1}} \mid \Pi_{\mathcal{R}\mathcal{L}^{n-1}}(-1) > 0,\ \delta < 1\right\} \tag{24}$$

(recall that for $(\tau_{\mathcal{L}}, \tau_{\mathcal{R}}, \delta) \in \mathcal{P}^{ex}_{\mathcal{R}\mathcal{L}^{n-1}}$, it holds that $\Pi_{\mathcal{R}\mathcal{L}^{n-1}}(1) > 0$). Its centre bifurcation boundary is

$$\chi_{\mathcal{R}\mathcal{L}^{n-1}} = \left\{(\tau_{\mathcal{L}}, \tau_{\mathcal{R}}, \delta) \in \mathcal{P}^{ex}_{\mathcal{R}\mathcal{L}^{n-1}} \mid \Pi_{\mathcal{R}\mathcal{L}^{n-1}}(-1) > 0,\ \delta = 1\right\} \tag{25}$$

and the flip bifurcation boundary is

$$\vartheta_{\mathcal{R}\mathcal{L}^{n-1}} = \left\{(\tau_{\mathcal{L}}, \tau_{\mathcal{R}}, \delta) \in \mathcal{P}^{ex}_{\mathcal{R}\mathcal{L}^{n-1}} \mid \Pi_{\mathcal{R}\mathcal{L}^{n-1}}(-1) = 0,\ \delta < 1\right\} \tag{26}$$

It is worth noting that if we fix $\delta = 0$ and relax the condition $\tau_{\mathcal{R}} > -1 - \delta$ (in fact, for the considered family we use only the condition $|\tau_{\mathcal{L}}| < 1 + \delta$), then from (21) and (24) we get the well-known stability regions of cycles $\mathcal{O}_{\mathcal{R}\mathcal{L}^{n-1}}$, $n \geqslant 3$, of the skew tent map:

$$\mathcal{P}^{att}_{\mathcal{R}\mathcal{L}^{n-1}} = \left\{(\tau_{\mathcal{L}}, \tau_{\mathcal{R}}) \mid -\frac{1}{\tau_{\mathcal{L}}^{n-1}} < \tau_{\mathcal{R}} < \frac{\tau_{\mathcal{L}}^{n-1} - 1}{\tau_{\mathcal{L}}^{n-2}(1 - \tau_{\mathcal{L}})}\right\} \tag{27}$$

bounded by the border collision bifurcation curves $\xi^{n-1}_{\mathcal{R}\mathcal{L}^{n-1}}$ and flip bifurcation curves $\vartheta_{\mathcal{R}\mathcal{L}^{n-1}}$ (see, e.g. [34]).

As for a cycle $\mathcal{O}_{\mathcal{R}^2\mathcal{L}^{n-2}}$ with the points $(\bar{x}_0, \bar{y}_0) \in D_{\mathcal{R}}$, $(\bar{x}_1, \bar{y}_1) \in D_{\mathcal{R}}$, $(\bar{x}_i, \bar{y}_i) \in D_{\mathcal{L}}$, $i = 2, \ldots, n-1$, its existence region $\mathcal{P}^{ex}_{\mathcal{R}^2\mathcal{L}^{n-2}}$ has the same border collision bifurcation boundaries as in (23a), $\xi^0_{\mathcal{R}^2\mathcal{L}^{n-2}} \equiv \xi^{n-1}_{\mathcal{R}\mathcal{L}^{n-1}}$ (relevant in case $0 < \delta < 1$), and $\xi^1_{\mathcal{R}^2\mathcal{L}^{n-2}} \equiv \xi^1_{\mathcal{R}\mathcal{L}^{n-1}}$ (valid for $\delta > 1$). The inequalities corresponding to these boundaries are the same as in (21), however, to satisfy the conditions $\bar{x}_0 \geqslant 0$, $\bar{x}_1 \geqslant 0$, it must hold that $\Pi_{\mathcal{R}^2\mathcal{L}^{n-2}}(1) < 0$, that is, the cycle $\mathcal{O}_{\mathcal{R}^2\mathcal{L}^{n-2}}$ is a saddle in its complete existence region that can be defined as follows:

$$\begin{aligned}\mathcal{P}^{ex}_{\mathcal{R}^2\mathcal{L}^{n-2}} = \{(\tau_{\mathcal{L}}, \tau_{\mathcal{R}}, \delta) \mid\ & (\tau_{\mathcal{R}} - \tau_{\mathcal{L}})(\delta \ell_{n-2} + 1) + \Pi_{\mathcal{R}\mathcal{L}^{n-1}}(1) \leqslant 0, \\ & (\tau_{\mathcal{R}} - \tau_{\mathcal{L}})(\ell_{n-2} + \delta^{n-1}) + \Pi_{\mathcal{R}\mathcal{L}^{n-1}}(1) \leqslant 0, \\ & \Pi_{\mathcal{R}^2\mathcal{L}^{n-2}}(1) < 0\}\end{aligned} \tag{28}$$

where the boundary

$$\tau_{\mathcal{R}^2\mathcal{L}^{n-2}} = \left\{(\tau_{\mathcal{L}}, \tau_{\mathcal{R}}, \delta) \in \mathcal{P}^{ex}_{\mathcal{R}^2\mathcal{L}^{n-2}} \mid \Pi_{\mathcal{R}^2\mathcal{L}^{n-2}}(1) = 0\right\} \tag{29}$$

corresponds to a degenerate transcritical bifurcation of $\mathcal{O}_{\mathcal{R}^2\mathcal{L}^{n-2}}$.

To summarize, we can state the following

Proposition 3.2: *Consider map F given in* (1a). *Let* $\delta_{\mathcal{L}} = \delta_{\mathcal{R}} = \delta$ *and* $(\tau_{\mathcal{L}}, \tau_{\mathcal{R}}, \delta) \in S$ *where S is given in* (4). *Map F has a cycle* $\mathcal{O}_{\mathcal{R}\mathcal{L}^{n-1}}$, $n \geqslant 5$, *for* $(\tau_{\mathcal{L}}, \tau_{\mathcal{R}}, \delta) \in \mathcal{P}^{ex}_{\mathcal{R}\mathcal{L}^{n-1}}$, *where the region* $\mathcal{P}^{ex}_{\mathcal{R}\mathcal{L}^{n-1}}$ *is defined in* (21), *and a complementary cycle* $\mathcal{O}_{\mathcal{R}^2\mathcal{L}^{n-2}}$, $n \geqslant 5$, *for* $(\tau_{\mathcal{L}}, \tau_{\mathcal{R}}, \delta) \in \mathcal{P}^{ex}_{\mathcal{R}^2\mathcal{L}^{n-2}}$, *where the region* $\mathcal{P}^{ex}_{\mathcal{R}^2\mathcal{L}^{n-2}}$ *is defined in* (28). *These regions have the same border collision bifurcation boundary* $\xi^0_{\mathcal{R}^2\mathcal{L}^{n-2}} \equiv \xi^{n-1}_{\mathcal{R}\mathcal{L}^{n-1}}$ *given in* (23b), *and different transcritical bifurcation boundaries:* $\tau_{\mathcal{R}\mathcal{L}^{n-1}}$ *given in* (22) *for the region* $\mathcal{P}^{ex}_{\mathcal{R}\mathcal{L}^{n-1}}$, *and* $\tau_{\mathcal{R}^2\mathcal{L}^{n-2}}$ *given in* (29) *for the region* $\mathcal{P}^{ex}_{\mathcal{R}^2\mathcal{L}^{n-2}}$. *The cycle* $\mathcal{O}_{\mathcal{R}^2\mathcal{L}^{n-2}}$ *is a saddle in its complete existence region, while* $\mathcal{O}_{\mathcal{R}\mathcal{L}^{n-1}}$ *is attracting in the region* $\mathcal{P}^{att}_{\mathcal{R}\mathcal{L}^{n-1}}$ *given in* (24). *This region has the degenerate flip bifurcation boundary* $\vartheta_{\mathcal{R}\mathcal{L}^{n-1}}$ *given in* (26).

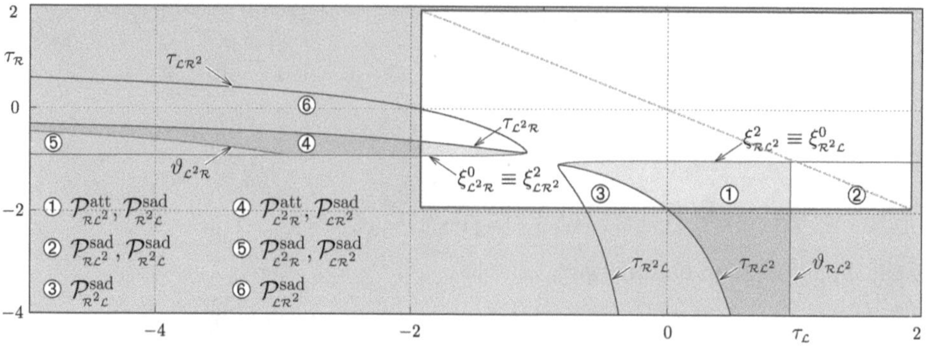

Figure 5. Existence and stability regions of 3-cycles $\mathcal{O}_{\mathcal{LR}^2}$ and $\mathcal{O}_{\mathcal{RL}2}$ of map F in the $(\tau_{\mathcal{L}}, \tau_{\mathcal{R}})$-parameter plane for $\delta = 0.92$. Here the window corresponding to the domain S is highlighted, and the line $\tau_{\mathcal{L}} = -\tau_{\mathcal{R}}$ associated with the Lozi map is shown in orange.

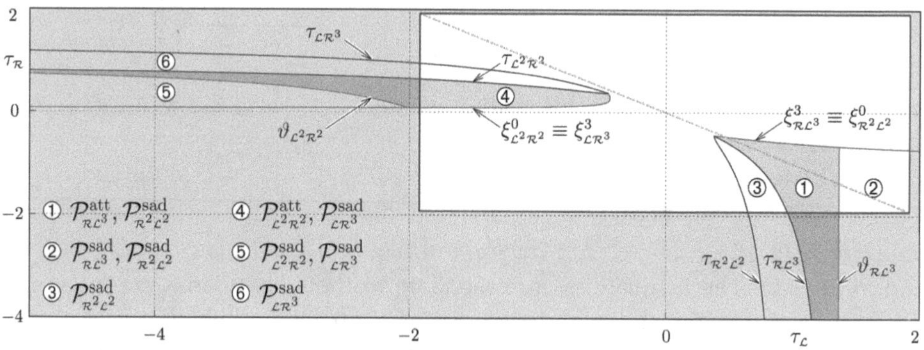

Figure 6. Existence and stability regions of 4-cycles $\mathcal{O}_{\mathcal{LR}^3}$, $\mathcal{O}_{\mathcal{L}^2\mathcal{R}^2}$ and $\mathcal{O}_{\mathcal{RL}^3}$ of map F in the $(\tau_{\mathcal{L}}, \tau_{\mathcal{R}})$-parameter plane for $\delta = 0.92$. Here the window corresponding to the domain S is highlighted, and the line $\tau_{\mathcal{L}} = -\tau_{\mathcal{R}}$ associated with the Lozi map is shown in orange.

4. Examples of the main periodicity regions

Let us illustrate and comment on the results obtained in the previous section.

First, in Figures 5–7 we show the existence and stability regions of the 3-, 4- and 5-cycles of map F in the $(\tau_{\mathcal{L}}, \tau_{\mathcal{R}})$-parameter plane for $\delta = 0.92$, where all the boundaries are marked, associated with border collision, degenerate transcritical and degenerate flip bifurcations of the related cycles. As mentioned above, for $0 < \delta < 1$, the curves $\xi^{n-1}_{\mathcal{LR}^{n-1}} \equiv \xi^0_{\mathcal{L}^2\mathcal{R}^{n-2}}$ and $\xi^{n-1}_{\mathcal{RL}^{n-1}} \equiv \xi^0_{\mathcal{R}^2\mathcal{L}^{n-2}}$ are acting as border collision bifurcation boundaries of the existence regions. In these figures, the highlighted window corresponds to the stability domain S, where $|\tau_{\mathcal{L}}| < 1 + \delta$, $|\tau_{\mathcal{R}}| < 1 + \delta$, and the stability regions of the 3-, 4- and 5-cycles are shown in different colours. Note that the bifurcation curves extend outside S, more precisely, the curves related to the first family of cycles, namely, $\mathcal{O}_{\mathcal{LR}^{n-1}}$ and $\mathcal{O}_{\mathcal{L}^2\mathcal{R}^{n-2}}$, are also valid for $\tau_{\mathcal{L}} < -1 - \delta$ (in fact, in our consideration of this family we used only the inequality $|\tau_{\mathcal{R}}| < 1 + \delta$), while the bifurcation curves related to the second family of cycles, $\mathcal{O}_{\mathcal{RL}^{n-1}}$ and $\mathcal{O}_{\mathcal{R}^2\mathcal{L}^{n-2}}$, extend to the region $\tau_{\mathcal{R}} < -1 - \delta$.

In Figure 5, the existence and stability regions of the 3-cycles $\mathcal{O}_{\mathcal{LR}^2}$ and $\mathcal{O}_{\mathcal{L}^2\mathcal{R}}$ are presented. Given that these cycles belong to both families, they both can be stable or saddle.

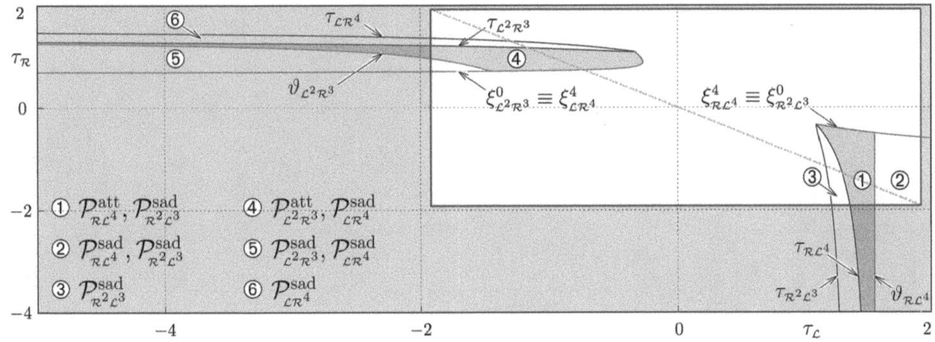

Figure 7. Existence and stability regions of 5-cycles $\mathcal{O}_{\mathcal{L}\mathcal{R}^4}$, $\mathcal{O}_{\mathcal{L}^2\mathcal{R}^3}$, $\mathcal{O}_{\mathcal{R}\mathcal{L}^4}$ and $\mathcal{O}_{\mathcal{R}^2\mathcal{L}^3}$ of map F in the $(\tau_{\mathcal{L}}, \tau_{\mathcal{R}})$-parameter plane for $\delta = 0.92$. Here the window corresponding to the domain S is highlighted, and the line $\tau_{\mathcal{L}} = -\tau_{\mathcal{R}}$ associated with the Lozi map is shown in orange.

To plot the boundaries of the corresponding regions, we treat the cycles $\mathcal{O}_{\mathcal{L}\mathcal{R}^2}$ and $\mathcal{O}_{\mathcal{L}^2\mathcal{R}}$ as those belonging to the first family for $\tau_{\mathcal{R}} > \tau_{\mathcal{L}}$, and $\mathcal{O}_{\mathcal{R}^2\mathcal{L}} \equiv \mathcal{O}_{\mathcal{L}\mathcal{R}^2}$ and $\mathcal{O}_{\mathcal{R}\mathcal{L}^2} \equiv \mathcal{O}_{\mathcal{L}^2\mathcal{R}}$ as those belonging to the second family for $\tau_{\mathcal{R}} < \tau_{\mathcal{L}}$, substituting $n = 3$ to the generic formulas obtained for these families in the previous section. Note however, that the Lozi map cannot have attracting 3-cycles (see [35]), and in fact, the regions $\mathcal{P}^{att}_{\mathcal{R}\mathcal{L}^2}$ and $\mathcal{P}^{att}_{\mathcal{L}\mathcal{R}^2}$ have no intersection with the line $\tau_{\mathcal{L}} = -\tau_{\mathcal{R}}$ (see also Figure 8). Figure 6 presents the existence and stability regions of the 4-cycles $\mathcal{O}_{\mathcal{L}\mathcal{R}^3}$, $\mathcal{O}_{\mathcal{L}^2\mathcal{R}^2}$ and $\mathcal{O}_{\mathcal{R}\mathcal{L}^3}$. Similar to the case of the 3-cycles, to plot the boundaries of the 4-periodicity regions, we treat the cycles $\mathcal{O}_{\mathcal{L}\mathcal{R}^3}$ and $\mathcal{O}_{\mathcal{L}^2\mathcal{R}^2}$ as those belonging to the first family for $\tau_{\mathcal{R}} > \tau_{\mathcal{L}}$, and $\mathcal{O}_{\mathcal{R}^2\mathcal{L}^2} \equiv \mathcal{O}_{\mathcal{L}^2\mathcal{R}^2}$ and $\mathcal{O}_{\mathcal{R}\mathcal{L}^3}$ as those belonging to the second family for $\tau_{\mathcal{R}} < \tau_{\mathcal{L}}$, substituting $n = 4$ to the generic formulas obtained in the previous section. As one can see in Figure 6, among the stability regions $\mathcal{P}^{att}_{\mathcal{R}\mathcal{L}^3}$ and $\mathcal{P}^{att}_{\mathcal{L}^2\mathcal{R}^2}$ only the region $\mathcal{P}^{att}_{\mathcal{R}\mathcal{L}^3}$ intersects the line $\tau_{\mathcal{L}} = -\tau_{\mathcal{R}}$.

The existence and stability regions associated with 5-cycles $\mathcal{O}_{\mathcal{L}\mathcal{R}^4}$ and $\mathcal{O}_{\mathcal{L}^2\mathcal{R}^3}$, as well as $\mathcal{O}_{\mathcal{R}\mathcal{L}^4}$ and $\mathcal{O}_{\mathcal{R}^2\mathcal{L}^3}$, are shown in Figure 7. All these regions are present in the bifurcation structure of the Lozi map.

In Figure 8, in the $(\tau_{\mathcal{L}}, \tau_{\mathcal{R}})$-parameter plane for $\delta = 0.7$ in (a), and $\delta = 0.8$ in (b), $\delta = 0.9$ in (c), and $\delta = 1$ in (d), the stability regions of the cycles $\mathcal{O}_{\mathcal{L}^2\mathcal{R}^{n-2}}$ and $\mathcal{O}_{\mathcal{R}\mathcal{L}^{n-1}}$, $n = 3, \ldots, 10$, are shown in the stability domain S. The family of periodicity regions $\mathcal{P}^{att}_{\mathcal{L}^2\mathcal{R}^{n-2}}$ is located above the diagonal $\tau_{\mathcal{L}} = \tau_{\mathcal{R}}$, while the family of periodicity regions $\mathcal{P}^{att}_{\mathcal{R}\mathcal{L}^{n-1}}$ is located below this diagonal.

It is easy to check that for $\delta = 1$, it holds that $\xi^{n-1}_{\sigma} \equiv \xi^1_{\sigma}$ for both families, with $\sigma = \mathcal{L}\mathcal{R}^{n-1}$ or $\sigma = \mathcal{R}\mathcal{L}^{n-1}$. So, we have that for $\delta < 1$, an existence region $\mathcal{P}_{\sigma}$ is bounded by ξ^{n-1}_{σ}, for $\delta = 1$ it is $\xi^{n-1}_{\sigma} \equiv \xi^1_{\sigma}$, and for $\delta > 1$ the region $\mathcal{P}_{\sigma}$ is bounded by ξ^1_{σ}. Note also that at $\delta = 1$ (see Figure 8(d)), the cusp points of the curves $\xi^{n-1}_{\mathcal{L}\mathcal{R}^{n-1}}$ belong to the diagonal $\tau_{\mathcal{L}} = \tau_{\mathcal{R}}$, at which map F is linear. These points are determined by $\tau_{\mathcal{L}} = \tau_{\mathcal{R}} = 2\cos(2\pi/n)$ and correspond to the rotation matrix $\mathcal{A}_{\mathcal{L}} = \mathcal{A}_{\mathcal{R}}$ with rotation number $1/n$, $n \geqslant 3$. Finally, in Figure 9, the stability regions associated with attracting cycles $\mathcal{O}_{\mathcal{L}^2\mathcal{R}^{n-2}}$ and $\mathcal{O}_{\mathcal{R}\mathcal{L}^{n-1}}$, $n = 3, \ldots, 10$, are shown in the $(\tau_{\mathcal{L}}, \tau_{\mathcal{R}}, \delta)$-parameter space on two horizontal cross-sections, for $\delta = 0.8$ and $\delta = 1$, and a vertical cross-section for $\tau_{\mathcal{L}} = -\tau_{\mathcal{R}}$, so that one can see how the main periodicity regions of the Lozi map are incorporated into the parameter space of the border collision normal form F.

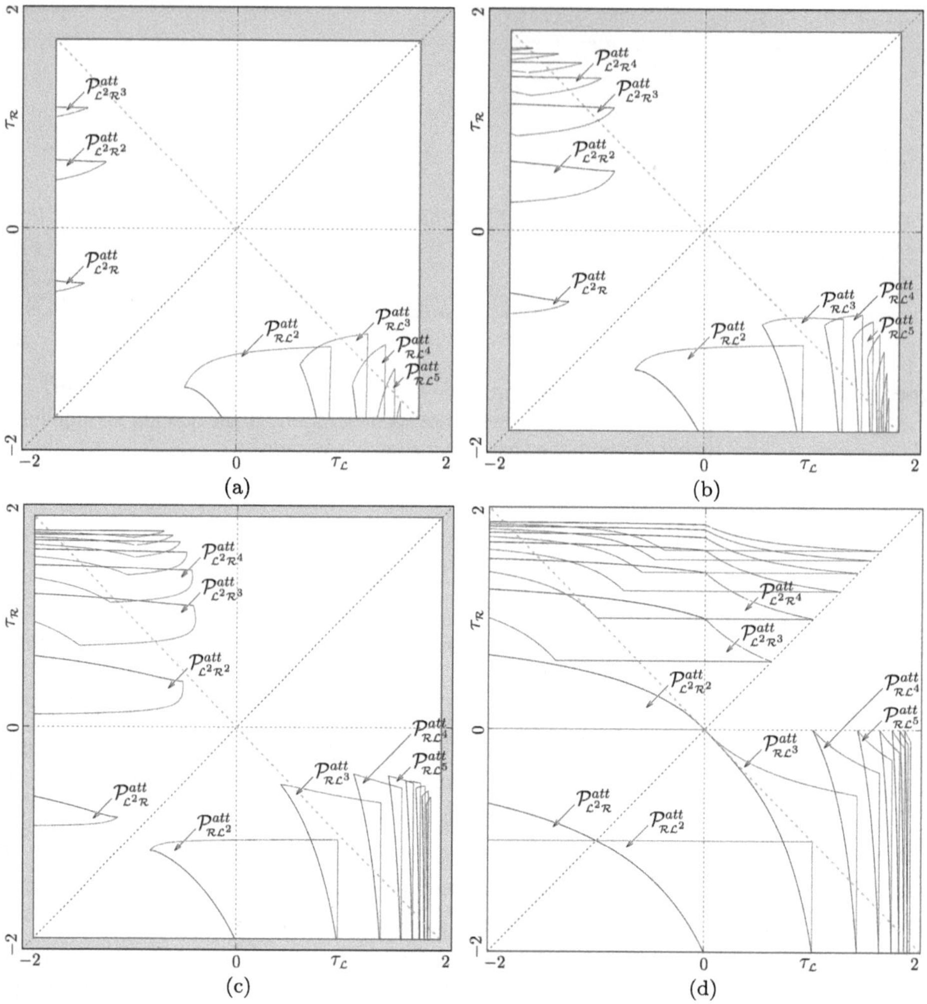

Figure 8. Stability regions $\mathcal{P}^{att}_{\mathcal{L}^2\mathcal{R}^{n-2}}$ and $\mathcal{P}^{att}_{\mathcal{R}\mathcal{L}^{n-1}}$, $n = 3, \ldots, 10$, in the $(\tau_{\mathcal{L}}, \tau_{\mathcal{R}})$-parameter plane, where the window corresponding to the domain S is highlighted, and the line $\tau_{\mathcal{L}} = -\tau_{\mathcal{R}}$ associated with the Lozi map is shown in orange. Here $\delta = 0.7$ in (a), $\delta = 0.8$ in (b), $\delta = 0.9$ in (c), and $\delta = 1$ in (d).

5. Conclusions

We considered the 2D border collision normal form F in case when the determinants of the Jacobian matrices in two partitions are equal. Since the same property holds also for the Lozi map, which is a special case of map F, we were able to compare the bifurcation structures in these maps. We focussed on the part of the parameter space of F near the centre bifurcation boundary of the stability domain of its fixed point. At this boundary both determinants equal 1 and map F is conservative. This property helps to explain why the bifurcation structure near the centre bifurcation boundary is so intricate both in map F and in the Lozi map. In the present paper, we obtained analytical representation of the boundaries of the existence and stability regions associated with two families of cycles with rotation number $1/n$, $n \geqslant 3$, having one or two points in one partition and all other points

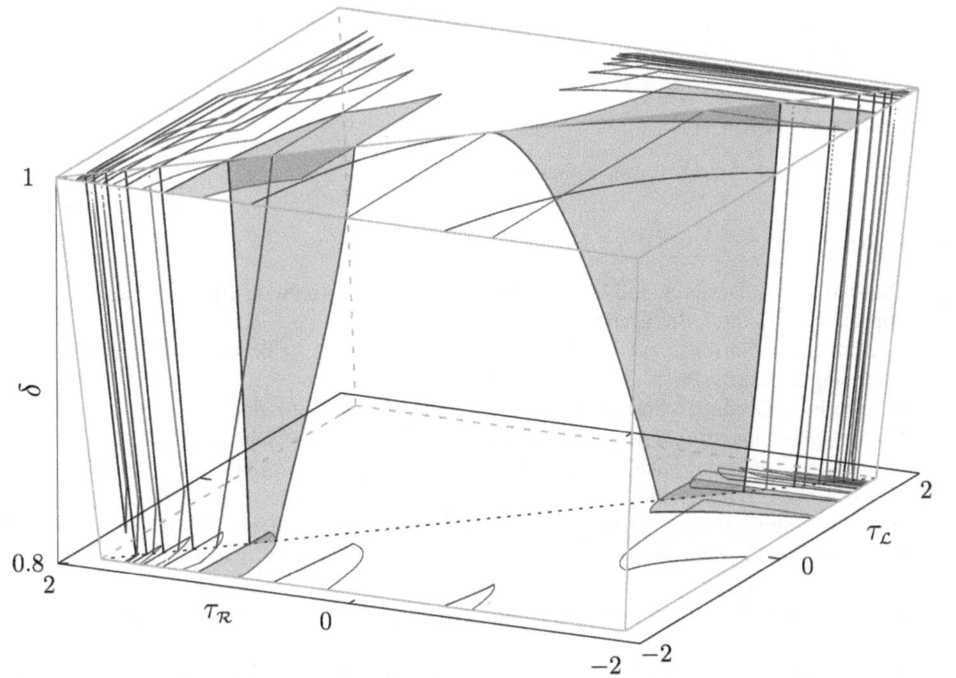

Figure 9. Map F for $\delta_\mathcal{L} = \delta_\mathcal{R} = \delta$. Three cross-sections of the $(\tau_\mathcal{L}, \tau_\mathcal{R}, \delta)$-parameter space: horizontal cross-sections correspond to $\delta = 0.8$, $\delta = 1$, and vertical cross-section is related to $\tau_\mathcal{L} = -\tau_\mathcal{R}$, i.e. to the Lozi map. The stability regions of the 4-cycle $\mathcal{O}_{\mathcal{R}^2\mathcal{L}^2}$ and 5-cycle $\mathcal{O}_{\mathcal{L}^2\mathcal{R}^3}$ are shown in blue and pink, respectively.

in the other partition. These regions are the largest among n-periodicity regions related to other rotation numbers. Their boundaries are associated with border collision, degenerate transcritical, degenerate flip and centre bifurcations of the related cycles. As a future work, it would be interesting to investigate other periodicity regions, in particular, those nested into the main periodicity regions.

Notes

1. The meaning of the notion of *normal form* used here is different from that in standard bifurcation theory where it is related to locally topologically equivalent maps. The border collision normal form can be seen as a piecewise linear approximation of the piecewise smooth map in the neighbourhood of a border crossing fixed point.
2. In the considered case, the unique fixed point of the map F is a focus, and points of n-cycles we are dealing with are located around this fixed point. Simplifying, the rotation number of this cycle is an irreducible fraction m/n, $m < n$, where m is the number of rotations of the trajectory around the fixed point in n iterations. For a rigorous definition see [33].

Acknowledgments

I. Sushko is grateful to the University of Urbino for its hospitality during her stay as a visiting professor as part of a project to support Ukrainian researchers during the Russian invasion of Ukraine. The work of V. Avrutin was supported by the German Research Foundation within the scope of the project 'Generic bifurcation structures in piecewise-smooth maps with extremely high number of borders in theory and applications for power converter systems – 2'.

Disclosure statement

No potential conflict of interest was reported by the author(s).

References

[1] D. Aharonov, R. Devaney, and U. Elias, The dynamics of a piecewise linear map and its smooth approximation, Int. J. Bif. Chaos 7 (1997), pp. 351–372.
[2] V. Avrutin, L. Gardini, and M. Schanz, On a special type of border-collision bifurcations occurring at infinity, Phys. D 239 (2010), pp. 1083–1094.
[3] V. Avrutin, L. Gardini, I. Sushko, and F. Tramontana, *Continuous and Discontinuous Piecewise-Smooth One-Dimensional Maps: Invariant Sets and Bifurcation Structures*, Nonlinear Science Series A, Vol. 95, World Scientific, 2019. https://doi.org/10.1142/8285.
[4] V. Avrutin, M. Schanz, and S. Banerjee, Occurrence of multiple attractor bifurcations in the two-dimensional piecewise linear normal form map, Nonlinear Dyn. 67 (2011), pp. 293–307.
[5] V. Avrutin, Zh.T. Zhusubaliyev, A. Saha, S. Banerjee, L. Gardini, and I. Sushko, Dangerous bifurcations revisited, Int. J. Bifurcat. Chaos 26 (2017), Article ID 1630040.
[6] M. Aziz-Alaoui, C. Robert, and C. Grebogi, Dynamics of a Hénon–Lozi-type map, Chaos Solit. Fractals 12 (2001), pp. 2323–2341.
[7] S. Banerjee and C. Grebogi, Border collision bifurcation in two-dimensional piecewise smooth maps, Phys. Rev. E 59 (1999), pp. 4052–4061.
[8] S. Banerjee and G.C. Verghese, Nonlinear Phenomena in Power Electronics – Attractors, Bifurcations, Chaos, and Nonlinear Control, IEEE Press, 2001.
[9] S. Banerjee, J.A. Yorke, and C. Grebogi, Robust Chaos, Phys. Rev. Lett. 80 (1998), pp. 3049–3052.
[10] G. Bischi, C. Chiarella, M. Kopel, and F. Szidarovszky, Nonlinear Oligopolies: Stability and Bifurcations, Springer Science & Business Media, 2010.
[11] V. Botella-Soler, J.M. Castelo, J.A. Oteo, and J. Ros, Bifurcations in the Lozi map, J. Phys. A: Math. Theor. 44 (2011), Article ID 305101.
[12] B. Brogliato, Nonsmooth Mechanics, Communications and Control Engineering, Springer, 1999.
[13] P. Collet and Y. Levy, *Ergodic Properties of the Lozi Mappings*, The Theory of Chaotic Attractors, Springer, 1984, pp. 222–242.
[14] M. di Bernardo, C.J. Budd, A.R. Champneys, and P. Kowalczyk, Piecewise-Smooth Dynamical Systems: Theory and Applications, Applied Mathematical Sciences Vol. 163, Springer, 2008.
[15] M. Dutta, H. Nusse, E. Ott, J.A. Yorke, and G. Yuan, Multiple attractor bifurcations: A source of unpredictability in piecewise smooth systems, Phys. Rev. Lett. 83 (1999), pp. 4281–4284.
[16] Z. Elhadj, Lozi Mappings: Theory and Applications, CRC Press, 2013.
[17] A. Ganguli and S. Banerjee, Dangerous bifurcation at border collision: When does it occur? Phys. Rev. E 71 (2005), Article ID 057202.
[18] L. Gardini, V. Avrutin, and M. Schanz, *Connection between bifurcations on the Poincaré equator and dangerous bifurcations*, in *Iteration Theory*, A. Sharkovsky and I. Sushko, eds., Grazer Math. Ber., 2009, pp. 53–72.
[19] L. Gardini and W. Tikjha, Role of the virtual fixed point in the center bifurcations in a family of piecewise linear maps, Int. J. Bif. Chaos 29 (2019), Article ID 1930041.
[20] M. Hassouneh, E. Abed, and H. & Nusse, Robust dangerous border-collision bifurcations in piecewise smooth systems, Phys. Rev. E 92 (2004), Article ID 070201.
[21] M. Hénon, A two-dimensional mapping with a strange attractor, Commun. Math. Phys. 50 (1976), pp. 69–77.
[22] Y. Ishii, Towards a kneading theory for Lozi mappings. II: Monotonicity of the topological entropy and hausdorff dimension of attractors, Commun. Math. Phys. 190 (1997), pp. 375–394.

[23] R. Leine and H. Nijmeijer, Dynamics and Bifurcations of Non-smooth Mechanical Systems, Vol. 18, Springer Science & Business Media, 2013.

[24] H. Li, K. Li, M. Chen, and B. Bao, Coexisting infinite orbits in an area-preserving Lozi map, Entropy 22 (2020), p. 1119. https://doi.org/10.3390/e22101119.

[25] R. Lozi, Un attracteur étrange (?) du type attracteur de Hénon, J. De Phys. Colloque 39 (1978), pp. C5–9.

[26] M. Misiurewicz, Strange attractors for the Lozi mappings, Ann. N. Y. Acad. Sci. 357 (1980), pp. 348–358.

[27] M. Misiurewicz and S. Štimac, Symbolic dynamics for Lozi maps, Nonlinearity 29 (2016), pp. 3031–3046.

[28] E. Mosekilde and J.L. Laugesen, Nonlinear dynamic phenomena in the beer model, Syst. Dyn. Rev. 23 (2007), pp. 229–252.

[29] H.E. Nusse and J.A. Yorke, Border-collision bifurcations including 'period two to period three' bifurcation for piecewise smooth systems, Phy. D 57 (1992), pp. 39–57.

[30] H.E. Nusse and J.A. Yorke, Border-collision bifurcations for piecewise smooth one dimensional maps, Int. J. Bif. Chaos 5 (1995), pp. 189–207.

[31] D. Simpson, Sequences of periodic solutions and infinitely many coexisting attractors in the border-collision normal form, Int. J. Bif. Chaos 24 (2014), Article ID 1430018.

[32] D. Simpson, *Bifurcations in Piecewise-Smooth Continuous Systems*, Nonlinear Science A, Vol. 70, World Scientific, 2010.

[33] D. Simpson and J. Meiss, Neimark–Sacker bifurcations in planar, piecewise-smooth, continuous maps, SIAM J. Appl. Dyn. Syst. 7 (2008), pp. 795–824.

[34] I. Sushko, V. Avrutin, and L. Gardini, Bifurcation structure in the skew tent map and its application as a border collision normal form, J. Differ. Equ. Appl. 22 (2016), pp. 1040–1087.

[35] I. Sushko, V. Avrutin, and L. Gardini, Center bifurcation in the Lozi map, Int. J. Bif. Chaos 31 (2021), Article ID 2130046.

[36] I. Sushko and L. Gardini, Center bifurcation for a two-dimensional border-collision normal form, Int. J. Bif. Chaos 18 (2008), pp. 1029–1050.

[37] I. Sushko and L. Gardini, Degenerate bifurcations and border collisions in piecewise smooth 1D and 2D maps, Int. J. Bif. Chaos 20 (2010), pp. 2045–2070. https://doi.org/10.1142/S0218127410026927.

[38] Zh. Zhusubaliyev and E. Mosekilde, *Bifurcations and Chaos in Piecewise-Smooth Dynamical Systems*, Nonlinear Science A, Vol. 44, World Scientific, 2003.

[39] Zh. Zhusubaliyev, E. Mosekilde, S.M. Maity, S. Mohanan, and S. Banerjee, Border collision route to quasiperiodicity: Numerical investigation and experimental confirmation, Chaos 16 (2006), Article ID 023122.

Appendix

For the map F given in (1a), the existence region of a cycle $\mathcal{O}_{\mathcal{LR}^{n-1}}$ with rotation number $1/n$, $n \geqslant 3$, and points $(x_0', y_0') \in D_{\mathcal{L}}$ and $(x_i', y_i') \in D_{\mathcal{R}}$, $i = 1, \dots, n-1$, can be determined from the following inequalities that have to be satisfied simultaneously:

$$
\begin{aligned}
x_0' &= \frac{\Pi_{\mathcal{R}^n}(1)}{\Pi_{\mathcal{LR}^{n-1}}(1)\Pi_{\mathcal{R}}(1)} \leqslant 0 \\
x_1' &= \frac{\tau_{\mathcal{L}} - \tau_{\mathcal{R}} + (\tau_{\mathcal{L}}\delta_{\mathcal{R}} - \tau_{\mathcal{R}}\delta_{\mathcal{L}})a_{n-2} + (\delta_{\mathcal{L}} - \delta_{\mathcal{R}})\delta_{\mathcal{R}}a_{n-3} + \Pi_{\mathcal{LR}^{n-1}}(1)}{\Pi_{\mathcal{LR}^{n-1}}(1)\Pi_{\mathcal{R}}(1)} \geqslant 0 \\
x_{n-1}' &= \frac{(\tau_{\mathcal{L}} - \tau_{\mathcal{R}})a_{n-2} + (\tau_{\mathcal{L}}\delta_{\mathcal{R}} - \tau_{\mathcal{R}}\delta_{\mathcal{L}})\delta_{\mathcal{R}}^{n-2} + (\delta_{\mathcal{R}} - \delta_{\mathcal{L}})a_{n-3} + \Pi_{\mathcal{LR}^{n-1}}(1)}{\Pi_{\mathcal{LR}^{n-1}}(1)\Pi_{\mathcal{R}}(1)} \geqslant 0
\end{aligned}
\tag{A1}
$$

where

$$
a_k = \tau_{\mathcal{R}} a_{k-1} - \delta_{\mathcal{R}} a_{k-2} \quad \text{with} \quad a_0 = 1,\ a_1 = \tau_{\mathcal{R}},\ k = 2, \dots,
$$

and

$$\Pi_{\mathcal{R}}(\lambda) = \lambda^2 - \tau_{\mathcal{R}}\lambda + \delta_{\mathcal{R}}$$

$$\Pi_{\mathcal{R}^n}(\lambda) = \lambda^2 - (a_n - \delta_{\mathcal{R}}a_{n-2})\lambda + \delta_{\mathcal{R}}$$

$$\Pi_{\mathcal{L}\mathcal{R}^{n-1}}(\lambda) = \lambda^2 - (\tau_{\mathcal{L}}a_{n-1} - (\delta_{\mathcal{L}} + \delta_{\mathcal{R}})a_{n-2})\lambda + \delta_{\mathcal{L}}\delta_{\mathcal{R}}^{n-1}$$

For $\Pi_{\mathcal{L}\mathcal{R}^{n-1}}(1) \neq 0$, $\Pi_{\mathcal{R}}(1) \neq 0$, the border collision bifurcation conditions are

$$\tau_{\mathcal{L}} - \tau_{\mathcal{R}} + (\tau_{\mathcal{L}}\delta_{\mathcal{R}} - \tau_{\mathcal{R}}\delta_{\mathcal{L}})a_{n-2} + (\delta_{\mathcal{L}} - \delta_{\mathcal{R}})\delta_{\mathcal{R}}a_{n-3} + \Pi_{\mathcal{L}\mathcal{R}^{n-1}}(1) = 0$$

$$(\tau_{\mathcal{L}} - \tau_{\mathcal{R}})a_{n-2} + (\tau_{\mathcal{L}}\delta_{\mathcal{R}} - \tau_{\mathcal{R}}\delta_{\mathcal{L}})\delta_{\mathcal{R}}^{n-2} + (\delta_{\mathcal{R}} - \delta_{\mathcal{L}})a_{n-3} + \Pi_{\mathcal{L}\mathcal{R}^{n-1}}(1) = 0$$

while degenerate transcritical, degenerate flip and centre bifurcation conditions are, respectively,

$$\Pi_{\mathcal{L}\mathcal{R}^{n-1}}(1) = 1 - \tau_{\mathcal{L}}a_{n-1} + (\delta_{\mathcal{L}} + \delta_{\mathcal{R}})a_{n-2} + \delta_{\mathcal{L}}\delta_{\mathcal{R}}^{n-1} = 0$$

$$\Pi_{\mathcal{L}\mathcal{R}^{n-1}}(-1) = 1 + \tau_{\mathcal{L}}a_{n-1} - (\delta_{\mathcal{L}} + \delta_{\mathcal{R}})a_{n-2} + \delta_{\mathcal{L}}\delta_{\mathcal{R}}^{n-1} = 0$$

$$\delta_{\mathcal{L}}\delta_{\mathcal{R}}^{n-1} = 1$$

For the complementary cycles $\mathcal{O}_{\mathcal{L}^2\mathcal{R}^{n-2}}$, the border collision bifurcation boundaries are determined by the same equations as those for the cycles $\mathcal{O}_{\mathcal{L}\mathcal{R}^{n-1}}$, while degenerated transcritical, degererate flip and centre bifurcations, respectively, by

$$\Pi_{\mathcal{L}^2\mathcal{R}^{n-2}}(1) = 1 - \delta_{\mathcal{L}}\delta_{\mathcal{R}}a_{n-4} + \tau_{\mathcal{L}}(\delta_{\mathcal{R}} + \delta_{\mathcal{R}})a_{n-3} - (\tau_{\mathcal{L}}^2 - \delta_{\mathcal{L}})a_{n-2} + \delta_{\mathcal{L}}^2\delta_{\mathcal{R}}^{n-2} = 0$$

$$\Pi_{\mathcal{L}^2\mathcal{R}^{n-2}}(-1) = 1 + \delta_{\mathcal{L}}\delta_{\mathcal{R}}a_{n-4} - \tau_{\mathcal{L}}(\delta_{\mathcal{R}} + \delta_{\mathcal{R}})a_{n-3} + (\tau_{\mathcal{L}}^2 - \delta_{\mathcal{L}})a_{n-2}) + \delta_{\mathcal{L}}^2\delta_{\mathcal{R}}^{n-2} = 0$$

$$\delta_{\mathcal{R}}^2\delta_{\mathcal{L}}^{n-2} = 1$$

Depending on the specific parameter setting, not all the bifurcation conditions can be satisfied for an actual cycle.

For the family of cycles $\mathcal{O}_{\mathcal{R}\mathcal{L}^{n-1}}$, $\mathcal{O}_{\mathcal{R}^2\mathcal{R}^{n-2}}$, with rotation number $1/n$, $n \geqslant 3$, the same results can be obtained from those presented above by interchanging $\mathcal{L} \leftrightarrow \mathcal{R}$, in particular, $\tau_{\mathcal{L}} \leftrightarrow \tau_{\mathcal{R}}$, $\delta_{\mathcal{L}} \leftrightarrow \delta_{\mathcal{R}}$, as well as inverting the inequalities in (A1).

A higher-dimensional generalization of the Lozi map: bifurcations and dynamics

Shakir Bilal and Ramakrishna Ramaswamy

ABSTRACT

We generalize the two-dimensional Lozi map in order to systematically obtain piecewise continuous maps in three and higher dimensions. Similar to higher dimensional generalizations of the related Hénon map, these higher dimensional Lozi maps support hyperchaotic dynamics. We carry out a bifurcation analysis and investigate the dynamics through both numerical and analytical means. The analysis is extended to a sequence of approximations that smooth the discontinuity of the derivatives in the Lozi map.

1. Introduction

The behaviour of low-dimensional nonlinear iterative maps and flows has been extensively studied and characterized over the past few decades, particularly with reference to the creation of chaotic dynamics [5,11,13–15]. The various scenarios or routes to chaos in such systems are by now fairly well known [2,15,16,18]. Similar exploration of the properties of higher dimensional dynamical systems – for instance, the dynamics of attractors with more than one positive Lyapunov exponent and the bifurcations through which they have been created – has not been studied in as much detail even in relatively simple systems [1,4,9,14,21].

Piecewise smooth mappings occur in applications ranging from power electronics to medicine [5,9]. Yet the linear and piecewise-linear mappings are among the simplest examples of iterative dynamical systems exhibiting complex dynamics. The so-called Lozi map [14] is analogous to the quadratic Hénon mapping [11] but has the advantage that more extensive analysis is possible [7]. The mapping itself is only piecewise continuous, and this introduces some additional features that need to be understood more clearly [8,12,19,20]. Indeed, specific bifurcation phenomenon such as border-collision bifurcations can only occur in piecewise smooth dynamical systems [8,20].

Our interest in the present paper is the generalization of the Lozi map to higher dimensions. One motivation is to compare this piecewise continuous system to a similar high-dimensional Hénon mapping [6]. Of the different ways in which this can be done, we choose to extend the map to d-dimensions by incorporating time-delay feedback while ensuring that the system remains an endomorphism in the absence of dissipation. The dissipation is introduced at the kth step, $k < d$, and this also ensures that the map is a diffeomorphism. The system can therefore have k-positive Lyapunov exponents, and we examine the transition to high-dimensional chaos as a function of parameters, characterizing the different bifurcations that can occur. An intermediate 'smooth' approximation [3] of the piecewise map is also investigated vis-á-vis bifurcations for comparison.

In the next Section, the generalized Lozi map is described and a detailed analysis of the local bifurcations of the elementary fixed points is presented. The emergence of chaotic and hyperchaotic attractors and the global bifurcations that arise are discussed in Section 3. Section 4 is devoted to the analysis of smooth approximations to the map. The paper concludes with a discussion and summary in Section 5.

2. The generalized Lozi map

The two-dimensional Lozi map [14] is given by

$$\begin{aligned} x_{n+1} &= 1 - (1-\nu)y_n - a|x_n| \\ y_{n+1} &= x_n \,. \end{aligned} \tag{1}$$

This map is a modification of the quadratic Hénon mapping, with the parameters ν and a tuning the dissipation and nonlinearity, respectively. Since the map is piecewise linear, it lends itself to extensive analysis, some of which has been recently summarized [9].

Rewriting Equation (1) as a difference delay equation, one has

$$x_n = 1 - a|x_{n-1}| - (1-\nu)x_{n-2}, \tag{2}$$

which suggests a natural generalization to higher dimensions,

$$x_n = 1 - a|x_{n-k}| - (1-\nu)x_{n-d}. \tag{3}$$

Here d and k are integers such that $k < d$, and we take $0 \le \nu \le 2$. The mapping is conservative when ν is 0 or 2 and is dissipative otherwise. For $\nu = 1$, the map reduces to a k-dimensional endomorphism, while for $\nu \neq 1$ the map is a d-dimensional diffeomorphism. In the next section, we analyse the implications of different choices of d and k for this map.

2.1. The base maps and q -degeneracy

The integers d and k are either co-prime or share a common factor q. When k and d have a common factor q, it is easy to see that all the eigenvalues of the Jacobian are q-fold degenerate, and this leads to a q-fold degeneracy in the Lyapunov exponents. It, therefore, suffices to examine the co-prime pair (d, k) since these form the *base* for all other values of $k < d$, leading to what we call as the *base-maps*, as can be seen by the following argument. For the $n = qm$th iterate, the substitution $x_{q\cdot}$ by $\xi_{\cdot}$ gives

$$\begin{aligned} x_{qm} &= 1 - a|x_{qm-qk'}| - (1-\nu)x_{qm-qd'} \\ &\downarrow \\ \xi_m &= 1 - a|\xi_{m-k'}| - (1-\nu)\xi_{m-d'}. \end{aligned} \tag{4}$$

The maps Equation (3) and Equation (4) differ in that there are q hidden variables within each ξ_m. Thus the Jacobian can be separated into q identical blocks, giving rise to a q-degenerate (d', k') map.

2.2. Fixed points: stability

The generalized map, Equation (3), can be rewritten as a d-dimensional iteration,

$$\begin{aligned} x_{n+1}^{(1)} &= 1 - a|x_n^{(k)}| - (1-\nu)x_n^{(d)} \\ x_{n+1}^{(2)} &= x_n^{(1)} \\ &\vdots \\ x_{n+1}^{(d)} &= x_n^{(d-1)} \end{aligned} \tag{5}$$

The fixed points of the mapping are those for which $\{x_{n+1}^{(1)}, \ldots, x_{n+1}^{(d)}\} = \{x_n^{(1)}, \ldots, x_n^{(d)}\}$. Solving, we find

$$\begin{aligned} x^{(1)*} &= x^{(2)*} = \cdots = x^{(d)*} = x_{\pm} \\ \text{with } x_{\pm} &= \frac{1}{2 - \nu \pm a}, \end{aligned} \tag{6}$$

of these two fixed points $x_{\pm}, x_{-}$ is always unstable. The matrix elements of the Jacobian of the map in Equation (3) are given by

$$[\mathbf{J}]_{ij} = \begin{cases} 1 & j = i-1, \quad d \geq i \geq 2 \\ \pm a & i = 1, \quad j = k \\ -(1-\nu) & i = 1, \quad j = d \\ 0 & \text{otherwise} \end{cases}, \tag{7}$$

and it is straightforward to obtain the stability conditions on the fixed point x_+ (for arbitrary d and k) from the characteristic polynomial $P(\lambda)$,

$$P_{1,2}(\lambda) = \lambda^d \pm a\lambda^{d-k} + (1-\nu). \tag{8}$$

If the number of real roots of the polynomials $P_{1,2}$ with real part greater than +1 (or smaller than -1) is $\sigma_{1,2}^{+}$ $(\sigma_{1,2}^{-})$, then according to Feigin's classification of border-collision bifurcations in piecewise smooth maps [8], a non-smooth fold bifurcation occurs when $\sigma_1^+ + \sigma_2^+$ is odd. If $\sigma_1^- + \sigma_2^-$ is odd, on the other hand, a non-smooth flip bifurcation occurs [8].

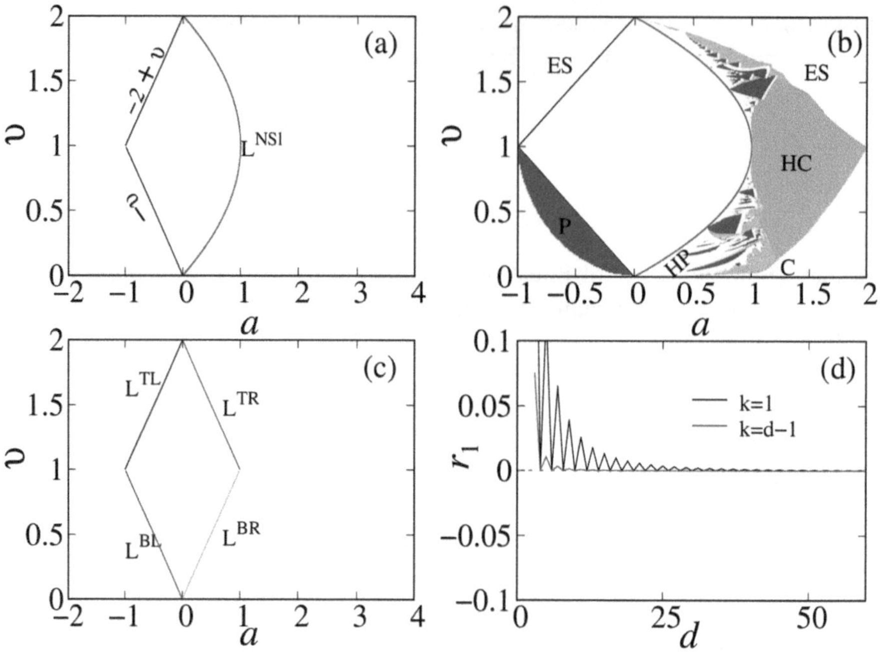

Figure 1. (a) Region of period-1 dynamics for $d = 3$, $k = 2$; see Equation (9). (b)The organization of a–ν plane into unbounded dynamics (ES) and bounded dynamics: hyperchaotic (HC,orange), chaotic (C, green), periodic (P,blue). White regions sandwiched between period one dynamics and hyperchaotic/chaotic/periodic corresponds to high period dynamics (HP)). (c) The region of period one dynamics converges asymptotically (i.e. for $d \to \infty$) and is contained within the curves $L^{TL}, L^{TR}, L^{BL}, L^{BR}$. (d) The convergence to the curves TR, BL, BR is captured by the distance r_1 of the largest root from the unit circle as a function of dimension d on these curves. L^{NSl} indicates the Neimark–Sacker-like boundary, and the other labels L^{TR}, L^{TL}, L^{BR} and L^{BL} are short for top left, right, and bottom left, right.

The characteristic of the Neimark–Sacker bifurcation in smooth dynamical systems is that a pair of complex eigenvalues cross the unit circle. This theory has been extended to piecewise smooth maps only recently [19] and although some of the features are common, there are major differences [19]. We find from numerical estimation of eigenvalues of $P_{1,2}$ for the base maps that a Neimark–Sacker-like [19] bifurcation occurs via a border-collision if an odd number of pairs of complex eigenvalues cross the unit circle. In particular, for the base map with $d = 3, k = 2$ we find that the stability region of the fixed point x_+ is bounded by the curves as shown in Figure 1(a):

$$L^{\text{ns-fold}}(d) \;:\; a = -2 + \nu$$

$$L^{\text{NSl}}(d = 3) \;:\; a = \nu(2 - \nu).$$

$$L^{\text{ns-flip}} \;:\; a = -\nu.$$

If the polynomials produce their largest roots with absolute values less than one, the fixed point x_+ is stable and contributes to the period-one region in the a–ν parameter space. The stability of fixed point is lost as either a high period/quasi-periodic or a period-two orbit is stabilized via a bisecting bifurcation, where a continuum of orbits exists at the bifurcation point [7]. Although the equivalence between a border collision bifurcation and bisection bifurcation has been argued [7] at the period-1–period-2 and period-2–period-4 boundary [7], similar results are not available for the generalized Lozi map studied here. However, we concur based on numerical evidence that bisecting bifurcations are the norm in this family of maps even for bifurcation from fixed point to quasiperiodic/high periodic orbits. Since these bifurcation clearly show that high period orbits hit the border $x = 0$ at bifurcation points, we attribute this distinct feature of the generalized version of the Lozi map (3) to bisecting bifurcations accompanied by a border-collisions in the map.

2.3. Bifurcations diagrams

Bifurcation diagram in the two-dimensional a–ν parameter space for $d = 3$, $k = 2$ is shown in Figure 1(b). A typical feature of a–ν parameter space is the existence of hyper-chaotic (HC), chaotic (C), high-period (HP) (which could include quasiperiodic curves in the parameter space), and periodic regions (tounge like features). These features are shared by other members (with different d and k values) of this generalized Lozi map. Representative bifurcation diagrams as a function of a for different (d, k) combinations are shown in Figure 2 and the corresponding orbital characteristics are shown via Lyapunov exponents in Figure 3. In each case, ν is different but fixed. An interesting feature of these bifurcation diagrams is that for $0 < \nu < 1$ some base maps exhibit bounded dynamics as the parameter a is decreased to the left of the period one boundary. Typically these were found to be a non-smooth flip bifurcation below $a < -\nu$ when $d = 3$, $k = 2$ and non-smooth Neimark–Sacker-like bifurcations for $d = 4$, $k = 3$. Such feature are typical

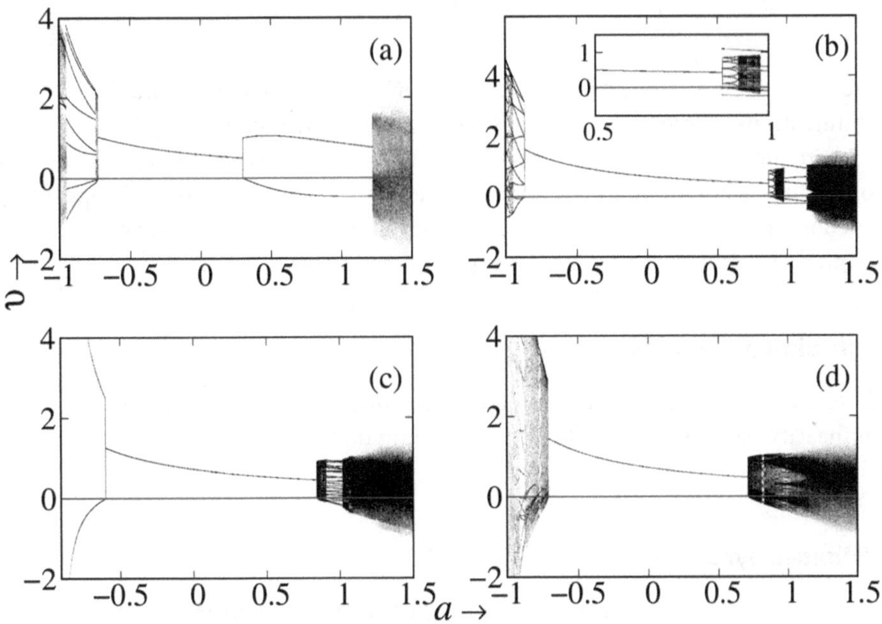

Figure 2. Bifurcation diagrams as a function of the nonlinearity a for different base maps (a) $d = 3, k = 1$, $\nu = 0.3$ (b) $d = 4, k = 1, \nu = 0.5$, (c) $d = 3, k = 2, \nu = 0.6$, (d) $d = 4, k = 3, \nu = 0.6$. The boundaries of stable period-1 dynamics are where high period orbits are created/destroyed via the bisection bifurcation that accompanied by border-collisions. The red line indicates the border at $x = 0$.

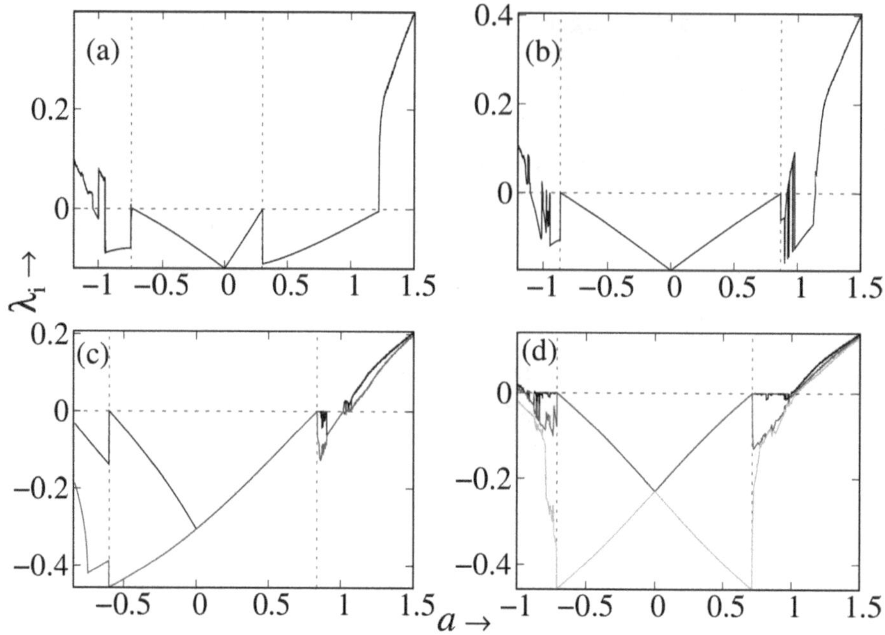

Figure 3. (Color online) k-largest Lyapunov exponents for the systems corresponding to the bifurcation diagrams shown in Figure 2. The different base maps are (a) $d = 3, k = 1, \nu = 0.3$, (b) $d = 4, k = 1, \nu = 0.5$, (c) $d = 3$, $k = 2$, $\nu = 0.6$, and (d) $d = 4$, $k = 3$, $\nu = 0.6$. The boundaries of the stable period-1 dynamics are indicated by dashed vertical lines; the boundaries are border-collision bifurcation points of different orbits depdening on d and k. The largest Lyapunov exponent is coloured black, the second largest in red (applicable only in (c) and (d)) and the third largest in green (applicable only in (d)).

of these maps even across different combination of dimensionality parameters (d,k). In these bifurcations, numerical evidence suggests that a form of bisecting bifurcation takes place, where the high period orbit hits the border at $x = 0$. We should mention that such phenomenon was not found for a similarly generalized Hénon map [6] and appears to be a result of border-collision bifurcations. It is important to note that the theory of bifurcations in smooth dynamical systems does not explain these features [12].

3. High-dimensional dynamics

In this section, we examine the bifurcations starting from the fixed point as a function of nonlinearity parameter for different embedding dimensions d and the endomorphism dimension k.

3.1. Bounded dynamics

In the limit $d \to \infty$, the stability region of period-1 motion converges to a region shown in Figure 1(c): the boundaries are the following curves,

$$L^{TL} \; : \; a = -2 + \nu$$

$$L^{TR} \; : \; a = 2 - \nu$$

$$L^{BL} \; : a = -\nu$$
$$L^{BR} \; : a = \nu, \tag{9}$$

in the $a - \nu$ parameter space. These curves can be understood from the properties of the characteristic polynomials Equation (8) in the limit of $d \to \infty$. The distance r_1 between the leading root of the characteristic polynomial Equation (8) and the unit circle on the curve L^{TL}, in the extreme case of $k = 1$ and $k = d-1$, are shown in Figure 1(d): r_1 approaches zero as the dimension d is increased. Similar behaviour of r_1 is also observed on L^{BR}, L^{BL}, T^{LR} and for $1 < k < d-1$.

3.2. *Hyperchaos*

The map Equation (3) exhibits at most k positive Lyapunov exponents as nonlinearity parameter a is varied: this is due to the fact that the piecewise-smoothness in the map occurs at the kth previous iteration step (therefore the map remains piecewise-smooth). The stretching and folding that is responsible for introducing sensitivity to initial conditions in the map [17], occurs in k directions, and this results in the maximum k number of possible positive Lyapunov exponents; see Figure 4(a,b). Additionally, for $\nu = 1$, the map is a k-dimensional endomorphism (and is not invertible) with k-fold degenerate Lyapunov exponents, i.e. all of these LEs are identical and become positive at the same value of the nonlinearity parameter a as can be seen in Figure 4(c). Embedding the k-dimensional endomorphism in a d-dimensional space does not change this behaviour. However, when the map is made diffeomorphic by enabling the contraction/dissipation parameter ($\nu \neq 1$ & $0 \leq \nu \leq 2$) the degeneracy in the Lyapunov exponents is lifted, although the maximum possible number of positive Lyapunov exponents is still k as can be seen in Figure 4(d).

Route to chaos is observed via the quasiperiodic and also via finite period-doubling route, as seen in the Lyapunov spectra Figure 3 and Figure 4. The period-doubling cascade terminates after a few doublings, leading to chaos. On the other hand, chaos and hyperchaos transition is smooth, since the first k-largest Lyapunov exponents behave smoothly as they hierarchically become positive at different values of the nonlinear parameter a. This typically means that the map Equation (3) can be written as a hierarchy of chaotically driven maps at subsequent transitions to higher chaos [10], this is similar to the chaos hyperchaos transition in the generalized Hénon map [6].

4. Smooth approximations

In this section, we analyse a smooth approximation of the generalized Lozi map (3). We replace the modulus function $|\cdot|$ in Equation (2) with a smooth function $S_\epsilon(\cdot)$:

$$x_n = 1 - aS_\epsilon(x_{n-k}) - (1-\nu)x_{n-d}$$
$$S_\epsilon(x_{n-k}) = \begin{cases} x_{n-k}^2/2\epsilon + \epsilon/2 & \text{if } |x_{n-k}| \leq \epsilon \\ |x_{n-k}| & \text{if } |x_{n-k}| \geq \epsilon \end{cases}, \tag{10}$$

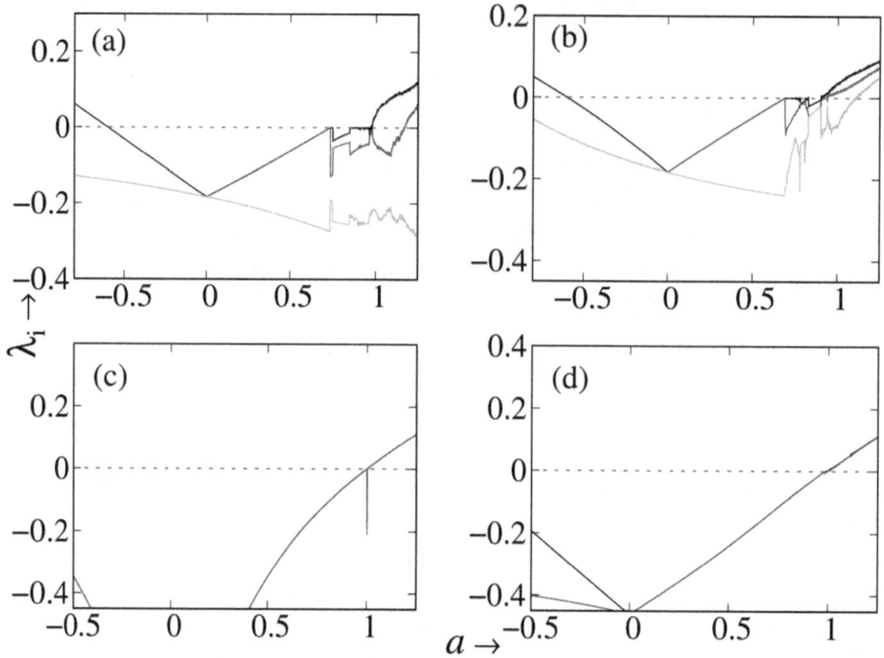

Figure 4. The Lyapunov exponent spectrum ($\lambda_i, i = 1, \ldots, k+1$) as a function of nonlinearity parameter for $d = 5$. (a) and (b) demonstrate that for a given dimension d the number of positive Lyapunov exponents are governed by the parameter k. We keep $\nu = 1.4$ and plot all the Lyapunov exponents for $k = 2$ in (a) and $k = 4$ in (b). (c) and (d) demonstrate the breaking of degeneracy in the Lyapunov spectrum observed in the endomorphism $\nu = 1$ by making the map diffeomorphic $\nu = 1.1$ (see text for explanation). The largest Lyapunov exponent is coloured black, the second largest in red and the third largest in green (applicable in (a) and (b)).

where $0 < \epsilon < 1$. The function $S_\epsilon(x_{n-k})$ extends the smooth approximation applied to the two-dimensional Lozi map [3] to our high-dimensional generalization of the Lozi map and removes the discontinuity in the slope at $x_{n-k} = 0$.

The fixed points of the new map in Equation (10) are given by:

$$x_{\pm} = \begin{cases} \dfrac{1}{2 - \nu \pm a} & \text{if } |x| > \epsilon \\ \dfrac{\epsilon}{a}\left(-(2-\nu) \pm \sqrt{(2-\nu)^2 + \dfrac{2a}{\epsilon} - a^2}\right) & \text{if } |x| \leq \epsilon \end{cases} \tag{11}$$

the first set of these fixed points are similar to those of Lozi map (see Equation (6)) and lose stability by colliding with one of the borders located at $\pm\epsilon$ as nonlinearity parameter a is varied. The new orbits that appear following this border-collision bifurcation depend on the delay parameters d and k, although the maximum number of positive Lyapunov exponents is still limited to k. Assuming fixed values of the dissipation parameter and ϵ, the variation in nonlinearity parameter can take iterations of the map also inside the region $|x| < \epsilon$ then subsequent bifurcations are no longer only due to border collision: borders

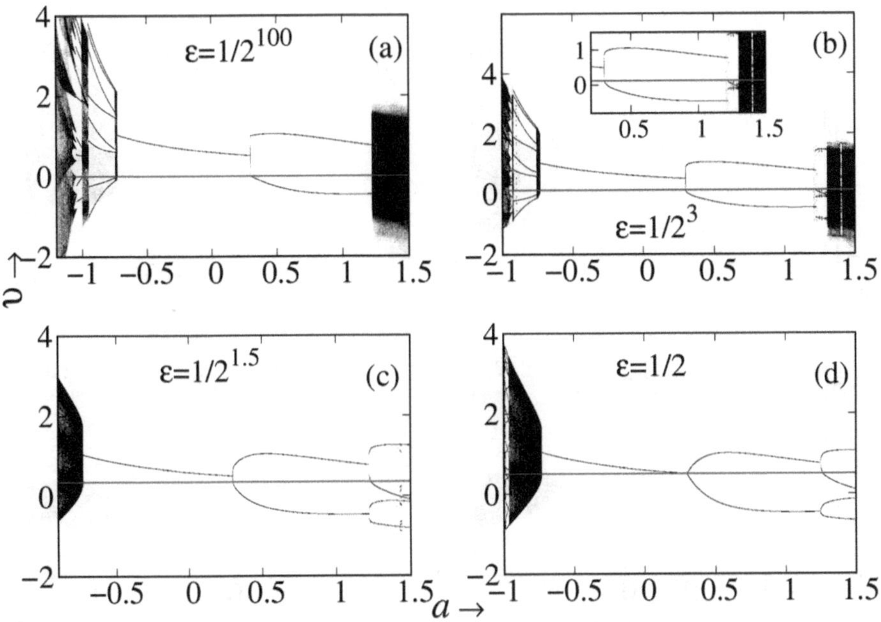

Figure 5. The bifurcation diagram for $d = 3$, $k = 1$ for different values of ϵ. As ϵ is increased from (a) to (d) characteristics of smooth bifurcations emerge in the form of period-doubling.

at $\pm\epsilon$ have well defined first derivatives and once an orbit enters the region $|x| < \epsilon$ the dynamics is also governed by the smooth approximation.

The case of $d = 2$ and $k = 1$ was illustrated in [3], where it was observed that the period-doubling route to chaos is achieved for finite values of ϵ (note that period-doubling route to chaos is absent for $\epsilon = 0$).

For $d > 2$ bifurcation scenarios easily understood by writing the smoothness parameter as $\epsilon = 1/2^{n_\epsilon}, n_\epsilon \geq 0$. For the case of $d = 3$, $k = 1$ the loss of stability of the fixed point as nonlinearity parameter is varied is dependent on n_ϵ: border-collision bifurcations gives way to period–doubling bifurcations as n_ϵ decreases Figure 5(a–d). Similar observations are made when a Neimark–Sacker type bifurcation is involved in the Lozi map with $d = 3$, $k = 2$, where border-collision bifurcations Figure 6(a) give way to smooth bifurcations Figure 6(d). In the Hénon maps the chaotic dynamics follows directly after the Neimark–Sacker bifurcation via a crisis-like transition [6]. Thus the effect of $\epsilon > 0$ is to introduce bifurcations which are a mix of border collisions and smooth bifurcations.

5. Discussion and summary

In this paper, we have introduced a generalized time-delayed Lozi map with nonlinear feedback from k earlier steps and linear feedback from d earlier steps. This simple feedback process governs the dimensionality of the maps. The parameter k ($< d$) determines the number of positive Lyapunov exponents. Furthermore, the family of maps formed as a result of different combinations of dimensionality parameters are classified into base maps when d and k are co-prime. All other maps reducible to these base maps exhibit

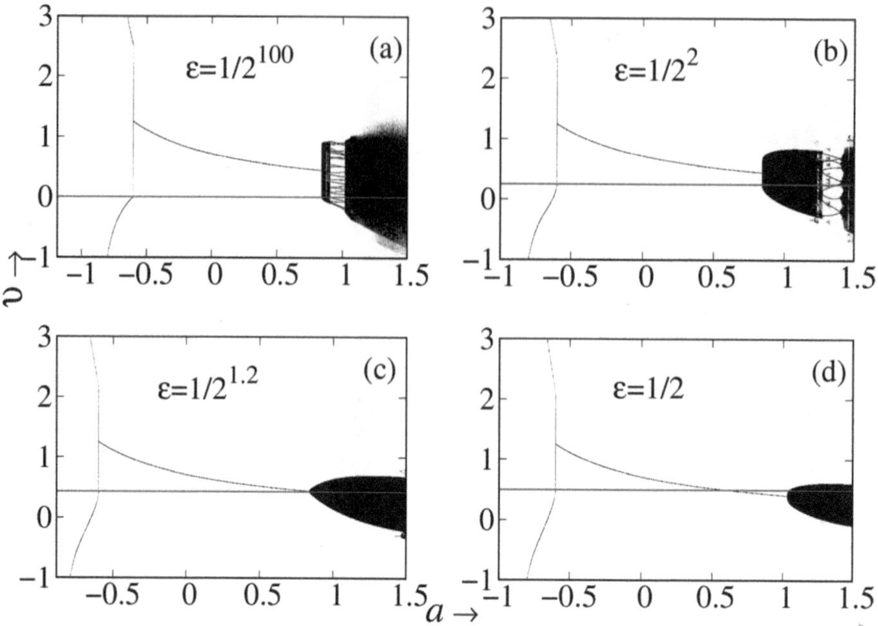

Figure 6. The bifurcation diagrams for $d = 3, k = 2$ for different values of ϵ. As ϵ is increased from (a) to (d) characteristics of smooth bifurcations emerge in the form of period–doubling. The figures (d)–(a) also indicate emergence of border-collision bifurcation beginning from smooth period-doubling bifurcations in (d).

a q-fold degenerate Lyapunov spectrum when d and k share a common factor q. Bifurcation analysis was performed in a limited region of the parameter space. In particular, fixed point dynamics is stabilized in regions of the parameters space surrounded by other invariant sets created via the non-smooth fold, non-smooth flip and the non-smooth Neimark–Sacker-like bifurcations which are often accompanied by border collisions. Analytic forms were determined for some of these boundaries, and the non-smooth flip and NS-like bifurcation curves were found to depend on the dimension d. With increasing dimension, the region of period-one dynamics was found to converge in the parameter plane.

The dynamics evolve abruptly from regular to chaos due to piecewise nature of the map. Subsequent transitions from chaos to hyperchaos, however, are smooth as indicated by Lyapunov spectrum: the dimension of the attractor changes smoothly if there are no abrupt transitions in Lyapunov spectrum [6,10].

A smooth approximation of the map enabled the analysis of the bifurcations vis-á-vis further comparing some of the bifurcations to the generalized the Hénon map. It showed that some of the bifurcations observed persist on both the piecewise Lozi and Hénon map. Further exploration of a more general unified mapping is a project for future work. Another possible project for a future work is the analysis of conservative limit in these class of maps: orbits in the conservative limit are only possible when $d = 2k$; therefore, in the conservative limit, hyperchaotic orbits are indeed possible with $k-$positive Lyapunov exponents. The exploration of the conservative limit of this map could be a task for future work.

Acknowledgments

S. B. would like to thank UGC (Govt of India) for support. S. B. and R. R. designed the study. S. B. performed the analysis. S. B. and R. R. wrote the manuscript. All authors read and approved the final manuscript.

Disclosure statement

The authors declare no conflict of interest.

Funding

R. R. is supported by the DST (Govt of India) through the JC Bose fellowship.

ORCID

Shakir Bilal http://orcid.org/0000-0003-4425-0805

References

[1] D.J. Albers and J.C. Sprott, *Routes to chaos in high-dimensional dynamical systems: A qualitative numerical study*, Phys. D 223 (2006), pp. 194–207.
[2] V.I. Arnold, *Small denominators. I. On the mappings of the circumference onto itself*, Trans. Am. Math. Soc. 2nd Ser. 46 (1965), pp. 213–284.
[3] M.A. Aziz-Alaoui, C. Robert and C. Grebogi, *Dynamics of a Hénon–Lozi-type map*, Chaos Solit. Fractal 12 (2001), pp. 2323–2341.
[4] G. Baier and M. Klein, *Maximum hyperchaos in generalized Hénon maps*, Phys. Lett. A 151 (1990), pp. 281–284.
[5] S. Banerjee and C. Grebogi, *Border collision bifurcations in two-dimensional piecewise smooth maps*, Phys. Lett. A 59 (1999), pp. 4052–4061.
[6] S. Bilal and R. Ramaswamy, *The generalized time-delayed Henon map: bifurcations and dynamics*, Int. J. Bifurc. Chaos. 23 (2013), pp. 1350045–7.
[7] V. Botella-Soler, J.M. Castelo, J.A. Oteo and J. Ros, *Bifurcations in the Lozi map*, J. Phys. A: Math. Theor. 44 (2011), pp. 305101–14.
[8] M. di Bernardo, C.J. Budd, A.R. Chumpney and P. Kowalczyk, Piecewise-Smooth Dynamical Systems:Theory and Applications, Springer-Verlag, London, 2008.
[9] Z. Elhadj, Lozi Mappings: Theory and Applications, CRC Press, Boca Raton, FL, 2014.
[10] M.A. Harrison and Y-C. Lai, *Bifurcation to high-dimensional chaos*, Int. J. Bifurc. Chaos. 10 (2000), pp. 1471–1483.
[11] M. Hénon, *A two-dimensional mapping with a strange attractor*, Commun. Math. Phys. 50 (1976), pp. 69–77.
[12] Y.A. Kuznetsov, Elements of Applied Bifurcation Theory (2/e), Springer Verlag, New York, 1998.
[13] T.Y. Lee and J.A. Yorke, *Period three implies chaos*, Amer. Math. Monthly. 82 (1975), pp. 985–992.
[14] R. Lozi, *Un attracteur étrange du type attracteur de Hénon*, J. Phys. Colloq. 39 (1978), pp. C5C9C5-9 -C5-10.
[15] R.M. May, *Simple mathematical models with very complicated dynamics*, Nature 261 (1976), pp. 459–467.
[16] S. Newhouse, D. Ruelle and F. Takens, *Occurrence of strange axiom A attractors near quasi periodic flows on T^m, $m \geq 3$*, Commun. Math. Phys. 64 (1978), pp. 35–40.
[17] E. Ott, Chaos in Dynamical Systems, Cambridge University Press, New York, 1993.
[18] D. Rand, S. Ostlund, J. Sethna and E.D. Siggia, *Universal transition from quasiperiodicity to chaos in dissipative systems*, Phys. Rev. Lett. 49 (1982), pp. 132–135.

[19] D.J.W. Simpson and J.D. Meiss, *Neimark-Sacker bifurcations in planar, piecewise-smooth, continuous maps*, SIAM J. Appl. Dyn. Sys. 7 (2008), pp. 795–824.
[20] D.J.W. Simpson and J.D. Meiss, *Aspects of bifurcation theory for piecewise-smooth, continuous systems*, Phys. D 241 (2012), pp. 1861–1868.
[21] J.C. Sprott, *High-dimensional dynamics in the delayed Hénon map*, Elec. J. Theor. Phys. 3 (2006), pp. 19–35.

Existence proofs for strange attractors in piecewise-smooth nonlinear Lozi-Hénon and Belykh maps

Vladimir N. Belykh, Nikita V. Barabash and Dina A. Grechko

ABSTRACT
Extensive research has previously been carried out on the existence of strange attractors in 2D piecewise linear maps, including the renowned Lozi map. However, the rigorous analysis of strange attractors in 2D nonlinear maps remains an underdeveloped area of study. In this paper, we introduce a 2D map with a single piecewise-smooth nonlinear function that is suitable for analytic studies of its strange attractor. This 2D piecewise-smooth nonlinear map represents a broad range of chaotic maps, including a hybrid Lozi-Hénon map and the Belykh map. To prove the existence of a strange attractor in the 2D map, we (i) construct its trapping region that contains all limit sets and (ii) demonstrate that the invariant set's trajectories have negative and positive Lyapunov exponents. We develop an invariant cone approach to establish the latter property, which involves constructing expanding and contracting cones as bounds for the eigenvectors of the variational equations along the chaotic trajectories. We apply our approach to analyze the chaotic hybrid Lozi-Hénon map, the original Lozi map, and the Belykh map.

1. Introduction

Strange attractors are fascinating and complex dynamical objects that can emerge in dynamical systems exhibiting chaos [22]. While rigorous analytic proofs of the existence of strange attractors in continuous-time dynamical systems are rare and typically limited to piecewise-smooth systems [11, 12], chaotic discrete-time maps are more analytically tractable. The piecewise-linear Lozi [32] and Belykh maps [13, 19] as well as the smooth nonlinear Hénon [29] are prime examples of 2D maps with chaotic attractors whose properties have been analytically studied (see [1, 8, 10, 14, 15, 17, 20, 21, 24–28, 33–35, 49] and references therein). Yet, a complete rigorous analysis of the Hénon map and other smooth nonlinear 2D maps remains challenging.

In this paper, we introduce a general 2D map that contains a single piecewise-smooth nonlinear function and is amenable to analytical investigations of its strange attractor. This map represents a wide range of chaotic maps, including a hybrid Lozi-Hénon map, and can be viewed as a proxy for studying the dynamics of smooth nonlinear 2D maps.

This general 2-D map $f : \mathbb{R}^2 \to \mathbb{R}^2$ has the form [10, 19]:

$$f : \begin{array}{l} \bar{x} = x + y + g(x), \\ \bar{y} = \lambda \left(y + g(x) \right), \end{array} \tag{1}$$

where $(\bar{x}, \bar{y})$ indicates the subsequent iterate (x_{k+1}, y_{k+1}) of (x_k, y_k) under the action of f, the function $g(x)$ is piecewise-smooth, and the parameter $\lambda \in (0, 1)$.

It is worth noticing that the Lozi and Hénon maps that are written in the standard form

$$\begin{aligned} \bar{x} &= y + 1 - aU(x), \\ \bar{y} &= bx, \end{aligned} \tag{2}$$

where the function $U(x) = |x|$ represents the Lozi map [32], and $U(x) = x^2$ corresponds to the Hénon map [29] can be cast into the form of the map (1). This can be done through the change of variables

$$u = x, \quad v = y - bx \tag{3}$$

which yields the map

$$\begin{aligned} \bar{u} &= v + u + (-u + bu + 1 - aU(u)), \\ \bar{v} &= -bv - b(-u + bu + 1 - aU(u)). \end{aligned} \tag{4}$$

Returning to the original variables (x, y) instead of (u, v) in (4) and denoting

$$g(x) = 1 - (1 + \lambda)x - aU(x), \quad \lambda = -b, \tag{5}$$

we obtain the general map (1). Note that the Hénon map with the smooth nonlinear function $U(x) = x^2$ does not belong to the assumed class of piecewise-smooth functions $g(x)$. Instead, we introduce a piecewise-smooth nonlinear map that we call the hybrid Lozi-Hénon map (1)–(5) which is determined by the function

$$U(x) = l|x| + (1 - l)x^2, \tag{6}$$

where the parameter $l \in (0, 1]$. The function (6) yields the Hénon map in the limit $l \to 0$, and the Lozi map for $l = 1$.

The general 2D map (1) also represents the chaotic Belykh map [13, 16] with the periodic function

$$g(x) = a(x - 1), \quad x(\mod 2). \tag{7}$$

In the following, we will analyze the 2D map (1) and derive sufficient conditions for the existence of strange attractors in the original Lozi map, hybrid Lozi-Hénon map

(1)–(5)–(6), and Belykh map (1)–(7). Toward this goal, for each of the three maps, we will construct a trapping region that contains all system's attractors and then prove that all trajectories confined inside the trapping region have negative and positive Lyapunov exponents. We introduce an invariant cone method to establish the latter property, which involves constructing expanding and contracting cones as bounds for the eigenvectors of Df along the chaotic trajectories.

Sufficient conditions for the existence of strange attractors in the 2D map (1) were previously obtained for a class of piecewise-smooth nonlinear functions f under some moderately restrictive conditions on the derivative of f [10]. These conditions included (i) the existence of a trapping region and the maximal attractor $\mathcal{A}$ and (ii) weak hyperbolicity of $f|_{\mathcal{A}}$. The construction of the trapping region was performed through the analysis of the mutual arrangement of stable and unstable manifolds of saddle fixed points. The weak hyperbolicity property [10] was proven by constructing expanding and contracting invariant cones. However, the hybrid Lozi-Hénon map (1)–(5)–(6) does not belong to the class of functions f studied in [10], requiring a non-trivial extension of the approach from [10] to proving the existence and hyperbolicity of the hybrid Lozi-Hénon map. Similarly, although existence proofs have been established for various 2D piecewise linear maps [21, 23, 25, 28, 32–35, 37, 39], extending these proofs to 2D piecewise nonlinear maps may not be straightforward and could present challenges.

In this paper, we address this gap by deriving sufficient conditions regarding the existence and hyperbolicity of an attractor in the Lozi-Hénon map (1)–(5)–(6). A noteworthy aspect of our approach is the derivation of upper and lower bounds for the Lyapunov exponents of all trajectories confined within the trapping domain, instead of isolating a specific chaotic trajectory from an infinite set of possible solutions and estimating its Lyapunov exponents. Extensive research was previously conducted on the orientation-preserving case of the Lozi map, as outlined in the introduction of [28] and in [21, 39]. Nonetheless, we will analyze the Lozi map in the form of (1)–(5) with $U(x) = |x|$ to illustrate the application of our approach and the direct construction of stable and unstable manifolds that enclose a trapping region. In particular, we will demonstrate that our approach can significantly simplify the rigorous analysis of the Lozi map. We will use the Belykh map as an additional example to demonstrate the effectiveness of our approach in analyzing the dynamical properties and structure of strange attractors.

The paper is structured as follows. Section 2 provides the definition and terminology used in our study. In Section 3, we analyze the 2D map (1) with the function $g(x)$ satisfying some general conditions and prove the existence of a trapping region and hyperbolicity of an attractor confined within it. In Sections 4, 5, and 6, we apply our approach to the Lozi, Lozi-Hénon, and Belykh maps, respectively. In Section 7, we establish a connection between 2D maps and physical models and demonstrate how the Belykh map arises from a model of phase-locked loop control systems. Finally, Section 8 contains concluding remarks.

2. Main definitions

We begin by adapting the standard definitions of a trapping region, the maximal attractor, and hyperbolicity, which were originally introduced for smooth maps [7, 46], to our piecewise-smooth map (1).

Definition 2.1: Let $G \subset \mathbb{R}^2$ be a compact domain that serves as a trapping region for the map f, such that $fG \subset G$. Then, the set

$$\mathcal{A} = \bigcap_{n=1}^{\infty} f^n G$$

is said to be the maximal (topological) attractor.

This attractor is an invariant set of full orbits under forward and backward iterates of any point of $\mathcal{A}$.

Definition 2.2: Maximal attractor $\mathcal{A}$ comprised only of unstable full trajectories is called a strange attractor [5, 18].

This 'physical' definition of a strange attractor as a maximal attractor composed only of unstable entire trajectories implies no more than 'stable instability.' However, for smooth maps, the hyperbolicity condition of an attractor, as defined by Anosov [4, 7], Katok [30], and others, ensures structural stability, mixing, and the existence of Sinai-Bowen-Ruelle (SBR) measures – the attributes of robust chaos. Strange attractors are classified into hyperbolic attractors and singular hyperbolic attractors. Hyperbolic attractors are structurally stable, while singular hyperbolic attractors undergo bifurcations of homoclinic and periodic orbits, such as in the Lorenz system [2, 11, 47, 48], or border-collision bifurcations in piecewise-smooth maps [23, 25–28, 37]. The use of hyperbolicity in chaos analysis of piecewise-smooth maps requires definitions for sets of discontinuities of f and Df. In the following, we employ a sufficient condition for the existence of a hyperbolic attractor based on the general definition of hyperbolicity of a compact set that is invariant under f [1, 4, 7, 30].

Let DF be the Jacobian matrix of a map F generating the linear map

$$\bar{\mathbf{u}} = L\mathbf{u}, \quad \mathbf{u} = \begin{pmatrix} u_1 \\ u_2 \end{pmatrix} \tag{8}$$

in the tangent space.

Definition 2.3: The cones $K^u(k_1^u, k_2^u)$, $K^u(k_1^s, k_2^s)$

$$K^u = \left\{ k_1^u < \frac{u_2}{u_1} < k_2^u \right\}, \quad K^s = \left\{ k_1^s < \frac{u_1}{u_2} < k_2^s \right\} \tag{9}$$

are said to be invariant if

$$DfK^u \subset K^u, \quad Df^{-1}K^s \subset K^s.$$

The cone K^u (K^s) is expanding (contracting) if $|L\mathbf{u}| > q|\mathbf{u}|$ ($|L^{-1}\mathbf{u}| > p|\mathbf{u}|$), where $q > 1$, $p > 1$.

Let $f : M \to M$ be a smooth map, where M is a $2D$ smooth manifold [1]. Let Λ be a compact invariant set of f. If, for any point $x \in \Lambda$, there exist an expanding cone K^u and a contracting cone K^s, then the set Λ is said to be hyperbolic. If Λ is an attractor $\mathcal{A}$, then it is a hyperbolic attractor. We will modify these definitions to fit the piecewise-smooth scenario, and present the following.

Definition 2.4: Consider the piecewise-smooth map f defined on $G_+ \subset \mathbb{R}^2$, and let S_0 be the set of points where the derivative Df is discontinuous and the function f may be discontinuous as well. Define $S = \bigcup_{i=0}^{\infty} f^{-i} S_0$, and let $G = G_+ \setminus S$ be a trapping region of f. Then, the intersection $\mathcal{A} = \text{closure} \bigcap_{i=0}^{\infty} f^i G$ is an attractor of f. If every orbit of G is hyperbolic, then $\mathcal{A}$ is a singular hyperbolic attractor.

This definition is similar to the definition of singular hyperbolic attractors for flows [36, 42] which include examples such as the Lorenz attractor.

3. Hyperbolicity conditions

In this section, we investigate the hyperbolic properties of the map (1) where the function $g(x)$ has the derivative $g'(x) \triangleq d(x)$ that is bounded away from zero.

Let I^0 be a finite set of points where the derivative changes sign, and let I^- and I^+ be sets of alternating intervals satisfying:

$$d(x) > 0,\ x \in I^+, \quad d(x) < 0,\ x \in I^-.$$

Denote $G^{\pm} = \{x, y | x \in I^{\pm}, y \in \mathbb{R}^1\}$, $G = G^+ \cup G^-$. The linearization of the map (1), Df, yields the map

$$Df : \bar{\mathbf{u}} = L(d)\mathbf{u},$$

where

$$\mathbf{u} = \begin{pmatrix} u_1 \\ u_2 \end{pmatrix}, \quad \bar{\mathbf{u}} = \begin{pmatrix} \bar{u}_1 \\ \bar{u}_2 \end{pmatrix}, \quad L(d) = \begin{pmatrix} 1 + d(x) & 1 \\ \lambda d(x) & \lambda \end{pmatrix}. \tag{10}$$

The eigenvalues $\mu_{1,2}(d)$ and the eigenvectors $\mathbf{V}_{1,2}(x) = (1, \alpha_{1,2}(d))$ of the matrix $L(d)$ have the form

$$\begin{aligned} \mu_{1,2}(d) &= \frac{1}{2}\left(1 + d(x) + \lambda \pm \sqrt{(1 + d(x) + \lambda)^2 - 4\lambda}\right), \\ \alpha_{1,2}(d) &= \frac{1}{2}\left(-1 - d(x) + \lambda \pm \sqrt{(1 + d(x) + \lambda)^2 - 4\lambda}\right). \end{aligned} \tag{11}$$

We refer to the values $\alpha_{1,2}(d)$ as the slopes. The eigenvalues (11) satisfy the following conditions

$$\begin{aligned} &\mu_1(d) > 1, \quad 0 < \mu_2(d) < 1, &&\text{for } d(x) > 0, \\ &-1 < \mu_1(d) < 0, \quad \mu_2(d) < -1, &&\text{for } d(x) < -2(1 + \lambda). \end{aligned} \tag{12}$$

The boundaries of the eigenvectors' slopes are defined by the bounds of the functions $\alpha_{1,2}(d)$. The derivatives of the slopes $\alpha_{1,2}(d)$ with respect to $d(x)$

$$\alpha'_{1,2}(d) = \frac{1}{2}\left(-1 \pm \frac{1 + d(x) + \lambda}{\sqrt{(1 + d(x) + \lambda)^2 - 4\lambda}}\right) \tag{13}$$

indicate the direction of rotation of the eigenvectors.

By using (11) and (13), we determine the following intervals over which the functions $\alpha_{1,2}$ have a monotonic dependence on $d(x)$:

$$\begin{array}{lll} \alpha_1'(d) > 0, & \alpha_1(d) \in (0, \lambda), & \text{for } d(x) \in (0, \infty), \\ \alpha_2'(d) < 0, & \alpha_2(d) \in (-\infty, -1+\lambda), & \\ \alpha_1'(d) < 0, & \alpha_1(d) \in (1+\lambda, \infty), & \text{for } d(x) \in (-\infty, -2(1+\lambda)). \\ \alpha_2'(d) > 0, & \alpha_2(d) \in (\lambda, 2\lambda), & \end{array} \tag{14}$$

From (11), (12) we conclude that the eigenvectors indicate

$$\begin{array}{llll} \mathbf{V}_1|_{G^+} & \text{and} & \mathbf{V}_2|_{G^-} & \text{– unstable directions,} \\ \mathbf{V}_2|_{G^+} & \text{and} & \mathbf{V}_1|_{G^-} & \text{– stable directions.} \end{array} \tag{15}$$

Theorem 3.1: *Let the conditions*

$$\begin{array}{ll} d(x) \geq \varepsilon > 0 & \textit{for } x \in I^+, \\ d(x) < d^- = -2(1+2\lambda) & \textit{for } x \in I^-, \end{array} \tag{16}$$

hold. Then, the map (1) *has invariant cones* (9), *where*

$$k_1^u = \alpha_1(\varepsilon), \quad k_2^u = \alpha_2(d^-); \quad k_1^s = \frac{1}{\alpha_2(\varepsilon)}, \quad k_2^s = \frac{1}{\alpha_1(d^-)}. \tag{17}$$

The expansion and contraction in the cones K^u and K^s are defined by the inequalities

$$|L\mathbf{u}| > q|\mathbf{u}|, \quad \mathbf{u} \in K^u, \, q > 1; \quad |L^{-1}\mathbf{u}| > p|\mathbf{u}|, \quad \mathbf{u} \in K^s, \, p > 1, \tag{18}$$

and the map (1) *is hyperbolic.*

Proof: The family of all possible eigenvectors $\mathbf{V}^u$ and $\mathbf{V}^s$ in the region G under the conditions of the Theorem 3.1 is bounded by the following boundary vectors (see Figure 1)

$$\mathbf{V}_{bot}^u = (1, k_1^u), \quad \mathbf{V}_{up}^u = (1, k_2^u),$$
$$\mathbf{V}_{left}^s = \left(-1, -\frac{1}{k_1^s}\right), \quad \mathbf{V}_{right}^s = \left(1, \frac{1}{k_2^s}\right).$$

Therefore, we conclude, that the images of the unstable boundary vectors, LV_{bot}^u and LV_{top}^u lie within the cone K^u and the preimages $L^{-1}V_{left}^s$ and $L^{-1}V_{right}^s$ lie within the cone K^s. Therefore, the cones (9), (17) are invariant under the condition (16).

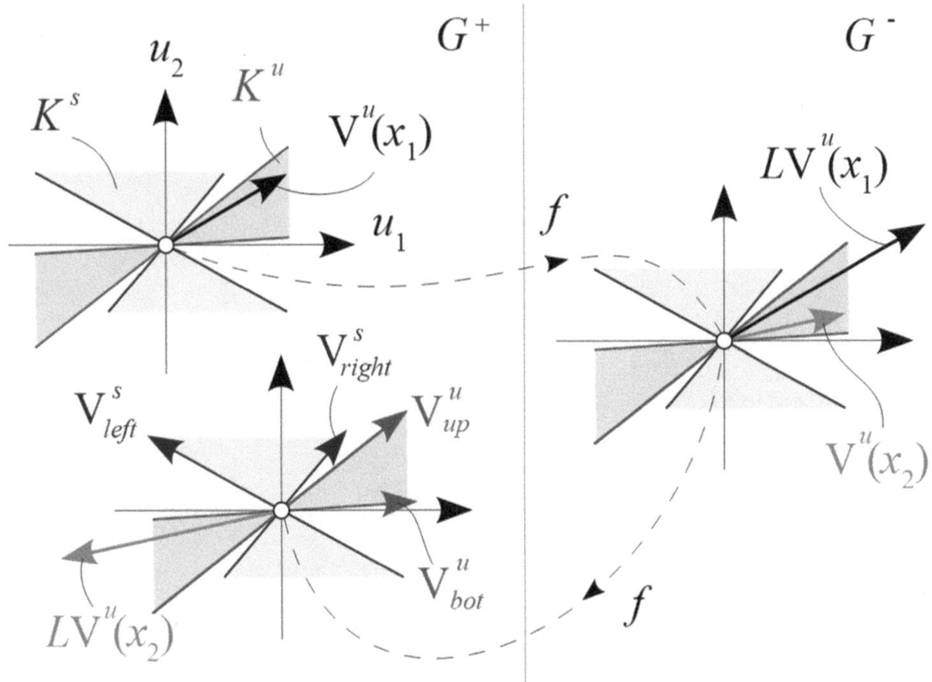

Figure 1. Unstable K^u and stable K^s invariant cones (red and blue triangles, respectively). Boundary vectors of the cones are $\mathbf{V}^u_{up}$, $\mathbf{V}^u_{bot}$ (red arrows) for K^u, and $\mathbf{V}^s_{left}$, $\mathbf{V}^s_{right}$ (blue arrows) for K^s. In the depicted case, while the point (x_1, y_1) is mapped by f (dashed black curves) from G^+ to G^-, its eigenvector $\mathbf{V}^u(x_1) \in K^u$ (black bold arrow) expands under linear map Df. In the next step, point (x_2, y_2) returns to G^+ under the action of f, and its unstable eigenvector $\mathbf{V}^u(x_2) \in K^u$ (gray bold arrow) expands under L rotated by π (due to negative multipliers, see the Equation (11)). All eigenvectors in the stable cone K^s are contracted [not shown].

To obtain conditions for the expansion in (18) of any vector $\mathbf{u}$ in the cone K^u under the condition (16), we first prove the inequality:

$$|L\mathbf{u}| > Q(k)|\mathbf{u}|, \quad \mathbf{u} \in K^u, \tag{19}$$

where $\mathbf{u} = \binom{1}{k}$, $L\mathbf{u} = \binom{1+d+k}{\lambda d+\lambda k}$, $k \in (k_1^u, k_2^u)$, and the function $Q(k) = \sqrt{1 + \frac{\lambda^2 k^2}{1+k^2}}$. The inequality (19) is expressed as follows

$$(1+\lambda^2)d^2 + 2(1+k+\lambda^2 k)d > 0. \tag{20}$$

It is valid in the parameter region defined by

$$(d(x) > 0) \cup \left(d(x) < d_h = -2\left(k + \frac{1}{1+\lambda^2}\right)\right). \tag{21}$$

The inequality (20) is valid for the first condition in (16) because $d > 0$. The second condition in (21) is compatible with the condition $d < d^-$ because the inequality

$$-2(1+2\lambda) < -2\left(k + \frac{1}{1+\lambda^2}\right), \quad k \in (k_1^u, k_2^u)$$

holds. Therefore, the cone K^u is expanding with the factor $q = \min_{\mathbf{u}\in K^u} Q(k) = Q(k_1^u) > 1$.

To determine the contraction condition in (18), we start by considering the inequality

$$|L^{-1}\mathbf{u}| > P(k)|\mathbf{u}|, \quad \mathbf{u} \in K^s, \tag{22}$$

where $\mathbf{u} = \binom{k}{1}$, $L^{-1} = \begin{pmatrix} 1 & -\frac{1}{\lambda} \\ -d & \frac{1+d}{\lambda} \end{pmatrix}$, $k \in (k_1^s, k_2^s)$, and the function

$$P(k) = \sqrt{1+\delta}, \quad \delta = \frac{2 - 2\lambda k - \lambda^2}{\lambda^2(1+k^2)}.$$

The inequality (22) takes the form

$$(1-\lambda k)d\,[(1-\lambda k)d + 2] > 0$$

that is valid for $\forall k \in (k_1^s, k_2^s)$ in the region

$$\{d > 0\} \cup \{d < -2(1+\lambda)\},$$

for which the conditions of the theorem are fulfilled. Therefore, the contraction condition in (18) is valid under the same conditions as the expansion condition. The factor p is given by

$$p = \min_{\mathbf{u}\in K^s} P(k) = \sqrt{1+\delta_0}, \quad \delta_0 = \frac{(1+\lambda)\left(2 - \lambda^2(1+\lambda)\right)}{\lambda^2\left(1 + (1+\lambda)^2\right)} > 0.$$

Thus, both inequalities in (18) are valid. ■

Corollary: *The value* ln q *is a lower bound on the maximum Lyapunov exponent, while the value* $(-\ln p)$ *is an upper bound on the minimum Lyapunov exponent.*

Figure 1 illustrates the first iterates of the map Df along a trajectory of the map f (Equation (1)) in the region G. According to Theorem 3.1, all trajectories in G are of saddle type.

By applying Theorem 3.1, we can conclude the following result.

Theorem 3.2: *Let G be a trapping region, $fG \subset G$, of the map* (1), *satisfying the conditions of Theorem* 3.1. *Then, the map f has a singular hyperbolic attractor.*

Proof: Since G is a trapping region, the map (1) has an attractor $\mathcal{A} = \bigcap_{n=1}^{\infty} f^n G$. According to Theorem 3.1, the attractor $\mathcal{A}$ is singular hyperbolic. The role of the set of points of discontinuity S_0 in Definition 2.4 is played by the set of lines, $S_0 = \{x, y | x \in I^0, y \in \mathbb{R}^1\}$. ■

4. Proving the existence of the Lozi attractor

In this section, we consider the Lozi map written in the form of the map (1) with the function

$$g(x) = 1 - a|x| - (1+\lambda)x, \quad \lambda = -b, \tag{23}$$

in the region $G^-(x > 0) \cup G^+(x < 0)$ for $\lambda \in (0, 1)$, i.e. in the orientation preserving case.

Although the Lozi map and a family of piecewise linear maps have been extensively studied in numerous papers, such as [21, 25–28, 33–35], our aim is to provide an example of the application of Theorem 3.2 to a simple piecewise linear map with only two pieces, $I^- = \{x|x > 0\}$ and $I^+ = \{x|x < 0\}$. As a case in point, we consider the Lozi map.

Theorem 4.1: *Let the parameters* λ *and* a *satisfy the following two conditions*

$$a > 1 + 3\lambda, \quad 0 < \lambda < 1, \tag{24}$$

and

$$H(\lambda, a) \triangleq a(2\lambda - a + 2)\sqrt{a^2 - 4\lambda} + a(2\lambda^2 - 6\lambda - a^2 + 2a) + 4\lambda^2 \geq 0. \tag{25}$$

Then, the Lozi map (1), (23) *has a singular hyperbolic attractor.*

Proof: The derivative of the function (23)

$$d(x) = -a \operatorname{sign} x - 1 - \lambda \tag{26}$$

satisfies the conditions (16) in the parameter region (24) for $0 < \varepsilon < 2\lambda$. Hence, the map (1)–(23) is singular hyperbolic. Note that in this case at each point of G the expanding cone K^u (9) contains only two constant eigenvectors $\mathbf{V}_1|_{x>0}$ and $\mathbf{V}_2|_{x<0}$, and the contracting cone K^s contains only two constant eigenvectors $\mathbf{V}_1|_{x<0}$ and $\mathbf{V}_2|_{x>0}$.

Therefore to prove the existence of the Lozi attractor, it remains to identify the part of the parameter region (24) that corresponds to the existence of a trapping region G_t such that $fG_t \subset G_t$. The map (1), (23) has two saddle fixed points $O_1(\frac{1}{1+\lambda-a}, 0)$ and $O_2(\frac{1}{1+\lambda+a}, 0)$ with multipliers $\mu_{1,2}$

$$\mu_{1,2} = \frac{1}{2}\left(-a \operatorname{sign} x \pm \sqrt{a^2 - 4\lambda}\right),$$

satisfying the conditions (12). The eigenvectors $\mathbf{V}_{1,2} = (1, \alpha_{1,2})$ are given by the slopes

$$\alpha_{1,2} = \frac{1}{2}(a \operatorname{sign} x + 2\lambda \pm \sqrt{a^2 - 4\lambda}).$$

Since the eigenvectors remain constant within each region, the map (1), (23) has two pairs of unstable and stable manifolds $W_{1,2}$ of the fixed points O_1 and O_2

$$W_{1,2}\big|_{x<0} : y = \alpha_{1,2}\left(x + \frac{1}{a-1-\lambda}\right),$$
$$W_{1,2}\big|_{x>0} : y = \alpha_{1,2}\left(x - \frac{1}{a+1+\lambda}\right). \tag{27}$$

We demonstrate the existence of the trapping region G_t by constructing a contour composed of the unstable and stable manifolds of the saddle O_1, along with an additional line segment that encloses the region G_t. To do so, we construct the continuation of the unstable manifold of O_1 as the first iterate of part (27). This part $W_1|_{x<0}$ intersects the discontinuity line $\{x = 0, y \in \mathbb{R}^1\}$ at the point $M_1(0, \frac{\alpha_1^-}{a-1-\lambda})$, where $\alpha_1^- = \alpha_1|_{x<0}$. Then, the continuation of $W_1|_{x<0}$ to the region $x > 0$ is the line segment M_1M_2, where the point $M_2 = f|_{x<0}M_1 = (x_2, y_2)$, $x_2 = \frac{\alpha_1^-}{a-1-\lambda} + 1$, $y_2 = \lambda x_2$. The further continuation of W_1 is the line segment M_2M_3, where $M_3 = f|_{x>0}(M_2)$, $M_3 = (x_3, y_3)$, $x_3 = y_2 + 1 - (a+\lambda)x_2$, $y_3 = \lambda(y_2 + 1 - (a+1+\lambda)x_2)$. For the parameter region (24), M_3 lies in the region $x < 0$ due to $x_3 = \lambda x_2 + 1 - (a+\lambda)x_2 = 1 - a - \frac{a\alpha_1^-}{a-1-\lambda} < 0$. We introduce the segment $I = \{x = x_3, \alpha_2(x_3 + \frac{1}{a-1-\lambda}) \le y \le y_3\}$, connecting M_3 and the stable manifold $W_2|_{x<0}$, and denote the part of $W_2|_{x<0}$ from the bottom end of I to the saddle O_1 as $I^* = \{x|\frac{1}{1+\lambda-a} < x < x_3\}$.

The area G_t bounded by $W_1|_{x<0} \cup M_1M_2 \cup M_2M_3 \cup I \cup W_2|_{I^*}$ is the trapping region, $fG_t \subset G_t$, where the point M_3 lies on or above the stable manifold $W_2|_{x<0}$. This requirement can be expressed as the inequality $\alpha_2(x_3 + \frac{1}{a-1-\lambda}) \le y_3$, which can be simplified to the condition (25) by substituting the expressions for the values of α_2, x_3, and y_3. Thus, the intersection of the parameter regions (24) and (25) corresponds to the Lozi attractor (see Figure 2).

■

Remark 4.1: If $H(\lambda, a) = 0$, the point M_3 lies on the stable manifold $W_2|_{x<0}$ and becomes a point on the homoclinic orbit. After this homoclinic bifurcation for $H(\lambda, a) < 0$, the point M_3 is located below $W_2|_{x<0}$, and therefore the area G_t is no longer a trapping region.

Figure 3 illustrates how the breakdown of the Lozi attractor can occur either due to the loss of hyperbolicity condition or via a homoclinic bifurcation.

5. Attractor of the nonlinear hybrid Lozi-Hénon map

In this section, we introduce an example of the piecewise-smooth nonlinear map in the form of the hybrid Lozi-Hénon map (1)–(5) with

$$U(x) = l|x| + (1 - l)x^2 \tag{28}$$

which yields the Lozi map for $l = 1$ and the Hénon map for $l = 0$. Thus, the function $g(x)$ determining the Lozi-Hénon map reads

$$g(x) = 1 - al|x| - (1 + \lambda)x - a(1 - l)x^2. \tag{29}$$

The map (1)–(29) has saddle points $O_{1,2}$ $(x_{1,2}^s, 0)$, where

$$x_{1,2}^s = \frac{1}{2a(1-l)}\left[-1 - \lambda \pm \left(al - \sqrt{(al \mp (1+\lambda))^2 + 4a(1-l)}\right)\right] \tag{30}$$

are zeros of the function $g(x)$. The manifolds of saddle O_1 are defined by some nonlinear functions

$$W_{1,2}\big|_{G^-} : y = \varphi_{1,2}(x - x_1^s), \quad x \in (x_1^s, 0),$$

where φ_1 (φ_2) corresponds to the unstable (stable) manifolds, $\varphi_{1,2}(0) = 0$.

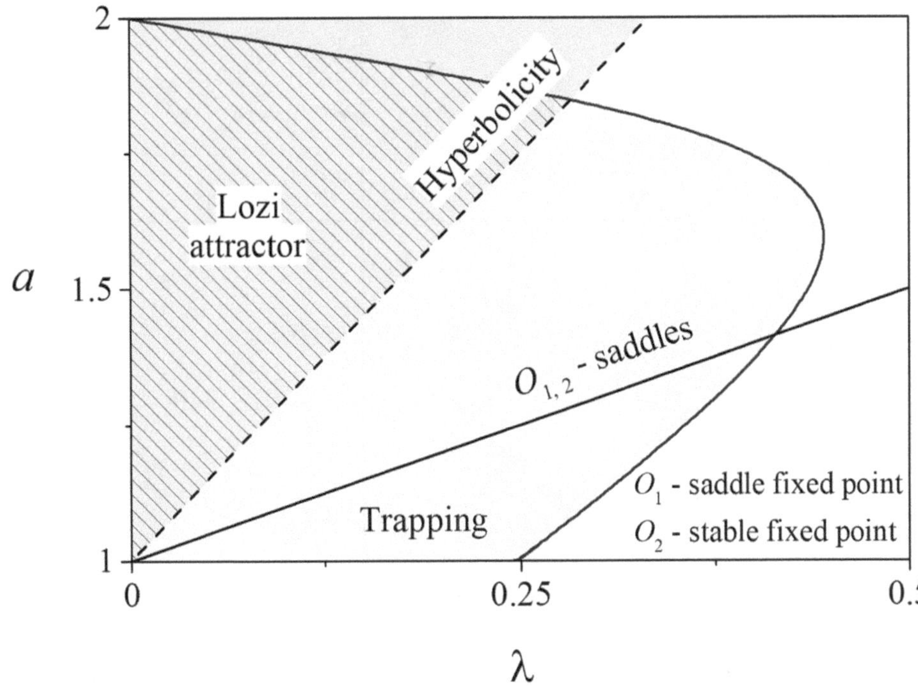

Figure 2. The parameter plane (λ, d) of the Lozi map (1), (23). The yellow region corresponds to the existence of the trapping region $G_t : fG_t \subset G_t$. This region is bounded by the bifurcation of a homoclinic orbit to saddle O_1 (blue curve), defined by the equation $H(\lambda, a) = 0$ in (25). The corresponding phase picture is displayed in Figure 3. On the solid dashed line, $a = 1 + \lambda$ the fixed point O_2 becomes a saddle. The dashed line $a = 1 + 3\lambda$ is the lower boundary of the parameter region (24) (grey region) corresponding to the hyperbolic phase space of the map. The intersection of the yellow and gray regions (dashed area) corresponds to the existence of the strange Lozi attractor.

Introduce the function $y = \psi(x)$ defined parametrically $\psi : \{x = t + g(t), y = \lambda g(t), t \in (x_1^s, 0)\}$, satisfying the conditions $\psi(x_1^s) = 0$, $\psi' = \frac{\lambda d(x(t))}{1+d(x(t))} > 0$ for $t \in [x_1^s, 0]$.

Lemma 5.1: *Manifolds $W_{1,2}$ of saddle O_1 are bounded by*

$$\begin{aligned} &\text{unstable separatrix } W_1 : \psi(x) < \varphi_1(x - x_1^s) < \lambda(x - x_1^s), \\ &\text{stable separatrix } W_2 : \varphi_2(x - x_1^s) < -g(x), \end{aligned} \quad \textit{for } x \in (x_1^s, 0). \tag{31}$$

Proof: The image of the region $R = \{x, y | x_1^s \le x \le 0, y \ge 0\}$ is $fR = \{x, y | \psi(x) \le y \le \lambda(x - x_1^s)\}$. The fixed point $O_1 \in \partial R$ and the derivative $\varphi_1'(0) = \alpha_1(d(x_1^s)) > 0$ (tangent to W_1 at O_1). Hence, the unstable manifold W_1 enters the region R and remains in it. The map $f|_R$ increases the coordinate x of each point in fR since for each point $(x_0, y) \in fR$, $x_0 + y + g > x_0$. Hence, W_1 intersects the line $\{x = 0, y > 0\}$, and the function φ_1 satisfies the condition (31). The image of the arc $= (y - g(x), x_1^s \le x \le x_2^s)$ is $f(arc) = \{y = 0, x_1^s \le x \le x_2^s\}$. Hence, we obtain the second condition in (31). ■

The unstable manifold W_1 intersects the line $\{x = 0, y \in \mathbb{R}^1\}$ at the point

$$M_1\left(0, y_1\right) : 0 < y_1 = \varphi_1(-x_1^s) < -\lambda x_1^s. \tag{32}$$

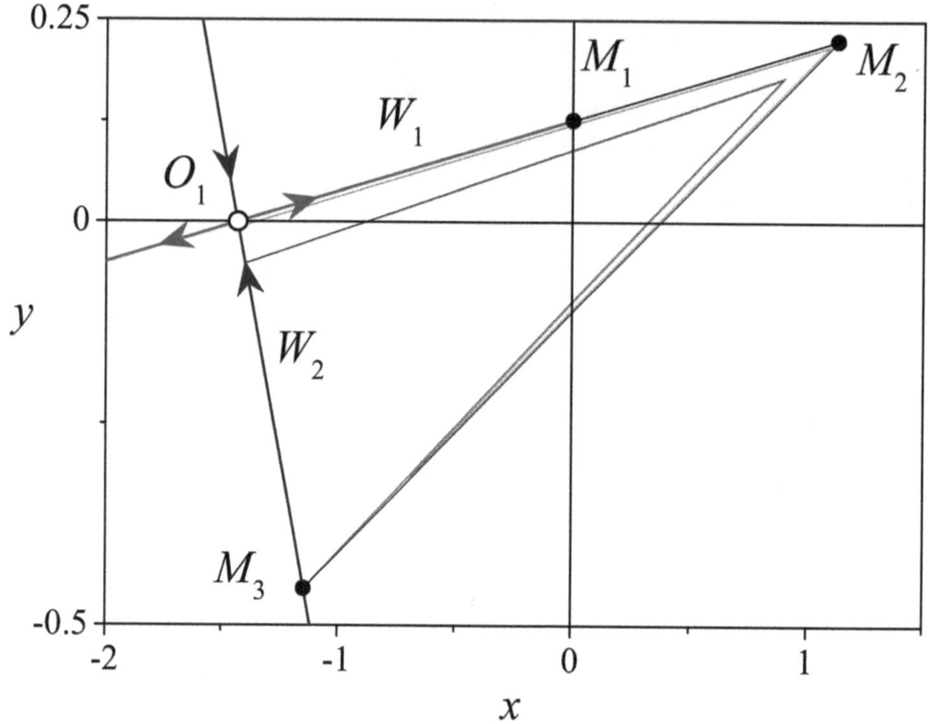

Figure 3. Formation of a homoclinic orbit in the Lozi map (1)–(23) for $\lambda = 0.2$ and $a = 1.905$. The red and blue lines represent the unstable and stable manifolds $W1, 2$ of saddle fixed point O_1, respectively. The black line represents the continuation of W_1, which forms the homoclinic orbit.

Theorem 5.2: *Let the following inequalities*

$$al > 1 + \lambda + \varepsilon, \ \ \varepsilon > 0, \ \ x < 0; \quad al > 1 + 3\lambda, \ \ x > 0. \tag{33}$$

$$y_3 > -g(x_3), \quad (x_3, y_3) = f^2 M_1. \tag{34}$$

hold. Then, the hybrid Lozi-Hénon map (1), (29) *has a singular hyperbolic attractor.*

Proof: The condition (33) removes the extrema of $g(x)$, where $g'(x) = 0$ and requires $g(x)$ to have a maximum at the kink point $(x = 0, y = 1)$ with a discontinuous derivative. As per Theorem 3.1, the map (1), (29) is singular hyperbolic under the condition (33).

To complete the proof, we need to derive sufficient conditions for the existence of a trapping region. This can be done as follows. The image $f|_{x<0}(M_1) = M_2(x_2, y_2)$ has the coordinates

$$x_2 = y_1 + 1, \quad y_2 = \lambda(y_1 + 1),$$

and the first piece of the continuation of the manifold $W_1|_{x<0}$ to the region $x > 0$ is the curve $\widehat{M_1M_2}$ connecting M_1 and M_2. The next piece is the curve M_2M_3 connecting M_2

with its image $M_3 = f|_{x>0}(M_2)$, $M_3(x_3, y_3)$, where

$$\begin{aligned} x_3 &= 1 - al(y_1+1) - a(1-l)(y_1+1)^2, \\ y_3 &= \lambda - \lambda(al+1)(y_1+1) - \lambda a(1-l)(y_1+1)^2. \end{aligned} \tag{35}$$

From (35), we obtain the inequalities

$$\begin{aligned} x_3 &< x_3|_{y_1=0} = 1 - a < 0, \\ y_3 &< y_3\big|_{y_1=0} = -\lambda a < 0. \end{aligned} \tag{36}$$

These inequalities indicate that M_3 belongs to $G^+(x<0)$ and that the part of the second piece of the manifold's continuation $\widehat{M_2M_3}$ intersects the region $x<0$.

Due to the second inequality in (31) and the condition (34) of Theorem 5.2, there exists an interval $I_t = \{x = x_3, y_4 < y < y_3\}$, where $y_4 = \varphi_2(x_3 - x_1^s)$. Denote the point $M_4 = (x_3, y_4)$. Then, the closed curve $O_1M_1M_2M_3M_4O_1$ bounds the area G_t. As the point M_4 lies on the stable manifold of O_1, then $f^iM_4 \to O_1$ for $i \to \infty$. The sequence of the curves $(f^iM_3, f^{i+1}M_3)$, $i = 0, 1, \ldots$ forms a continuation of the unstable manifold of saddle O_1. Hence, $fG_t \subset G_t$, G_t is a trapping region and $\bigcap_{i=0}^{\infty} f^iG_t$ is an attractor of the map (1), (29) under the conditions (33), (34). ■

Remark 5.1: The parameter domain (33)–(34) defines the lower bound of the parameter l, $l^*(\lambda, a)$, such that Theorem 5.2 is valid for $l^* < l \le 1$.

6. Analysis of the Belykh map with non-periodic and periodic functions

The Belykh map [13] and its multidimensional generalizations [6, 38, 40, 41] have been studied from an ergodic theory perspective to establish the existence of a continuous invariant measure [3, 13, 38, 40, 41, 43, 44]. In this section, we focus on the dynamical properties of the Belykh map and analyze its new characteristics for the map f with period-2 function $g(x)$, defined on the cylinder $\mathbb{S}^1 \times \mathbb{R}^1$. The connection between the Belykh map and the dynamics phase locked loop control systems will be detailed in Section 7.

The earlier version of the Belykh map was introduced in the form of the 2D map (1) with the period-2 sawtooth function [16, 19]:

$$g(x) = a(x - (2k-1)) \quad \text{for } x \in (2k-2, 2k),\ k \in \mathbb{Z}. \tag{37}$$

Note that (37) becomes (7) defined on the circle $x \in \mathbb{S}^1$. However, the standard Belykh map [13] is traditionally determined in the case of the non-periodic function, obtained from (37) by setting $k = 0$ and $k = 1$. This function $g(x)$ has the form

$$g(x) = \begin{cases} a(x+1), & x < 0, \\ a(x-1), & x \ge 0. \end{cases} \tag{38}$$

In the following, we will analyze the dynamics of the Belykh map with the non-periodic function (38) and the period-2 function (37) separately. Notably, we will demonstrate that the period-2 function allows for the chaotic attractor to exhibit oscillatory, rotating, and mixed oscillatory-rotating hyperbolic components, owing to the cylindrical phase space it provides.

6.1. Non-periodic function $g(x)$

The derivative of the function (38) $g'(x) \equiv d(x) \equiv a > 0, x \neq 0$ and the matrix L in (10) is constant. Denote $s \triangleq \operatorname{tr} L = 1 + a + \lambda$. Due to (11), the eigenvalues $\mu_1 > 1, 0 < \mu_2 < 1$ and the eigenvectors' $\mathbf{V}_{1,2} = (1, \alpha_{1,2})$ coordinates satisfy the conditions

$$0 < \alpha_1 < 1, \quad \alpha_2 < -(1 + \lambda) < 0.$$

To prevent the non-invertibility of the map, we impose the constraint $\mu_2 < \frac{1}{2}$.

The constant eigenvectors $\mathbf{V}_{1,2}$ of the map f are the same for all values of x, resulting in unstable Φ_c^u and stable Φ_c^s invariant foliations given by:

$$\begin{aligned} \Phi_c^u &= \{x, y \mid y = \alpha_1(x + c_1)\}, \\ \Phi_c^s &= \{x, y \mid y = \alpha_2(x + c_2)\}, \end{aligned} \tag{39}$$

where the constants $c_{1,2} \in \mathbb{R}^1$. The image of each line from $\Phi_c^u(\Phi_c^s)$ forms two axle shafts lying at $\Phi_c^u(\Phi_c^s)$ respectively. The cones K^u and K^s are unnecessary in the case of the map (1)–(38), as they contain only one constant vector everywhere and degenerate into lines.

The map (1)–(38) possesses two saddle-type fixed points $O_{1,2}(\pm 1, 0)$ with unstable $W_{1,2}^u$ and stable $W_{1,2}^s$ manifolds, which are depicted in Figure 4. These manifolds are part of the invariant foliations (39) and can be expressed as

$$\begin{aligned} W_{1,2}^u &= \{x, y \mid y = \alpha_1(x \pm 1)\}, \\ W_{1,2}^s &= \{x, y \mid y = \alpha_2(x \pm 1)\}. \end{aligned} \tag{40}$$

Furthermore, let $M_1(x_1, y_1) = W_1^u \cap \{x = 0\}$, where $x_1 = 0$ and $y_1 = \alpha_1$, and $M_2(x_2, y_2) = W_2^s \cap \{x > 0, y = \alpha_1(x + 1)\}$, where

$$x_2 = \frac{s - 2\lambda}{\sqrt{s^2 - 4\lambda}}, \quad y_2 = \alpha_1(x_2 + 1).$$

Theorem 6.1: *For*

$$a \leq 1 - \lambda \tag{41}$$

the map (1)–(38) *has a singular hyperbolic Belykh attractor.*

Proof: The image $f|_{x<0}(M_1) = (\bar{x}_1, \bar{y}_1) = (\alpha_1 + a, \lambda(\alpha_1 + a))$ does not reach the point M_2 when $\bar{x}_1 < x_2$, i.e. for

$$\alpha_2 + a < \frac{s - 2\lambda}{\sqrt{s^2 - 4\lambda}}. \tag{42}$$

Similarly, $f|_{x>0}(\bar{M}_1)$ does not reach the point $\bar{M}_2$ for (42). Inequality (42) can be rewritten as

$$(2 - s)(s + \sqrt{s^2 - 4\lambda}) > 0 \tag{43}$$

which is valid under the condition (41). Therefore, the parallelogram $P = O_1 M_2 O_2 \bar{M}_2$ is invariant, $fP \subset P$, and there exists an attractor $\mathcal{A} \subset P$.

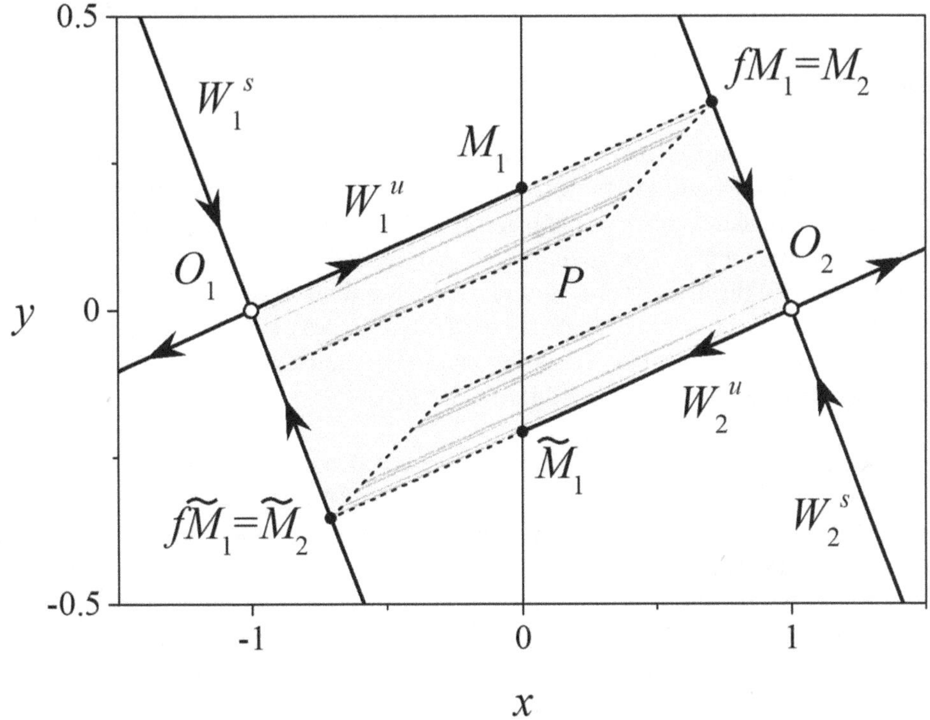

Figure 4. Phase portrait of the Belykh map (1), (38) for the bifurcational case $\lambda = a = 0.5$ (see condition (41)) when point M_1 maps to the stable manifold W_2^s such that $fM_1 = M_2$. This is a special case where the parallelogram P (yellow) forms a trapping region, i.e. $fP \subset P$. The parallelogram is mapped into two trapezoids (the areas bounded by the dashed lines): the upper $fP|_{x<0}$ and the bottom $fP|_{x>0}$. Numerical simulations reveal the singular hyperbolic Belykh attractor, represented by the grey points.

The foliations (39) intersect the domain P for $c_{1,2} \in [-1, 1]$, stretching ($\mu_1 > 1$) along Φ_c^u and shrinking ($\mu_2 < 1$) along Φ_c^s. Hence, every orbit of the attractor $\mathcal{A}$ is uniformly hyperbolic with Lyapunov exponents $h_1 = \ln \mu_1 > 0$ and $h_2 = \ln \mu_2 < 0$. ∎

6.2. *Periodic function g(x)*

In this subsection, we will demonstrate that by introducing periodicity in $g(x)$, the trapping region is converted into an annulus that slices off a portion of the cylindrical phase space. Consequently, chaotic trajectories inside the annulus can revolve around the cylinder, adding a rotational element to the oscillatory-type Belykh attractor that is absent in the non-periodic function case.

Lemma 6.2: *The annulus* $R = \{x, y \,|\, x \in \mathbb{S}^1, |y| < \frac{a}{1-\lambda}\}$ *is a trapping region of the map* (1)–(37).

Proof: The function $g(x) = a(x-1)$ is bounded on $\mathbb{S}^1$: $|g(x)| < a$ when $0 < x < 2$. Therefore, $|\bar{y}| < \lambda(|y| + a)$ and the variable $|y|$ decreases at each iteration for $\lambda(|y| + a) < |y|$, i.e. for $|y| \geq \frac{a}{1-\lambda}$. This implies that the annulus R is a trapping region. ∎

Theorem 6.3: *For $a > 1-\lambda$, $0 < \lambda < 1$ the map (1)–(37) has a singular hyperbolic attractor consisting of three components: oscillating $\mathcal{A}_o \subset P$, rotating $\mathcal{A}_r \subset R \setminus P$ and rotating with oscillations $\mathcal{A}_{ro} \subset R$.*

Proof: Lemma 6.2 implies that $fR \subset R$ and the map has the attractor $\mathcal{A}$. Due to the existence of foliations Φ_c^u and Φ_c^s, the attractor is singular hyperbolic. When $a > 1-\lambda$, the image of fM_1 overshoots the point M_2, and $fP \setminus P \neq \emptyset$ forms a triangle $\Delta_1 = M_2 fM_1 M_4$ (or trapezoid, see Fugure 5) for $y > 0$, and its symmetrical counterpart $\bar{\Delta}_1$ for $y < 0$. Starting from some k^*, the image $f^k\Delta$ begins to wrap around the cylinder and forms a rotating attractor $\mathcal{A}_r$, intersecting the triangle $\Delta_2 = M_1M_3M_2$. Clearly, $f^k\Delta_2 \subset P$ for $k \geq 1$, indicating that trajectories from Δ_2 reach the region P. Almost all trajectories $f^k\Delta_2$, where $k > k^*$, leave the region P and form the attractor $\mathcal{A}_{ro}$. ■

It is worth noting that for all parameter values $a > 0$ and $0 < \lambda < 1$, the map (1)–(37) has a singular hyperbolic attractor with only the oscillatory component, similar to the non-periodic case described in Theorem 6.1, even without imposing the conditions of Theorem 6.3. However, when the conditions of Theorem 6.3 are satisfied by the parameters a and λ, the oscillatory attractor transforms into a Cantor set of trajectories similar to the Smale horseshoe. Unlike the attractors $\mathcal{A}_r$ and $\mathcal{A}_{ro}$, the basin of this attractor has zero Lebesgue measure, making it unobservable in numerical simulations.

7. The Belykh map as a discrete-time model of phase-locked loop control systems

To establish a stronger link between the dynamics of 2D piecewise maps and physical systems, we briefly sketch the reasoning that led to the construction of the Belykh map [10, 19] from phase-locked loop (PLL) control systems [1, 9, 16, 31, 45].

The simplest form of a PLL model can be described by a system of ordinary differential equations:

$$\dot{x} = u + \omega_1 g(x), \quad \dot{u} = -\lambda_0 u + \omega_2 g(x). \tag{44}$$

Here, $\omega_{1,2}$ and λ_0 are positive parameters, and $g(x)$ is a periodic function resembling a sine wave that acts as an output of the PLL phase detector [31].

The continuous-time system (44) can be reduced to a discrete-time PLL model via the Euler discretization method which replaces the ordinary derivatives with the corresponding time differences so that

$$\begin{aligned} x_{i+1} &= x_i + hu_i + h\omega_1 g(x_i), \\ u_{i+1} &= u_i - h\lambda_0 u_i + h\omega_2 g(x_i). \end{aligned} \tag{45}$$

The following change of the variables and parameters in (45)

$$hu = y, \quad \lambda = 1 - \lambda_0 h, \quad h\omega_1 = a, \quad \omega_2 h^2 = \lambda a \tag{46}$$

yields the general 2D map (1). Choosing the periodic function $g(x)$ in the form (37), we obtain the Belykh map. Historically, to facilitate the analysis of the maps' chaotic properties such as the mixing and SBR measure, the Belykh map (1) with the non-periodic

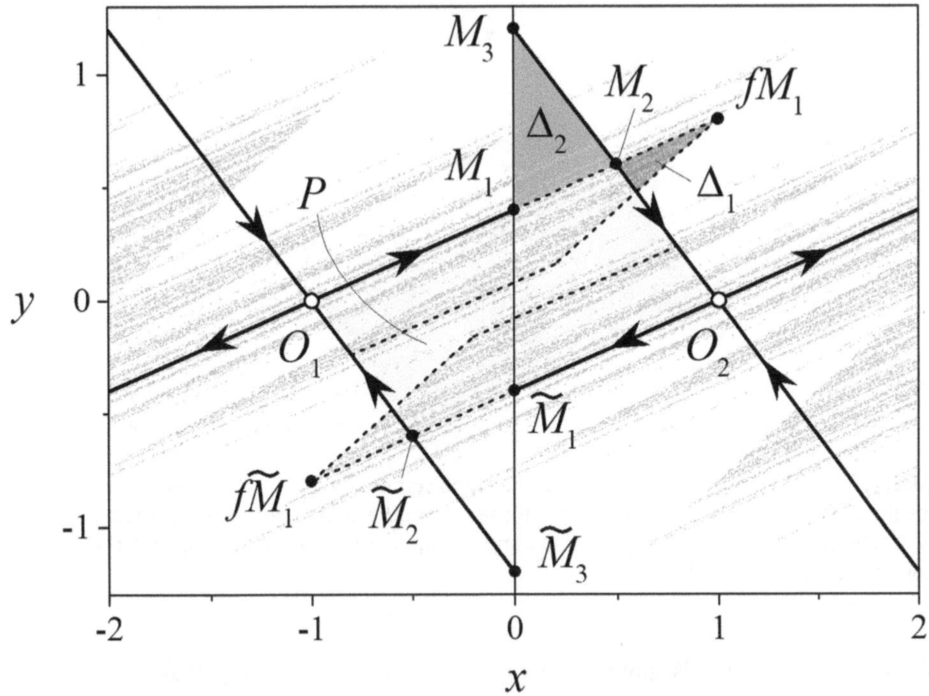

Figure 5. Phase portrait of the Belykh map (1) with periodic function $g(x)$ (37) under the conditions of Theorem 6.3 with parameters $\lambda = 0.8$, $a = 0.6$. The numerically calculated attractor, composed of $\mathcal{A}_o$, $\mathcal{A}_r$ and $\mathcal{A}_{ro}$ [not shown], are indicated as light gray dots. The parallelogram P (yellow) is mapped into two trapezoids (dashed contours): the upper $fP|_{x<0}$ and the bottom $fP|_{x>0}$. The red triangle Δ_1 acts as an exit gate from P for the trajectories of A_ro, and the blue triangle Δ_2 serves as an entry gate to P. The coordinates of the points are: $M_1(0, 0.4)$, $M_2(0.5, 0.6)$, $M_3(0, 1.2)$ $fM_1(1, 0.8)$, $\tilde{M}_1(0, -0.4)$, $\tilde{M}_2(-0.5, -0.6)$, and $f\tilde{M}_1(-1, -0.8)$. Points $\tilde{M}$ are odd symmetric to points M.

function (38) was transformed in the basis of eigenvectors into its conventional form [13]:

$$\begin{aligned} \bar{u}_1 &= \mu_1(u_1+1)+1, \quad \bar{u}_2 = \mu_2(u_2+1)+1, \quad v<0, \\ \bar{u}_1 &= \mu_1(u_1-1)-1, \quad \bar{u}_2 = \mu_2(u_2-1)-1, \quad v>0, \end{aligned} \tag{47}$$

where $v = u_1 + ku_2, 0 < k < 1$.

Of particular importance is the relationship between the dynamics of the Belykh map and its continuous-time phase-locked loop (PLL) counterpart, described by system (44). In contrast to the hyperbolic chaotic dynamics present in the Belykh map, the PLL model can only exhibit trivial dynamics. It is straightforward to prove that the system (44) has a globally stable equilibrium point at the origin. This can be achieved by constructing a positively defined Lyapunov function candidate

$$V = \frac{u^2}{2} - \omega_2 \int_0^x g(\xi)\, d\xi,$$

for the system (44) with the function $g(x)$ (37). The time derivative of V with respect to the system (44) is negative due to

$$\dot{V} = -\alpha u^2 - \omega_1\omega_2 g^2(x) < 0 \quad \text{for } x \neq 2k-1, \ u \neq 0.$$

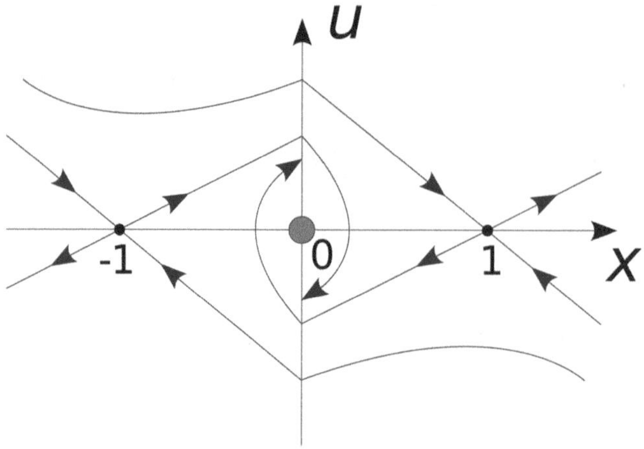

Figure 6. The phase portrait of the phase-locked loop system (44) for $\nu = 0$, where the red point indicates the globally stable equilibrium. The transition from the continuous-time PLL model (44) to the Belykh map (1)–(37) results in an explosive transformation of the equilibrium into the chaotic Belykh attractor (compare with Figure 5).

Therefore, system (44) is globally asymptotically stable, implying that all trajectories, except for those on the stable manifold of the saddle equilibrium points at $x = \pm 1, u = 0$, converge to the point ($x = u = 0$). It should be noted that the stable equilibrium ($x = u = 0$) is situated on the discontinuity boundary $x = 0$, which does not allow for sliding motions, and the trajectories near the boundary converge to the stable equilibrium (see Figure 6).

The x-amplitude of the Belykh attractor in the map (1)–(38) is determined by the x-coordinate of the point $f|_{x<0}(M_1) = (\alpha_1 + a, \lambda(\alpha_1 + a))$. Using (9) and (14), we can see that the order of this coordinate is $\alpha_1 + a \sim h$. Therefore, transitioning from the continuous-time PLL system (44)–(37) to the Belykh map (1)–(38) transforms the equilibrium point $(0, 0)$ of (44) into the chaotic Belykh attractor in the map (1)–(37). This explosive transition, which can occur when numerically integrating continuous-time piecewise-smooth systems using discrete-integration schemes, highlights the complexity of the dynamics and the potential for highly unexpected outcomes.

Moving beyond the realm of PLL models, it is worth noting that a Belykh-like attractor, exhibiting similar dynamical properties as that of the Belykh map with the non-periodic function (38), can be found in a 2D piecewise-linear map that models financial markets. Specifically, this attractor appears in the region of parameters R_4 in [24].

8. Conclusion

In this paper, we have given the existence proofs for strange attractors in a 2D map with arbitrary piecewise-smooth linear and nonlinear functions. We developed an approach to deriving upper bounds for trapping regions and an approach to constructing invariant cones for proving the hyperbolicity of the maps' limit set. We have introduced the Lozi-Hénon map as our primary example and demonstrated that it inherits the main chaotic dynamical properties of both the Lozi and Hénon maps, with a closer resemblance to the

Hénon map for l approaching 0 and to the Lozi map when l approaches 1. We also provided an alternative proof for the existence and hyperbolicity of the Lozi attractor using our approach, showing that it can significantly simplify the proof.

We also applied our approach to the Belykh map, which has degenerate invariant cones that form lines. By analyzing the stable and unstable manifolds of saddle fixed points, we showed that the proof of the existence of trapping regions can be simplified.

Taking a broader perspective, the analytically-tractable Lozi-Hénon map, which we introduced and analyzed in this paper, can serve, in the limit of l approaching 0, as an approximation of the smooth nonlinear Hénon map, which has proven to be analytically challenging to study in full detail. Our study offers a methodological approach to proving the existence of strange attractors in 2D piecewise-smooth nonlinear maps, which opens up new avenues for future research into their transitivity and the construction of their SBR measures.

Acknowledgments

We are thankful to René Lozi for the invitation to contribute to this special issue. We are also grateful to Igor Belykh for helpful discussions and suggestions.

Disclosure statement

No potential conflict of interest was reported by the author(s).

Funding

This work was supported by the Ministry of Science and Higher Education of the Russian Federation under Project No. 0729-2020-0036 and the Russian Science Foundation under Grant No. 22-21-00553.

References

[1] V. Afraimovich and S. Hsu, *Lectures on chaotic dynamical systems*, American Mathematical Societyand International Press, 2003.

[2] V. Afraimovich, V. Bykov, and L. Shilnikov, *On structurally unstable attracting limit sets of Lorenz attractor type*, Trudy Moskov. Mat. Ob-va 44 (1982), pp. 150–212.English transl. in Trans. Moscow Math. Soc., 2 (1983), pp. 153–216.

[3] V. Afraimovich, N. Chernov, and E. Sataev, *Statistical properties of 2-D generalized hyperbolic attractors*, Chaos 5(1) (1995), pp. 238–252.

[4] D. Anosov, *Geodesic flows on closed Riemann manifolds with negative curvature*, Trudy MIAN SSSR 92 (1967), pp. 3–210.

[5] D. Anosov and V. Belykh, *Attractor*, in *Great Russian Encyclopedia*, Volume 31, Great Russian Encyclopedia, Moscow, 2016. pp. 487.

[6] D.V. Anosov and J. Sinai, *Some smooth ergodic systems*, Uspekhi Mat. Nauk 22(137:5) (1967), pp. 107–172.

[7] D. Anosov, S. Aranson, V. Grines, R.V. Plykin, E.A. Sataev, A.V. Safonov, V.V. Solodov, A.N. Starkov, A.M. Stepin, and S.V. Shlyachkov, *Dynamical systems with hyperbolic behavior*, Itogi Nauki I Tekhniki. Ser. Modern Prob. Mat. Fundam. Directions, Dyn. Syst. - 9 66 (1991), pp. 5–242.

[8] N. Barabash and V. Belykh, *Chaotic driven maps: Non-stationary hyperbolic attractor and hyperchaos*, Eur. Phys. J. Spec. Top. 229(6) (2020), pp. 1071–1081.

[9] V. Belykh, Qualitative Methods of the Theory of Nonlinear Oscillations in Point Systems, Gorki University Press, Gorki, 1980. (in Russian).

[10] V. Belykh, *Chaotic and strange attractors of a two-dimensional map*, Mat. Sb. 186(3) (1995), pp. 311–326.
[11] V. Belykh, N. Barabash, and I. Belykh, *A Lorenz-type attractor in a piecewise-smooth system: Rigorous results*, Chaos 29(10) (2019), Article ID 103108.
[12] I. Belykh, R. Kuske, M. Porfiri, and D.J.W. Simpson, *Beyond the Bristol book: Advances and perspectives in non-smooth dynamics and applications*, Chaos 33 (2023), Article ID 010402.
[13] V. Belykh and I. Belykh, *Belykh map*, Scholarpedia 6(10) (2011), p. 5545.
[14] V. Belykh and D. Grechko, *Singular hyperbolic attractor of a multidimensional cylinder map*, Dyn. Syst. 8(4) (2018), pp. 373–383.
[15] V.N. Belykh and B. Ukrainsky, *A discrete-time hybrid Lurie type system*, in *Dynamics and Control of Hybrid Mechanical Systems*, Volume 14 World Scientific Publishing Co Pte Ltd, Singapore, 2010. pp. 43–52.
[16] V.N. Belykh and I.L. Zheleznyak, *On the structure and dimension of a strange attractor of a piecewise linear map of the cylinder*, in *Methods of Qualitative Theory and Bifurcation Theory*, L. P. Shil'nikov, ed., Nizhnii Novgorod. Gos. Univ., Nizhnii Novgorod, 1991, pp. 18–25.
[17] V. Belykh , N. Komrakov , and B. Ukrainsky, *Hyperbolic attractors in a family of multidimensional maps with cusp-points*. Proceedings of International Conference 'Progress in Nonlinear Science' Dedicated to the 100th Anniversary of A. Andronov (2001), pp. 23–24.
[18] V. Belykh, *Strange attractor*, in *Great Russian Encyclopedia*, Volume 31, Great Russian Encyclopedia, Moscow, 2016. pp. 285–286.
[19] V. Belykh1976, *On models of phase synchronization systems and their study*. Gorky: State University Press. Dinamika System. 11 (1976), pp. 23–32. (In Russian)
[20] M. Benedicks and L. Carleson, *The dynamics of the Hénon map*, Ann. Math. 133 (1991), pp. 73–169.
[21] Y. Cao and Z. Liu, *Strange attractors in the orientation preserving Lozi map*, Chaos Solit. Fractals 9(11) (1998), pp. 1857–1863.
[22] J.-P. Eckmann and D. Ruelle, *Ergodic theory of chaos and strange attractors*, Rev. Mod. Phys. 57(3) (1985), p. 617.
[23] L. Gardini, D. Fournier-Prunaret, and P. Charge, *Border collision bifurcations in a two-dimensional piecewise smooth map from a simple switching circuit*, Chaos 21(2) (2011), pp. 023106-1–023106-16.
[24] L. Gardini, D. Radi, N. Schmitt, I. Sushko, and F. Westerhoff, *perception of fundamental values and financial market dynamics: Mathematical insights from a 2D piecewise linear map*, SIAM J. Appl. Dyn. Syst. 21(4) (2022), pp. 2314–2337.
[25] I. Ghosh and D. Simpson, *Robust Devaney chaos in the two-dimensional border-collision normal form*, Chaos 32(4) (2022), pp. 1–9.
[26] P. Glendinning, *Robust chaos revisited*, Eur. Phys. J. Spec. Top. 226(9) (2017), pp. 1721–1738.
[27] P. Glendinning and D. Simpson, *Robust chaos and the continuity of attractors*, Trans. Math. Appl. 4(1) (2020), Article ID tnaa002, pp. 1–15.
[28] P. Glendinning and D. Simpson, *A constructive approach to robust chaos using invariant manifolds and expanding cones*, Discrete Continuous Dyn. Syst. 41(7) (2021), pp. 3367–3387.
[29] M. Hénon, *A two-dimensional mappings with a strange attractor*, Commun. Math. Phys. 50(1) (1976), pp. 69–77.
[30] A. Katok and B. Hasselblatt, Introduction to the Modern Theory of Dynamical Systems, Factorial, Moscow, 1999.
[31] J. Kudrewicz and S. Wasowicz, Equations of Phase-Locked Loops: Dynamics on Circle, Torus and Cylinder, Vol. 59, World Scientific Publishing Co Pte Ltd, Singapore, 2007.
[32] R. Lozi, *Un attracteur de Hénon*, UJ. Phys. 39 (1978), pp. 9–10.
[33] M. Misiurewicz, *Strange attractors for the Lozi mappings*, Ann. N. Y. Acad. Sci. 357(1) (1980), pp. 348–358.
[34] M. Misiurewicz and S. Štimac, *Symbolic dynamics for Lozi maps*, Nonlinearity 29(10) (2016), pp. 3031–3046.
[35] M. Misiurewicz and S. Stimac, *Lozi-like maps*, Discrete Continuous Dyn. Syst. 38(6) (2018), pp. 2965–2985.

[36] C. Morales, M. Pacifico, and E. Pujals, *On C^1 robust singular transitive sets for three-dimensional flows*, C. R. Acad. Sci. Paris Ser. I Math. 326(1) (1998), pp. 81–86.
[37] H. Nusse, E. Ott, and J. Yorke, *Border-collision bifurcations: An explanation for observed bifurcation phenomena*, Phys. Rev. E 49(2) (1994), pp. 1073–1076.
[38] Y.B. Pesin, *Dynamical systems with generalized hyperbolic attractors: Hyperbolic, ergodic and topological properties*, Ergod. Theory Dyn. Syst. 12(1) (1992), pp. 123–151.
[39] M. Rychlik, *Invariant measures and the variational principle for Lozi mappings*, Ph.D. thesis, ProQuest LLC, Ann Arbor, MI; University of California, Berkeley, 1983.
[40] E. Sataev, *Invariant measures for hyperbolic maps with singularities*, Uspekhi Mat. Nauk 47(1) (1992), pp. 147–204.
[41] E. Sataev, *Ergodic properties of the Belykh map*, J. Math. Sci. 95 (1999), pp. 2564–2575.
[42] E. Sataev, *Some properties of singular hyperbolic attractors*, Sb. Math. 200(1) (2009), pp. 35–76.
[43] J. Schmeling, *A dimension formula for endomorphisms – the Belykh family*, Ergod. Theory Dyn. Syst. 18 (1998), pp. 1283–1309.
[44] J. Schmeling and S. Troubetzkoy, *Dimension and invertibility of hyperbolic endomorphisms with singularities*, Ergod. Theory Dyn. Syst. 18(5) (1998), pp. 1257–1282.
[45] V. Shakhildyan and L. Belyustina, Systems of Phase Synchronization, Radio i Svyaz, Moscow, 1982, 288 p. (in Russian).
[46] S. Smale, *Differentiable dynamical systems*, Bull. Am. Math. Soc. 73 (1967), pp. 747–817.
[47] D. Turaev and L. Shilnikov, *An example of a wild strange attractor*, Mat. Sb. 189(2) (1998), pp. 291–314.
[48] D. Turaev and L. Shilnikov, *Pseudo-hyperbolisity and the problem on periodic perturbations of Lorenz-like attractors*, Dokl. Math. 77(1) (2008), pp. 17–21.
[49] Zh. Zhusubaliyev and E. Mosekilde, Bifurcations and Chaos in Piecewise-Smooth Dynamical Systems, World Science Series, Nonlinear Sci. Ser. A Monogr. Treatises Vol. 44, World Scientific, Hackensack, NJ, 2003.

On the relationship between Lozi maps and max-type difference equations

Antonio Linero Bas and Daniel Nieves Roldán

ABSTRACT
In the present work we revise a transformation that links generalized Lozi maps with max-type difference equations. In this view, according to the technique of topological conjugation, we relate the dynamics of a concrete Lozi map with a complete uniparametric family of max-equations, and we apply this fact to investigate the dynamics of two particular families. Moreover, we present some numerical simulations related to the topic and, finally, we propose some open problems that look into the relationship established between generalized Lozi maps and max-equations.

1. Introduction

In 1963, the meteorologist E. N. Lorenz [15], studied a system of three first-order differential equations whose solutions stand out for being unstable with respect to small modifications, that is, minimal variations on the initial conditions of the system can produce a considerably different evolution of the terms. This property, which makes the trajectories wander in an apparently erratic manner, yields to what is known as a strange attractor.

The term *strange attractor* was coined by Ruelle and Takens in [19] while studying the generation of turbulence[1] in dissipative systems. In concrete, the analysed system appeared to have an attractor which was locally the product of a Cantor set and a piece of two-dimensional manifold. Such structure was named by the authors as *strange*. In an informal way, when an attractor is fractal, that is, its dimension is not an integer, we will called it strange attractor. In these cases, the dynamics on the attractor are said to be chaotic and they are sensitive to initial conditions (see [18]).

This phenomenon motivated the analysis of similar models that exhibit the same properties. For instance, in 1976, M. Hénon investigated the following two-dimensional mapping:

$$\begin{cases} x_{n+1} = y_n + 1 - ax_n^2 \\ y_{n+1} = bx_n \end{cases}, \tag{1}$$

where a, b are positive real numbers. In [13], he simulated such model in the particular case $a = 1.4$, $b = 0.3$. Depending on the selection of the initial conditions, the solution of Equation (1) either diverges to infinity or tends to a strange attractor, which appears to be the product of a one-dimensional manifold by a Cantor set. Later on, Benedicks and Carleson [5], proved analytically the existence of a strange attractor; they considered Equation (1) with $1 < a < 2$ and $b > 0$, for a small b and an a close to 2. In such case, there exists an hyperbolic fixed point[2] through which there are a stable and an unstable manifold. The closure of the unstable manifold is the attractor of the system.

In 1978, Lozi [16], exchanges the quadratic term in Hénon's map by considering the system of difference equation

$$\begin{cases} x_{n+1} = 1 - a|x_n| + y_n \\ y_{n+1} = bx_n \end{cases}, \tag{2}$$

where $a, b \in \mathbb{R}$. The numerical simulations developed for the particular case $a = 1.7$ and $b = 0.5$ suggested the existence of a strange attractor simpler than the one exhibited in Hénon map. In concrete, it seemed the product of parts of straight lines by a Cantor set. Lozi map gave raise to an abundant literature, as can be appreciated in the different chapters of this volume. Indeed, Lozi map was the first system for which the existence of a strange attractor was analytically established. In [17], for certain values of the parameters a and b, trapping regions[3] were found and the author showed the hyperbolic structure of the map, which enabled him to prove that the intersection of the images of the trapping regions is a strange attractor. In fact, the restrictions that the parameters must verify in order to have such attractor are

$$0 < b < 1; \quad a > b + 1; \quad 2a + b < 4; \quad b < \frac{a^2 - 1}{2a + 1}; \quad \sqrt{2}a > b + 2. \tag{3}$$

Then, the attractor can be constructed from the successive forward iteration of a trapping region which is the triangle with vertices I, $F(I)$ and $F^2(I)$, where $F(x, y) = (1 - a|x| + y, bx)$ and I is the point given by the intersection of the unstable manifold of the fixed point $(\frac{1}{a+1-b}, \frac{b}{a+1-b})$ with the horizontal axis, that is, $I = (\frac{2+a+\sqrt{a^2+4b}}{2(1+a-b)}, 0)$.

It is worth mentioning that in [3] the authors even characterize the basin of attraction[4] for the strange attractor of the Lozi map whose parameters a and b verify the conditions established in (3).

As an example of further research developed after Lozi map, we can highlight the particular case of Equation (2) with $a = -1$ and $b = -1$ that receives the name of *Gingerbreadman equation*:

$$\begin{cases} x_{n+1} = 1 + |x_n| + y_n \\ y_{n+1} = -x_n \end{cases}, \tag{4}$$

studied by Devaney in [8]. He proved that there exists a unique fixed point, $(1, 1)$, which is elliptic.[5] Furthermore, there exists a hexagon where every point except the fixed point is periodic of period 6. However, he showed that the fixed point is surrounded by infinitely many invariant polygons of arbitrarily large radius and the regions between any two of those consecutive polygons provide the equation of zones of instability. In this sense, Equation (4) is chaotic in certain regions and stable in others.

Moreover, in 1992, Crampin [7], studied the piecewise linear equation

$$x_{n+1} = |x_n| - x_{n-1}, \tag{5}$$

which is globally periodic of period 9.[6] Also, he observed that each one of the linear difference equations, $x_{n+1} = x_n - x_{n-1}$ and $x_{n+1} = -x_n - x_{n-1}$, are periodic of periods 6 and 3, respectively. As a consequence he proposed the study of the combination of two periodic linear difference equations through a piecewise linear equation. In this direction, in [4] Beardon *et al.* considered the difference equation

$$x_{n+1} = \alpha|x_n| + \beta x_n - x_{n-1}, \tag{6}$$

where α, β are real numbers, and studied for which values of the parameters α, β, Equation (6) was globally periodic. In fact, they were able to prove that the set of points (α, β) for which Equation (6) is globally periodic is unbounded and uncountable. For instance, they showed, [4, Theorem 1.5.], that for $p \geq 2$, if

$$\beta^2 - \alpha^2 = 2\left(1 + \cos\frac{\pi}{p}\right), \quad \beta < 0,$$

then Equation (6) is periodic with period $4p$.

Furthermore, in [1], the authors gave necessary conditions to assure periodicity of

$$x_{n+1} = \alpha|x_n| + \beta x_{n-1}, \tag{7}$$

where α, β are real numbers. In concrete, for $\alpha \notin \{0, 1\}$, it is necessary that $|\alpha| \in (0, 2) \setminus \mathbb{Q}$ and $\beta = -1$, for Equation (7) to be globally periodic.

In the literature, see for instance [12], Lozi map also appears in connection with a class of difference equations, so-called max-type equations. To have a general scope of the dynamics of different classes of max-type equations, consult the survey paper [14]. In the present paper we are interested in deepening in the relationship between Lozi maps and some class of max-type difference equations. In this sense, in Section 2, we will show well-known changes of variables transforming a generalized Lozi map into max-type difference equations when some additional conditions to the parameters are considered. We will clarify the casuistic concerning the values of an arbitrary positive value A, fixed in advance. These transformations allow us to relate the dynamics of concrete generalized Lozi maps with that of some one-parametric families of max-type difference equations. In this direction, it is interesting to point out that all the members of this family share the same dynamics as they are topologically conjugate[7] to the same generalized Lozi map. Afterwards, in Sections 3 and 4, we will apply this study in order to obtain some properties related to the dynamics of these families by interpreting them in the light of the conjugation of Lozi map and max-type equations. Then, Section 5 collects some simulations and display some particular dynamics which are the leitmotiv to present some associated problems. Finally, we will present some conclusions in Section 6.

2. The transformation

As it can be read in the introduction of [10], the idea of exchanging the quadratic term of Hénon map, see Equation (1), into the absolute value function, came to Lozi's mind when he embedded the shape of the area generated by Hénon map, bounded by two parabolas, into another area bounded by four line segments, which reminded him to the graph of the absolute value function. Thus, in 1978, [16], Lozi introduced the following system of difference equations

$$\begin{cases} x_{n+1} = 1 - a|x_n| + y_n \\ y_{n+1} = bx_n \end{cases}, \tag{8}$$

where a, b are positive real numbers. Such system, which can be reduced into the difference equation

$$x_{n+1} = 1 - a|x_n| + bx_{n-1}, \tag{9}$$

was called *Lozi map*.

In the Lozi's original paper, the mapping is presented under the form of Equation (8). Additionally, in different papers the Lozi map is also given by

$$\begin{cases} x_{n+1} = 1 - a|x_n| + by_n \\ y_{n+1} = x_n, \end{cases} \tag{10}$$

even, we can find the definition of a Lozi map as the bidimensional map

$$\begin{cases} x_{n+1} = y_n \\ y_{n+1} = 1 - a|y_n| + x_n \end{cases}. \tag{11}$$

It is easy to establish that the three formulations are topologically conjugate, and for the sake of completeness we proceed to show the conjugations. This observation will allow us to deal with one of them and to translate automatically its dynamics to the other ones. In our case, we will consider the system (11). We denote by F_1, F_2, F_3 the corresponding two-dimensional maps $F_j(x, y)$ associated to the systems (8), (10) and (11), respectively. To see that (8) and (10) are conjugate, consider the homeomorphism $\sigma(x, y) = (x, \frac{1}{b}y)$. Then, it is immediate to check that $F_1 = \sigma^{-1} \circ F_2 \circ \sigma$. On the other hand, consider the homeomorphism $\tau(x, y) = (y, x)$. It is immediate to see that $F_3 = \tau^{-1} \circ F_2 \circ \tau$. Since the topological conjugation is an equivalence relation, we conclude that the three systems (8), (10) and (11) are topologically conjugate.

A natural generalization of Equation (9) can be given by

$$y_{n+1} = \alpha|y_n| + \beta y_n + \gamma y_{n-1} + \delta, \tag{12}$$

where $\alpha, \beta, \gamma, \delta$ are real numbers with $\alpha \neq 0$, otherwise we would obtain a linear difference equation whose dynamics is very well known. In the sequel we will refer to Equation (12) as *generalized Lozi map*. For instance, in the literature (see [12]) we can find a version of such equation

$$y_{n+1} = \frac{k}{2}|y_n| + \left(\frac{k}{2} - l\right) y_n - my_{n-1} + \delta, \tag{13}$$

where $k, l, m, \delta \in \mathbb{Z}$.

Returning to Equation (12), firstly, notice that in the particular case $\alpha = -a$, $\beta = 0$, $\gamma = b$ and $\delta = 1$, we recover Equation (9).

Now, we are going to show the general changes of variables which transform the generalized Lozi equation described in (12) into a max-equation.

By means of a change of the form

$$y_n = \log_A(z_n^q) + p$$

(where p and q are, in principle, arbitrarily taken real numbers with $q \neq 0$, and $A > 0$ is also arbitrary, $A \neq 1$, with $z_n > 0$ for all $n \geq -1$), and the observation that $|z| = \max\{z, -z\}$, Equation (12) is transformed into (with $P = A^p$):

$$\log_A(z_{n+1}) = \frac{\alpha}{q} \max\{\log_A(z_n^q \cdot P), -\log_A(z_n^q \cdot P)\} + \log_A(z_n^\beta \cdot z_{n-1}^\gamma) + \frac{\beta p + \gamma p + \delta - p}{q}. \tag{14}$$

Next, bearing in mind that the logarithm function $\log_A x$ is increasing (decreasing) when $A > 1$ $(0 < A < 1)$, we take an arbitrary q holding $\frac{\alpha}{q} > 0$ $(\frac{\alpha}{q} < 0)$. Then we can exchange the maximum function and the logarithm as

$$\frac{\alpha}{q} \max\{\log_A(z_n^q \cdot P), -\log_A(z_n^q \cdot P)\} = \log_A \max\left\{z_n^\alpha \cdot P^{\alpha/q}, \frac{1}{z_n^\alpha \cdot P^{\alpha/q}}\right\},$$

for $A > 1$ and $\frac{\alpha}{q} > 0$; whereas

$$\begin{aligned}
&\frac{\alpha}{q} \max\left\{\log_A(z_n^q \cdot P), -\log_A(z_n^q \cdot P)\right\} \\
&= \min\left\{\log_A(z_n^\alpha \cdot P^{\alpha/q}, \log\left(\frac{1}{z_n^\alpha \cdot P^{\alpha/q}}\right)\right\} \\
&= \log_A\left(\max\left\{z_n^\alpha \cdot P^{\alpha/q}, \frac{1}{z_n^\alpha \cdot P^{\alpha/q}}\right\}\right),
\end{aligned}$$

for $0 < A < 1$ and $\frac{\alpha}{q} < 0$.

We can gather the above discussion in the following result.

Lemma 2.1: *Let*

$$y_{n+1} = \alpha|y_n| + \beta y_n + \gamma y_{n-1} + \delta$$

be a generalized Lozi map. Let p and q arbitrarily real numbers, and $A > 0$. Then, the equation is transformed into

$$\log_A(z_{n+1}) = \log_A\left(\frac{\max\left\{z_n^{2\alpha}, \frac{1}{A^{2\alpha p/q}}\right\}}{z_n^{\alpha-\beta} \cdot z_{n-1}^{-\gamma}}\right) + \frac{p(\alpha + \beta + \gamma - 1) + \delta}{q}, \tag{15}$$

if $A > 1$ and $\frac{\alpha}{q} > 0$; or $0 < A < 1$ and $\frac{\alpha}{q} < 0$.

Remark 2.1: Notice that, by Lemma 2.1, Equations (12) and (15) are topologically conjugate. Indeed, it suffices to consider the homeomorphism $\phi : (0,\infty)^2 \to \mathbb{R}^2$, $\phi(x,y) = (\alpha(x), \alpha(y))$, where $\alpha(u) = \log_A(u^q) + p$, thus $\alpha^{-1}(w) = A^{\frac{w-p}{q}}$. Then, if we denote by F the discrete dynamical system in $\mathbb{R}^2$ given by $F(x,y) = (y, \alpha|y| + \beta y + \gamma x + \delta)$, and by $\tilde{F}$ the system defined in $(0,\infty)^2$ as $\tilde{F}(x,y) = (y, \frac{\max\{y^{2\alpha}, A^{-\frac{2\alpha p}{q}}\}}{y^{\alpha-\beta}\cdot x^{-\gamma}} \cdot A^{\frac{p(\alpha+\beta+\gamma-1)+\delta}{q}})$, it is easily seen that $\tilde{F} = \phi^{-1} \circ F \circ \phi$.

Some consequences of Lemma 2.1 are the following:

Proposition 2.1: *If $\delta = 0$, then the generalized Lozi map*

$$y_{n+1} = \alpha|y_n| + \beta y_n + \gamma y_{n-1}$$

is topologically conjugate to the max-type equation

$$z_{n+1} = \frac{\max\{z_n^{2\alpha}, B\}}{z_n^{\alpha-\beta} \cdot z_{n-1}^{-\gamma}} \cdot B^{\frac{\alpha+\beta+\gamma-1}{-2\alpha}} \tag{16}$$

for all $B > 0$.

In particular, if $\alpha + \beta + \gamma - 1 = 0$, then the generalized Lozi map is topologically conjugate to the max-type equation

$$z_{n+1} = \frac{\max\{z_n^{2\alpha}, B\}}{z_n^{\alpha-\beta} \cdot z_{n-1}^{-\gamma}} \tag{17}$$

for all $B > 0$.

Proof: If $\delta = 0$ and $\alpha + \beta + \gamma - 1 \neq 0$, then (15) reads as follows

$$\log_A(z_{n+1}) = \log_A\left(\frac{\max\left\{z_n^{2\alpha}, \frac{1}{A^{2\alpha p/q}}\right\}}{z_n^{\alpha-\beta} \cdot z_{n-1}^{-\gamma}}\right) + \frac{p(\alpha+\beta+\gamma-1)}{q}.$$

If we take $p = 0$ in the above equation, we find

$$z_{n+1} = \frac{\max\{z_n^{2\alpha}, 1\}}{z_n^{\alpha-\beta} \cdot z_{n-1}^{-\gamma}}.$$

If, otherwise, $p \neq 0$, the equation presents this writing

$$z_{n+1} = \frac{\max\left\{z_n^{2\alpha}, \frac{1}{A^{\frac{2\alpha p}{q}}}\right\}}{z_n^{\alpha-\beta} \cdot z_{n-1}^{-\gamma}} \cdot A^{\frac{p(\alpha+\beta+\gamma-1)}{q}} = \frac{\max\left\{z_n^{2\alpha}, A^{\frac{-2\alpha p}{q}}\right\}}{z_n^{\alpha-\beta} \cdot z_{n-1}^{-\gamma}} \cdot A^{\left(\frac{-2\alpha p}{q}\right)\cdot\frac{\alpha+\beta+\gamma-1}{-2\alpha}}.$$

When $A > 1$ and $\frac{\alpha}{q} > 0$, since p is arbitrarily taken, then once fixed the value q, having the same sign as α, we have that $A^{-\frac{2\alpha p}{q}} \in (0,\infty)$. Therefore, if we write $B = A^{-\frac{2\alpha p}{q}}$, we obtain (16). We can proceed analogously for $0 < A < 1$ and $\frac{\alpha}{q} < 0$.

Moreover, if $\alpha + \beta + \gamma - 1 = 0$, we can choose the value p *ad libitum*, and we have

$$z_{n+1} = \max\left\{z_n^{2\alpha}, \frac{1}{A^{2\alpha p/q}}\right\} \cdot z_n^{\beta-\alpha} \cdot z_{n-1}^{\gamma} = \frac{\max\left\{z_n^{2\alpha}, \left(\frac{1}{A}\right)^{\frac{2\alpha p}{q}}\right\}}{z_n^{\alpha-\beta} \cdot z_{n-1}^{-\gamma}}.$$

Since the choice of p is arbitrary, if $A > 1$ and $\frac{\alpha}{q} > 0$, we know that the range of $(\frac{1}{A})^{\frac{2\alpha p}{q}}$ is $(0, \infty)$, and we obtain

$$z_{n+1} = \frac{\max\{z_n^{2\alpha}, B\}}{z_n^{\alpha-\beta} \cdot z_{n-1}^{-\gamma}}$$

for all $B > 0$. The same applies if $0 < A < 1$ and $\frac{\alpha}{q} < 0$. ■

Example 2.1: In particular, $y_{n+1} = \frac{3}{2}|y_n| + \frac{1}{2}y_n - y_{n-1}$ is topologically conjugate to any max-type equation

$$z_{n+1} = \frac{\max\{z_n^3, B\}}{z_n \cdot z_{n-1}}, \quad B > 0.$$

On the other hand, $y_{n+1} = |y_n| + y_n + y_{n-1}$ is topologically conjugate to any max-type equation

$$z_{n+1} = \frac{1}{B}\max\{z_n^2, B\} \cdot z_{n-1}, \quad B > 0.$$

Proposition 2.2: *Let us consider*

$$y_{n+1} = \alpha|y_n| + \beta y_n + \gamma y_{n-1} + \delta, \tag{18}$$

with $\delta \neq 0$. Then, either for $A > 1$ and $\frac{\alpha}{q} > 0$, or for $0 < A < 1$ and $\frac{\alpha}{q} < 0$, with $q \in \mathbb{R} \setminus \{0\}$, Equation (18) is topologically conjugate to

$$z_{n+1} = \frac{\max\{z_n^{2\alpha}, A^{\frac{-2\alpha p}{q}}\}}{z_n^{\alpha-\beta} \cdot z_{n-1}^{-\gamma}} \cdot A^{\frac{p(\alpha+\beta+\gamma-1)+\delta}{q}},$$

for all $p \in \mathbb{R}$. In particular, assuming that $\alpha + \beta + \gamma - 1 \neq 0$, Equation (18) is topologically conjugate to:

- $z_{n+1} = \frac{\max\{z_n^{2\alpha}, B\}}{z_n^{\alpha-\beta} \cdot z_{n-1}^{-\gamma}}$, *for all $B > 1$, if $\frac{\delta}{\alpha+\beta+\gamma-1} > 0$.*
- $z_{n+1} = \frac{\max\{z_n^{2\alpha}, C\}}{z_n^{\alpha-\beta} \cdot z_{n-1}^{-\gamma}}$, *for all $0 < C < 1$, if $\frac{\delta}{\alpha+\beta+\gamma-1} < 0$.*

Moreover, if $\alpha + \beta + \gamma - 1 = 0$, additionally we get that Equation (18) is topologically conjugate to:

- $z_{n+1} = \frac{\max\{z_n^{2\alpha}, 1\}}{z_n^{\alpha-\beta} \cdot z_{n-1}^{-\gamma}} \cdot B$, *for all $B > 1$, if $\frac{\delta}{\alpha} > 0$.*

- $z_{n+1} = \frac{\max\{z_n^{2\alpha},1\}}{z_n^{\alpha-\beta} \cdot z_{n-1}^{-\gamma}} \cdot C$, *for all* $0 < C < 1$, *if* $\frac{\delta}{\alpha} < 0$.

Proof: Firstly, by Lemma 2.1, it follows directly that Equation (18) is topologically conjugate to

$$z_{n+1} = \frac{\max\{z_n^{2\alpha}, A^{\frac{-2\alpha p}{q}}\}}{z_n^{\alpha-\beta} \cdot z_{n-1}^{-\gamma}} \cdot A^{\frac{p(\alpha+\beta+\gamma-1)+\delta}{q}}, \tag{19}$$

for $A > 1$ and $\frac{\alpha}{q} > 0$, or for $0 < A < 1$ and $\frac{\alpha}{q} < 0$, and for all $p \in \mathbb{R}$.

Now, if $\alpha + \beta + \gamma - 1 \neq 0$, by taking $p = \frac{-\delta}{\alpha+\beta+\gamma-1}$, (19) reduces to

$$z_{n+1} = \frac{\max\left\{z_n^{2\alpha}, A^{\frac{2\alpha\delta}{q(\alpha+\beta+\gamma-1)}}\right\}}{z_n^{\alpha-\beta} \cdot z_{n-1}^{-\gamma}}.$$

Now, depending on the sign of $\frac{\delta}{\alpha+\beta+\gamma-1}$ and taking into account if $A > 1$ or $0 < A < 1$, we deduce that the corresponding generalized Lozi map is topologically conjugate to:

- The equations $z_{n+1} = \frac{\max\{z_n^{2\alpha},B\}}{z_n^{\alpha-\beta} \cdot z_{n-1}^{-\gamma}}$ for all $B > 1$ when $\frac{\delta}{\alpha+\beta+\gamma-1} > 0$.
- The equations $z_{n+1} = \frac{\max\{z_n^{2\alpha},C\}}{z_n^{\alpha-\beta} \cdot z_{n-1}^{-\gamma}}$ for all $0 < C < 1$ when $\frac{\delta}{\alpha+\beta+\gamma-1} < 0$.

Finally, if $\alpha + \beta + \gamma - 1 = 0$, then, Equation (19) is converted into

$$z_{n+1} = \frac{\max\left\{z_n^{2\alpha}, A^{-\frac{2\alpha p}{q}}\right\}}{z_n^{\alpha-\beta} \cdot z_{n-1}^{-\gamma}} A^{\delta/q}.$$

So, if we take $p = 0$, we arrive to the difference equation

$$z_{n+1} = \frac{\max\{z_n^{2\alpha}, 1\}}{z_n^{\alpha-\beta} \cdot z_{n-1}^{-\gamma}} \cdot A^{\delta/q}.$$

Since A is arbitrary, depending on the interval where A belongs and the corresponding sign of q, we find:

- The equations $z_{n+1} = \frac{\max\{z_n^{2\alpha},1\}}{z_n^{\alpha-\beta} \cdot z_{n-1}^{-\gamma}} \cdot B$ for all $B > 1$ when $\frac{\delta}{\alpha} > 0$.
- The equations $z_{n+1} = \frac{\max\{z_n^{2\alpha},1\}}{z_n^{\alpha-\beta} \cdot z_{n-1}^{-\gamma}} \cdot C$ for all $0 < C < 1$ when $\frac{\delta}{\alpha} < 0$.

■

Example 2.2: For instance, $y_{n+1} = \frac{1}{2}|y_n| - \frac{1}{2}y_n - 2y_{n-1} + 1$ is topologically conjugate to

$$z_{n+1} = \frac{\max\{z_n, 1\}}{z_n \cdot z_{n-1}^2} \cdot B,$$

for all $B > 1$. □

Finally, in order to show some conditions that allow us to connect Equation (12) with max-type difference equations, we recover Equation (13) introduced at the beginning of the section, that is,

$$y_{n+1} = \frac{k}{2}|y_n| + \left(\frac{k}{2} - l\right) y_n - m y_{n-1} + \delta.$$

Obviously, the two formulations of Equations (12) and (13) are equivalent, since it suffices to take into account that the equalities $\frac{k}{2} = \alpha$, $\frac{k}{2} - l = \beta$, $-m = \gamma$ and $\delta = \delta$ define a bijective relation between the coefficients of (12) and (13). In this sense, we find:

- $\delta = k - 1 - l - m \Leftrightarrow \delta = \alpha + \beta + \gamma - 1$;
- $\delta = -(k - 1 - l - m) \Leftrightarrow \delta = -(\alpha + \beta + \gamma) + 1$;
- $\delta = 0$.

It should be emphasized that the above conditions $\delta = k - 1 - l - m$, $\delta = -(k - 1 - l - m)$ and $\delta = 0$ appear in [12] when applying the change of variables to Equation (13) in order to obtain a max-type difference equation, which is

$$x_{n+1} = \frac{\max\{x_n^k, A\}}{x_n^l \cdot x_{n-1}^m}. \tag{20}$$

In the light of our study, we can establish:

Corollary 2.1 ([12]): *Let $k, l, m, \delta \in \mathbb{R}$. Then, the generalized Lozi map $y_{n+1} = \frac{k}{2}|y_n| + (\frac{k}{2} - l)y_n - my_{n-1} + \delta$ is topologically conjugate to:*

- $x_{n+1} = \frac{\max\{x_n^k, A\}}{x_n^l \cdot x_{n-1}^m}$ *for all $A > 0$, if $\delta = 0$;*
- $x_{n+1} = \frac{\max\{x_n^k, A\}}{x_n^l \cdot x_{n-1}^m}$ *for all $A > 1$, if $\delta = k - 1 - l - m$;*
- $x_{n+1} = \frac{\max\{x_n^k, A\}}{x_n^l \cdot x_{n-1}^m}$ *for all $0 < A < 1$, if $\delta = -(k - 1 - l - m)$.*

The transformation that links Equation (13) with Equation (20) can be found in [12] or [11]. However, we have made a more general change of variables and the parameters are not restricted to the integer set $\mathbb{Z}$ as they are in the references cited, but they can be arbitrary real numbers. Furthermore, as far as we are concerned, it is the first time that the problem is treated from the point of view of topological conjugacies.

On the other hand, recall that we recover Lozi equation (9) from the generalized Lozi equation (12) if $\alpha = -a$, $\beta = 0$, $\gamma = b$ and $\delta = 1$. In this sense, from the above conditions, it yields that $\delta = \alpha + \beta + \gamma - 1$ implies $b{-}a = 2$; and $\delta = -(\alpha + \beta + \gamma) + 1$ implies $a = b$. So, this motivates to analyse in detail Lozi map in the particular cases $a = b$ and $b{-}a = 2$. Nevertheless, we will restrict our attention to case $a = b$.

3. A family of max-type difference equations

In the present section we will analyse a concrete family of max-type difference equations. Our purpose with such study is to illustrate that, thanks to the transformations developed in the previous section, the analysis of a concrete Lozi map is sufficient to know the behaviour of a whole family of equations.

In this direction, we will deal with the particular one-parametric family of max-type difference equations

$$x_{n+1} = \frac{\max\{x_n^3, A\}}{x_n x_{n-1}}, \quad A > 0,$$

which is topologically conjugate to the generalized Lozi map

$$y_{n+1} = \frac{3}{2}|y_n| + \frac{1}{2}y_n - y_{n-1}. \tag{21}$$

Notice that Equation (21) is a particular case of Equation (12) with $\delta = 0$ and $\alpha + \beta + \gamma = 1$. The dynamics of (21) is easily described by the following results. The proof of the first one is straightforward and will be omitted.

Proposition 3.1: *Equation* (21) *has infinitely many equilibrium points, namely,* $\mathcal{F} = \{\bar{x} : \bar{x} \geq 0\}$,

Theorem 3.1: *Every solution of Equation* (21) *which is not an equilibrium point diverges to* ∞.

Proof: We distinguish several cases depending on the values of the initial conditions.

Case 1: If the initial conditions (y_{-1}, y_0) of (21) hold $0 \leq y_{-1} < y_0$, then by induction it is easily seen that $y_n = y_{-1} + (n+1)(y_0 - y_{-1})$ for all $n \geq 1$, and, hence, $\lim_{n\to+\infty} y_n = +\infty$.

Case 2: Similarly, if the initial conditions (y_{-1}, y_0) satisfy $y_{-1} \leq 0 < y_0$ or $y_{-1} < 0 \leq y_0$, then $y_1 = 2y_0 - y_{-1} > y_0 > 0$, and we can apply Case 1 to affirm that the solution goes to $+\infty$ when n tends to $+\infty$.

Case 3: Again, the solution is unbounded if the initial conditions verify $y_{-1} \leq y_0 \leq 0$, $|y_{-1}| + |y_0| > 0$, since now we find $y_1 = -\frac{3}{2}y_0 + \frac{1}{2}y_0 - y_{-1} = -y_0 - y_{-1} > 0$ and we can apply Case 2.

Case 4: Suppose that $0 \leq y_0 < y_{-1}$. Consider the value $\varepsilon = y_{-1} - y_0$ and fix the smallest non negative integer N such that

$$y_0 - (N+1)\varepsilon < 0 \leq y_0 - N\varepsilon.$$

Then, it is a simple matter to check that

$$y_1 = 2y_0 - y_{-1} = y_0 - \varepsilon, \ldots, y_N = y_0 - N\varepsilon, y_{N+1} = y_0 + (N+1)\varepsilon,$$

and being $y_{N+1} < 0 \leq y_N$, we obtain

$$\begin{aligned} y_{N+2} &= \frac{3}{2}(-y_0 + (N+1)\varepsilon) + \frac{1}{2}(y_0 - (N+1)\varepsilon) - y_0 + N\varepsilon \\ &= -2y_0 + (2N+1)\varepsilon. \end{aligned}$$

Now, if $y_{N+2} \geq 0$, we finish because we have the pair of new initial conditions (y_{N+1}, y_{N+2}), with $y_{N+1} < 0 \leq y_{N+2}$ and we apply Case 2. But if $y_{N+2} < 0$, following the iteration we deduce that

$$\begin{aligned} y_{N+3} &= \frac{3}{2}(2y_0 - (2N+1)\varepsilon) + \frac{1}{2}(-2y_0 + (2N+1)\varepsilon) - (y_0 - (N+1)\varepsilon) \\ &= y_0 - N\varepsilon \geq 0, \end{aligned}$$

and we can apply Case 2 to the new initial conditions $y_{N+2} < 0 \leq y_{N+3}$.

Case 5: For the case $y_0 \leq y_{-1} \leq 0$, $|y_{-1}| + |y_0| > 0$, take into account that $y_1 = -y_0 - y_{-1} > 0$ and use Case 2 to the initial conditions $y_0 \leq 0 < y_1$.

Case 6: With respect to the last case of our discussion, we have $y_1 = -y_0 - y_{-1}$.

If $y_{-1} + y_0 < 0$, then $y_0 \leq 0 < y_1$ and we finish by a simple application of Case 2. If, on the contrary, $y_{-1} + y_0 \geq 0$, then $y_0 < 0$, $y_1 \leq 0$, and it suffices to apply Case 3 or Case 5 depending on whether $y_0 \leq y_1 \leq 0$ or $y_1 \leq y_0 \leq 0$, respectively. ■

Corollary 3.1: *Given an arbitrary value $A \in (0, +\infty)$, consider the max-type difference equation*

$$x_{n+1} = \frac{\max\{x_n^3, A\}}{x_n x_{n-1}}.$$

For any arbitrary positive initial conditions (x_{-1}, x_0), either $x_{-1} = x_0 = \bar{x}$ is an equilibrium point, or the solution generated by them diverges to infinity. The stationary solution $(\bar{x}, \bar{x}, \bar{x}, \ldots)$ appears for all the values $\bar{x} \geq \sqrt[3]{A}$.

In view of this result, a question that may be of some interest is to study the dynamics of generalized Lozi map in the case $\delta = 0$ and $\alpha + \beta + \gamma = 1$.

4. Lozi map for $a = b$

Now, we will focus on the Lozi map, in the particular case where the parameters verify $a = b$,

$$x_{n+1} = 1 - a|x_n| + ax_{n-1}. \tag{22}$$

For such equation it is straightforward to determine its fixed points and we will omit its proof.

Lemma 4.1: *The equilibrium points of Equation (22) are given by:*

(a) $\bar{x} = 1$ *if* $a \in (-\infty, \frac{1}{2}]$.
(b) $\bar{x} = 1$ *and* $\bar{x} = \frac{1}{1-2a}$ *if* $a \in (\frac{1}{2}, +\infty)$.

It is well-known (see [10, pages 181–182]) that if $|a| + |b| < 1$, then the Lozi map (9) has a unique fixed point which is globally attractor. In the case of $a = b$, such condition reads as $|a| < \frac{1}{2}$. Therefore,

Proposition 4.1: *Let $a \in (-\frac{1}{2}, \frac{1}{2})$. Then, all the solutions of Equation (22) converge to the equilibrium point $\bar{x} = 1$.*

In view of the fact that $a = \frac{1}{2}$ and $a = -\frac{1}{2}$ are boundary values for the condition $|a| < \frac{1}{2}$, the study of the difference equations

$$x_{n+1} = 1 - \frac{1}{2}|x_n| + \frac{1}{2}x_{n-1} \quad \text{and} \quad x_{n+1} = 1 + \frac{1}{2}|x_n| - \frac{1}{2}x_{n-1}$$

can merit to pay our attention.

In the case of $a = \frac{1}{2}$, it seems that all the solutions, other than periodic solutions, converge to a 2-periodic solution. We will prove it later.

Concerning periodic solutions of Lozi equation (22), we have to discard the interval of values $a \in (-\frac{1}{2}, \frac{1}{2})$ because in this segment of values we know that all the solutions converge to the global attractor $\bar{x} = 1$. For $|a| > \frac{1}{2}$, by a straightforward way, we find:

Lemma 4.2: *Assume that $|a| > \frac{1}{2}$. Then, the only initial conditions (x_{-1}, x_0) which generate 2-periodic solutions are given by:*

$$(x_{-1}, x_0) = \left(\frac{1}{2a^2 - 2a + 1}, \frac{1 - 2a}{2a^2 - 2a + 1}\right),$$

or

$$(x_{-1}, x_0) = \left(\frac{1 - 2a}{2a^2 - 2a + 1}, \frac{1}{2a^2 - 2a + 1}\right)$$

with $a > \frac{1}{2}$.

In this case, notice that the initial conditions have different signs.

Concerning the stability of these 2-periodic points for the Lozi map $F(x, y) = (y, 1 - a|y| + ax)$, we have

$$DF\left(\frac{1}{2a^2 - 2a + 1}, \frac{1 - 2a}{2a^2 - 2a + 1}\right) = \begin{pmatrix} 0 & 1 \\ a & a \end{pmatrix},$$

and

$$DF\left(\frac{1 - 2a}{2a^2 - 2a + 1}, \frac{1}{2a^2 - 2a + 1}\right) = \begin{pmatrix} 0 & 1 \\ a & -a \end{pmatrix}.$$

Therefore,

$$DF^2\left(\frac{1}{2a^2 - 2a + 1}, \frac{1 - 2a}{2a^2 - 2a + 1}\right) = \begin{pmatrix} 0 & 1 \\ a & a \end{pmatrix} \cdot \begin{pmatrix} 0 & 1 \\ a & -a \end{pmatrix} = \begin{pmatrix} a & -a \\ a^2 & a - a^2 \end{pmatrix}.$$

The corresponding eigenvalues are computed by the characteristic equation

$$(-a^2 + a - \lambda) \cdot (a - \lambda) + a^3 = 0,$$
$$\lambda^2 + (a^2 - 2a)\lambda + a^2 = 0,$$

whose roots are given by

$$\lambda_1 = \frac{2a - a^2 + a\sqrt{a^2 - 4a}}{2}, \quad \lambda_2 = \frac{2a - a^2 - a\sqrt{a^2 - 4a}}{2},$$

and these roots lie inside the unit circle if and only if the coefficients of the characteristic polynomial verify (consult [9, Th.2.37]):

$$|-a^2+2a| < 1+a^2 < 2,$$

which is obviously true for the range of values $a \in (\frac{1}{2}, 1)$. This means that these 2-periodic points are locally asymptotically stable[8] when $\frac{1}{2} < a < 1$.

4.1. Case $a = \frac{1}{2}$

We consider the parameter $a = \frac{1}{2}$ for the Lozi map

$$x_{n+1} = 1 - \frac{1}{2}|x_n| + \frac{1}{2}x_{n-1}. \tag{23}$$

The max-difference equation associated to Equation (23) is given by (use Proposition 2.2)

$$x_{n+1} = x_n^{1/2} \cdot x_{n-1}^{1/2} \cdot \max\left\{\frac{1}{x_n}, A\right\}, \quad 0 < A < 1. \tag{24}$$

The following results establish the fixed points and the 2-periodic orbits of Equation (23), respectively. The proof of the first one is immediate and will be omitted.

Lemma 4.3: *The unique equilibrium point of* (23) *is* $\bar{x} = 1$.

Lemma 4.4: *Suppose that the initial condition* $(x_{-1}, x_0) = (x, y)$ *generates a 2-periodic solution of Equation* (23). *Then,*

$$0 \le x, y \le 2; \quad x + y = 2; \quad x \ne y.$$

Proof: Assume that the pair (x, y) provides a 2-periodic sequence. From the corresponding iteration, we deduce that

$$\begin{cases} x = 1 - \frac{1}{2}|y| + \frac{1}{2}x \\ y = 1 - \frac{1}{2}|x| + \frac{1}{2}y \end{cases} . \tag{25}$$

Equivalently, $\frac{1}{2}x = 1 - \frac{1}{2}|y|$ and $\frac{1}{2}y = 1 - \frac{1}{2}|x|$. Now, if we subtract both equations, we deduce

$$x - y = |x| - |y|. \tag{26}$$

Next, we do a proof by cases.

- If $x, y \ge 0$, suppose that $x = \alpha \cdot y$, for some $\alpha \ge 0$. If $\alpha = 0$, $x = 0$, we deduce that $y = 2$, and it is immediate to check that the initial conditions $(x_{-1}, x_0) = (0, 2)$ generates a 2-periodic solution. If $\alpha > 0$, we find $\frac{1}{2}x = 1 - \frac{1}{2}\frac{x}{\alpha}$; or $x = \frac{2\alpha}{1+\alpha}$. Then, $y = \frac{x}{\alpha} = \frac{2}{1+\alpha}$.
  Notice that $x + y = 2$. It is easy to see that $(\frac{2\alpha}{1+\alpha}, \frac{2}{1+\alpha})$ generates a 2-periodic orbit.
- If $x \ge 0, y \le 0$, from (26) we deduce that $y = 0$, and therefore $x = 2$. The initial conditions are $(2, 0)$, already included in the above case. By symmetry, the case $x \le 0, y \ge 0$ leads us to the same conclusion.

- If $x \leq 0, y \leq 0$, from (26) we have $x - y = 0$, or $x = y$. Replacing this equality in (25) yields to $\frac{1}{2}x = 1 + \frac{1}{2}x$, which is impossible. ■

Now, we return to the max-equation, for instance through the change of variables $y_n = \log_A(\frac{A}{x_n^k}) = \log_A(Ax_n) = 1 - \log_A x_n$ (we take $p = 1, q = -1$), where (x_n) and (y_n) are the corresponding solutions of Equation (24) and the Lozi equation, respectively. It is direct to see then that the following result holds.

Corollary 4.1: *Suppose that the initial conditions $(x_{-1}, x_0) = (x, y)$ generate a 2-periodic solution of Equation* (24). *Then,*

$$x, y \in \left[A, \frac{1}{A}\right], \quad A \in (0,1) \text{ with } xy = 1, \ x \neq y.$$

Realize that we have included the condition $x \neq y$ in order to avoid the appearance of the equilibrium point $\bar{x} = 1$.

Next, we are going to prove that any initial condition (x_{-1}, x_0) generates by Equation (23) an asymptotically periodic orbit to some appropriate 2-periodic orbit $(x, y, x, y, \ldots)$ with $x + y = 2$.

Concerning the analysis of the stability for 2-periodic orbits of $x_{n+1} = 1 - \frac{1}{2}|x_n| + \frac{1}{2}x_{n-1}$, since $F(x, y) = (y, 1 - \frac{1}{2}|y| + \frac{1}{2}x)$, and the coordinates are positive, it is $DF(x, y) = \begin{pmatrix} 0 & 1 \\ \frac{1}{2} & -\frac{1}{2} \end{pmatrix}$, therefore

$$DF^2(x, y) = \begin{pmatrix} 0 & 1 \\ \frac{1}{2} & -\frac{1}{2} \end{pmatrix} \cdot \begin{pmatrix} 0 & 1 \\ \frac{1}{2} & -\frac{1}{2} \end{pmatrix} = \begin{pmatrix} \frac{1}{2} & -\frac{1}{2} \\ -\frac{1}{4} & \frac{3}{4} \end{pmatrix};$$

the corresponding eigenvalues are computed by the characteristic equation

$$\lambda^2 - \frac{5}{4}\lambda + \frac{1}{4} = 0,$$

whose roots are given by $\lambda_1 = 1$ and $\lambda_2 = \frac{1}{4}$. Since one of the eigenvalues of the 2-periodic point $(x, 2 - x)$ has modulo 1, it is necessary to analyse its stability by other means different to the linearization technique.

As an initial example, consider $(x_{-1}, x_0) = (0, 0)$. We try to guess the behaviour at large of its associated orbit $(x_{-1}, x_0, x_1, x_2, \ldots, x_n, \ldots)$. The first elements are given by

$$\left(0, 0, 1, \frac{1}{2}, \frac{5}{4}, \frac{5}{8}, \frac{21}{16}, \frac{21}{32}, \ldots\right).$$

It is a simpler matter to see that all the elements are positive because by induction it is easily seen that $x_n = \frac{1}{3}(-1)^{n+1} - \frac{4}{3}(\frac{1}{2})^{n+1} + 1$ for all $n \geq -1$. Notice that this expression is nothing but the expression for the solution of the non-homogenous linear difference equation $x_{n+1} = 1 - \frac{1}{2}x_n + \frac{1}{2}x_{n-1}$, when initial conditions $(x_{-1}, x_0) = (0, 0)$ are considered. Therefore, $\lim_{n\to+\infty} x_{2n} = \frac{2}{3}$, $\lim_{n\to+\infty} x_{2n+1} = \frac{4}{3}$, and the orbit is asymptotically periodic to the 2-periodic orbit $(\frac{4}{3}, \frac{2}{3}, \frac{4}{3}, \frac{2}{3}, \ldots)$.

The above computations for the orbit of $(0, 0)$ suggest to do a case study, by distinguishing the different zones in which the initial conditions can be located.

Returning to our problem on the convergence of the orbits generated under $x_{n+1} = 1 - \frac{1}{2}|x_n| + \frac{1}{2}x_{n-1}$ in an asymptotically periodic form, we present the following result on invariance of two triangles in the plane by $F(x,y) = (y, 1 - \frac{1}{2}|y| + \frac{1}{2}x)$.

Lemma 4.5: *Consider the triangles*

$$\Delta_\ell = \{(x,y) : 0 \le x, y \le 2,\ x + y \le 2\}$$

and

$$\Delta_u = \{(x,y) : 0 \le x, y \le 2,\ x + y \ge 2\}.$$

Then, $F(\Delta_\ell) \subset \Delta_\ell$ and $F(\Delta_u) \subset \Delta_u$, where $F(x,y) = (y, 1 - \frac{1}{2}|y| + \frac{1}{2}x)$.

Proof: Let $(x,y) \in \Delta_\ell$. Then, $F(x,y) = (y, 1 - \frac{1}{2}y + \frac{1}{2}x)$, with $0 \le y \le 2$,

$$1 - \frac{1}{2}y + \frac{1}{2}x \ge 1 - \frac{1}{2}y \ge 1 - 1 = 0,$$

$$1 - \frac{1}{2}y + \frac{1}{2}x \le 1 - \frac{1}{2}y + 1 \le 2 - \frac{1}{2}y \le 2,$$

and

$$y + \left(1 - \frac{1}{2}y + \frac{1}{2}x\right) = 1 + \frac{1}{2}x + \frac{1}{2}y \le 1 + \frac{x+y}{2} \le 1 + 1 = 2.$$

Therefore, $F(x,y) \in \Delta_\ell$.

The proof of the invariance of Δ_u by F is analogous; now,

$$y + \left(1 - \frac{1}{2}y + \frac{1}{2}x\right) = 1 + \frac{x+y}{2} \ge 2.$$

■

As a consequence of this lemma, if some iterate of the orbit by F lies in Δ_ℓ or Δ_u, then the orbit will remain indefinitely in these triangles. Since the coordinates of the iterates in these zones are always positive, in fact the Lozi map becomes in the linear system $F(x,y) = (y, 1 - \frac{1}{2}y + \frac{1}{2}x)$ and the solution will be explicitly known.

Proposition 4.2: *Let $(x,y) \in \Delta_\ell \cup \Delta_u$. Then, $(F^n(x,y))_{n\ge 0}$ is asymptotically periodic to some 2-periodic point (v,w).*

Proof: Due to the invariance of $\Delta_\ell \cup \Delta_u$, if we denote the n-th iterate of (x,y) under F by $F^n(x,y) = (x_n, y_n)$, $n \ge 0$, it is clear that $y_{n+1} = 1 - \frac{1}{2}y_n + \frac{1}{2}y_{n-1}$, with $y_{-1} = y$, $y_0 = 1 - \frac{1}{2}y + \frac{1}{2}x$. The general solution of the non-homogenous linear difference equation is given by $y_n = A(-1)^n + B(\frac{1}{2})^n + 1$, for arbitrary constants by $A, B \in \mathbb{R}$.

If we impose the initial conditions $y_{-1} = y$, $y_0 = 1 - \frac{1}{2}y + \frac{1}{2}x$, we obtain the solution

$$y_n = \frac{2y - x - 1}{3}(-1)^n + \frac{x + y - 2}{3}\left(\frac{1}{2}\right)^n + 1.$$

Since $y_{2n} \to \frac{2y-x-1}{3} + 1 = \frac{2y-x+2}{3}$ and $y_{2n+1} \to -\frac{2y-x-1}{3} + 1 = \frac{-2y+x+4}{3}$, we conclude that the orbit converge to the 2-periodic sequence

$$\left(\ldots, \frac{2y - x + 2}{3}, \frac{-2y + x + 4}{3}, \frac{2y - x + 2}{3}, \frac{-2y + x + 4}{3}, \ldots\right).$$

■

Bearing in mind this proposition, in order to prove that all the orbits converge to periodic solutions of period 2, it suffices to prove that for any initial condition (x, y) there exists a positive integer N such that $F^N(x, y) \in \Delta_\ell \cup \Delta_u$.

For this task, we are going to prove that for any square

$$C_{m,n} := [2m, 2m+2] \times [2n, 2n+2], \quad \text{for } m, n \in \mathbb{Z},$$

either there exists $N \in \mathbb{Z}$ such that $F^N(C_{m,n}) \subset C_{0,0}$ or $\cap_{p\geq 0} F^p(C_{m,n}) = \{(0, 2), (0, 2)\}$. In both cases, we will deduce that for any $(x, y) \in \mathbb{R}^2$, it holds that $(F^n(x, y))_{n\geq 0}$ is asymptotically periodic to some 2-periodic point (v, w).

To do it, we will use the following facts. In their proofs, we will use the notation

$$H_u = \{(x, y) : y \geq 0\} \quad \text{and} \quad H_\ell = \{(x, y) : y \leq 0\}.$$

Lemma 4.6: *The map F transforms the square $C_{m,n}$ into a parallelogram and the vertices of $C_{m,n}$ go to the vertices of that parallelogram.*

Proof: It is a consequence of the linearity of F and the fact of being located the square $C_{m,n}$ entirely on the upper half-plane H_u or in the lower half plane H_ℓ. ■

Lemma 4.7: *Let $A(C_{m,n})$ and $A(F(C_{m,n}))$ be the areas of the square $C_{m,n}$ and its image by F, respectively. Then $A(F(C_{m,n})) = \frac{1}{2}A(C_{m,n})$.*

Proof: In this case, $A(F(C_{m,n})) = \int\int_{C_{m,n}} |\frac{\partial(x,y)}{\partial(u,v)}| du\ dv$, where $|\frac{\partial(x,y)}{\partial(u,v)}|$ is the absolute value of the determinant $\begin{vmatrix} \frac{\partial x}{\partial u} & \frac{\partial x}{\partial v} \\ \frac{\partial y}{\partial u} & \frac{\partial y}{\partial v} \end{vmatrix}$, with $(x(u, v), y(u, v)) = (v, 1 - \frac{1}{2}v + \frac{1}{2}u)$ or $(x(u, v), y(u, v)) = (v, 1 + \frac{1}{2}v + \frac{1}{2}u)$ depending on whether the square is located or not in the upper half-plane H_u. In both cases, $\frac{\partial(x,y)}{\partial(u,v)} = \begin{vmatrix} \frac{\partial x}{\partial u} & \frac{\partial x}{\partial v} \\ \frac{\partial y}{\partial u} & \frac{\partial y}{\partial v} \end{vmatrix} = \begin{vmatrix} 0 & 1 \\ \frac{1}{2} & \pm\frac{1}{2} \end{vmatrix} = -\frac{1}{2}$, and the result follows. ■

Concerning the iteration of triangles located entirely on H_u or H_ℓ, again we have that any triangle $T \subset H_u$ (respectively, $T \subset H_\ell$) is transformed in a new triangle whose vertices are the vertices of T by the action of F, and additionally $A(F(T)) = \frac{1}{2}A(T)$.

From now on, our strategy consists in proving:

(a) The region R_1 surrounding $C_{0,0}$, including $C_{0,0}$ itself, is invariant, that is, $R_1 := \bigcup_{i,j\in\{-1,0,1\}} C_{i,j} = [-2, 4] \times [-2, 4]$ is invariant; after that, to prove that either the images $F^N(C_{i,j})$ are included in $C_{0,0}$ for some positive integer N or the intersection of images not included in $C_{0,0}$ converges to the 2-periodic orbit $\{(0, 2), (0, 2)\}$.
(b) The image of any square $C_{i,j}$ for $i \geq 2$ or $j \geq 2$ is finally entirely contained in R_1.

For (a), we need to use the previous lemmas concerning the contraction of areas and the transformation of triangles and parallelograms in the same type of figures whenever the geometric object is included in H_ℓ or H_u.

As a first step, we analyse the evolution of the segment

$$S_{\varepsilon,0} = \{(x,0) : 2 \le x \le 2+\varepsilon\}, \quad \varepsilon \in [0,1].$$

Lemma 4.8: *Given $\varepsilon \in [0,1]$, consider $F^n(S_{\varepsilon,0}), n \ge 0$. Then, there exists N such that $F^N(S_\varepsilon) \subset C_{0,0}$.*

Proof: It suffices to study the evolution of the endpoints of $S_{\varepsilon,0}$. Realize that $F^n(S_{\varepsilon,0})$ are lines lying entirely in either H_ℓ or H_u. Then,

$$\begin{aligned}(2+\varepsilon,0) &\xrightarrow{F} \left(0, 2+\frac{\varepsilon}{2}\right) \xrightarrow{F} \left(2+\frac{\varepsilon}{2}, -\frac{\varepsilon}{4}\right) \xrightarrow{F} \left(-\frac{\varepsilon}{4}, 2+\frac{\varepsilon}{8}\right) \xrightarrow{F} \left(2+\frac{\varepsilon}{8}, -\frac{3}{16}\varepsilon\right)\\ &\xrightarrow{F} \left(-\frac{3}{16}\varepsilon, 2-\frac{\varepsilon}{32}\right) \xrightarrow{F} \left(2-\frac{\varepsilon}{2}, -\frac{5}{64}\varepsilon\right) \xrightarrow{F} \left(-\frac{5}{64}\varepsilon, 2-\frac{7}{128}\varepsilon\right)\\ &\xrightarrow{F} \left(2-\frac{7}{128}\varepsilon, -\frac{3}{256}\varepsilon\right) \xrightarrow{F} \left(-\frac{3}{256}\varepsilon, 2-\frac{17}{512}\varepsilon\right) \xrightarrow{F} \left(2-\frac{17}{512}\varepsilon, \frac{11}{1024}\varepsilon\right).\end{aligned}$$

At this point, we notice that $F^{10}(2+\varepsilon,0) = (2-\frac{17}{512}\varepsilon, \frac{11}{1024}\varepsilon) \in C_{0,0}$, which ends the proof. ■

In a similar way, we will keep studying the behaviour of the lines

$$S_{\varepsilon,2} = \{(x,2) : -\varepsilon \le x \le 0\}, \quad \varepsilon \in [0,1].$$

Lemma 4.9: *Given $\varepsilon \in [0,1]$, consider $F^n(S_{\varepsilon,2}), n \ge 0$. Then, there exists N such that $F^N(S_{\varepsilon,2}) \subset C_{0,0}$.*

Proof: In this case, the iterates of the endpoint $(-\varepsilon, 2)$ are

$$\begin{aligned}(-\varepsilon,2) &\xrightarrow{F} \left(2, -\frac{\varepsilon}{2}\right) \xrightarrow{F} \left(-\frac{\varepsilon}{2}, 2-\frac{\varepsilon}{4}\right) \xrightarrow{F} \left(2-\frac{\varepsilon}{4}, -\frac{\varepsilon}{8}\right)\\ &\xrightarrow{F} \left(-\frac{\varepsilon}{8}, 2-\frac{3}{16}\varepsilon\right) \xrightarrow{F} \left(2-\frac{3}{16}\varepsilon, \frac{\varepsilon}{32}\right).\end{aligned}$$

Since $(2-\frac{3}{16}\varepsilon, \frac{\varepsilon}{32}) \in C_{0,0}$, we finish the proof. ■

As a second step, we are going to analyse the behaviour of F under appropriate neighbourhoods of $(2,0)$ and $(0,2)$. The proof is straightforward and we omit it. Instead, we encourage the reader to follow the reasoning via Figure 1.

Lemma 4.10: *Let $\varepsilon \in [0,1]$. Consider the squares $Q_{1,\varepsilon} = [2, 2+\varepsilon] \times [0,\varepsilon]$, $Q_{2,\varepsilon} = [2, 2+\varepsilon] \times [-\varepsilon, 0]$ and $Q_{3,\varepsilon} = [2-\varepsilon, 2] \times [-\varepsilon, 0]$. Then:*

(a.1) *$F(Q_{1,\varepsilon}) \subset C_{0,0} \cup T_{1,\varepsilon}$, where $T_{1,\varepsilon}$ is the triangle with vertices $(0,2)$, $(0, 2+\frac{\varepsilon}{2})$, $(\varepsilon, 2)$.*
(a.2) *$F(Q_{2,\varepsilon})$ is the parallelogram with vertices $(0,2)$, $(0,2+\frac{\varepsilon}{2})$, $(-\varepsilon,2)$, $(-\varepsilon, 2-\frac{\varepsilon}{2})$.*
(a.3) *$F(Q_{3,\varepsilon})$ is the parallelogram with vertices $(0,2)$, $(0,2-\frac{\varepsilon}{2})$, $(-\varepsilon, 2-\frac{\varepsilon}{2})$, $(-\varepsilon, 2-\varepsilon)$.*

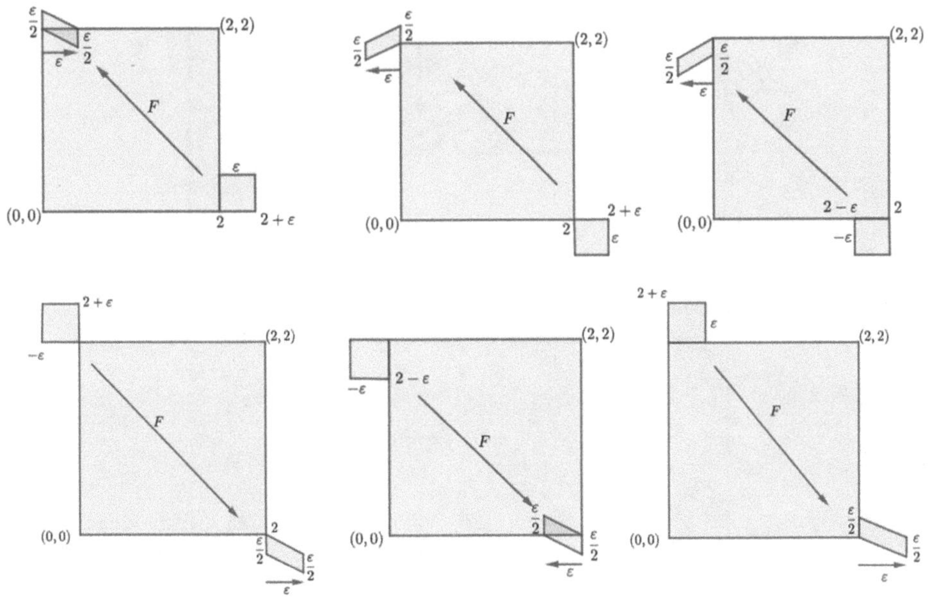

Figure 1. Evolution of neighbourhoods $Q_{j,\varepsilon}$ of (2, 0) (top) and neighbourhoods $W_{j,\varepsilon}(0,2)$ (bottom), $j \in \{1,2,3\}$, in Lemma 4.10.

Consider the squares $W_{1,\varepsilon} = [-\varepsilon, 0] \times [2, 2+\varepsilon]$, $W_{2,\varepsilon} = [-\varepsilon, 0] \times [2-\varepsilon, 2]$ *and* $W_{3,\varepsilon} = [0, \varepsilon] \times [2, 2+\varepsilon]$. *Then:*

(b.1) $F(W_{1,\varepsilon})$ *is the parallelogram with vertices* $(2,0)$, $(2, -\frac{\varepsilon}{2})$, $(2+\varepsilon, -\varepsilon)$, $(2+\varepsilon, -\frac{\varepsilon}{2})$.
(b.2) $F(W_{2,\varepsilon}) \subset C_{0,0} \cup T_{2,\varepsilon}$, *where* $T_{2,\varepsilon}$ *is the triangle with vertices* $(2-\varepsilon, 0)$, $(2,0)$, $(2, -\frac{\varepsilon}{2})$.
(b.3) $F(W_{3,\varepsilon})$ *is the parallelogram with vertices* $(2,0)$, $(2, \frac{\varepsilon}{2})$, $(2+\varepsilon, 0)$, $(2+\varepsilon, -\frac{\varepsilon}{2})$.

We are now in a position to prove that all the orbits of the region $R_1 = [-2, 4] \times [-2, 4]$ enter the square $C_{0,0}$.

Proposition 4.3: *For any point* (x, y) *in the region* R_1, *its orbit* $(F^n(x,y))_n$ *eventually enters in the square* $C_{0,0}$.

Proof: Taking into account that $R_1 = \bigcup_{-1 \le i,j \le 1} C_{i,j}$, we proceed by cases. We only present the proof for some of them, the rest is left to the care of the reader.

If $(x,y) \in C_{1,1}$, we consider that $F(C_{1,1}) \subset C_{1,0}$ is the parallelogram having vertices $(2,1)$, $(2,2)$, $(4,1)$, $(4,0)$, and the second iterate $F^2(C_{1,1})$ is the parallelogram with vertices $(1, \frac{5}{2})$, $(2,1)$, $(1, \frac{3}{2})$, $(0,3)$. Notice that $F^2(C_{1,1}) \subset C_{0,0} \cup W_{3,\varepsilon} \cup \tilde{T}$ with $\varepsilon = 1$ and $\tilde{T}$ the triangle of vertices $(1, \frac{5}{2})$, $(\frac{4}{3}, 2)$ and $(1,2)$; moreover, $F^2(\tilde{T}) \subset C_{0,0} \cup W_{3,\varepsilon}$, since the vertices of $\tilde{T}$ by F^2 go to $(\frac{1}{4}, \frac{17}{8})$, $(\frac{2}{3}, \frac{5}{3})$ and $(\frac{1}{2}, \frac{7}{4})$. From here, by Lemmas 4.8 – 4.10, we deduce that the images of $W_{3,\varepsilon}$ by F either enter to $C_{0,0}$ or either the intersection of such images narrow to segments of type $S_{\varepsilon,0}$ or $S_{\varepsilon,2}$. In any case, we derive the statement of the proposition.

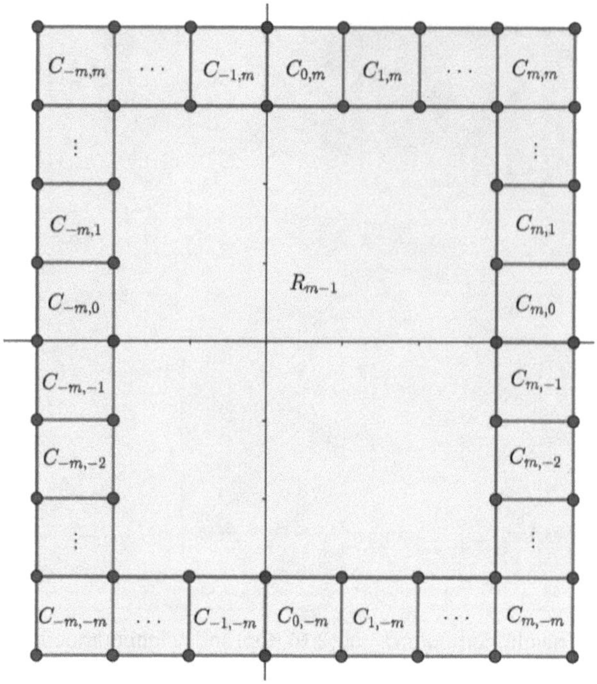

Figure 2. The squares $C_{m,n}$, $C_{n,m}$, $C_{-m,n}$ and $C_{n,-m}$ with $n \in [-m, m]$.

For the square $C_{1,0}$, we have that its image $F(C_{1,0})$ is the parallelogram with vertices $(0, 2)$, $(0, 3)$, $(2, 2)$, $(2, 1)$ and consequently $F^2(C_{1,0})$ is a new parallelogram having vertices $(2, 0)$, $(3, -\frac{1}{2})$, $(2, 1)$, $(1, \frac{3}{2})$; since $F^2(C_{1,0}) \subset C_{0,0} \cup Q_{1,\varepsilon} \cup Q_{2,\varepsilon}$, with $\varepsilon = 1$, a similar reasoning to that carried out in the former case gives us the desired conclusion on the enter of the orbit into $C_{0,0}$.

For the square $C_{-1,-1} = [-2, 0] \times [-2, 0]$, we find that $F(C_{-1,-1})$ is a parallelogram with vertices $(-2, -1)$, $(-2, 0)$, $(0, 1)$, $(0, 0)$; we decompose the figure into two triangles $T_1 \cup T_2$, with vertices $(-2, -1)$, $(-2, 0)$, $(0, 0)$ for T_1 and $(-2, 0)$, $(0, 0)$, $(0, 1)$ for T_2; then $F(T_2) \subset C_{0,0}$ while $F(T_1)$ is sent to the new triangle T_3 with vertices $(-1, -\frac{1}{2})$, $(0, 0)$, $(0, 1)$. Since T_3 intersects both half-planes H_ℓ and H_u, we need to decompose it as $T_4 \cup T_5$; in this case, consider that T_4 has vertices $(-1, -\frac{1}{2})$, $(-\frac{2}{3}, 0)$, $(0, 0)$, so $F(T_4)$ has vertices $(-\frac{1}{2}, \frac{1}{4})$, $(0, \frac{2}{3})$, $(0, 1)$ and it is easily seen that, in fact, $F^2(T_4) \subset C_{0,0}$; with respect to the triangle T_5, with vertices $(-\frac{2}{3}, 0)$, $(0, 0)$, $(0, 1)$, it is a simple matter to see that $F(T_5) \subset C_{0,0}$. ■

For (b) we need to control the evolution of the images corresponding to the squares $C_{i,j}$. We want to generalize the process by induction in the different levels $2, 3, 4, \ldots$ To this purpose, recall that the vertices of the squares are sent to the vertices of the new parallelograms.

Suppose that $m \geq 2$. If we denote $R_t = \cup_{-t \leq i,j \leq t} C_{i,j}$, we assume that the orbits of points of R_t by F eventually enter into the region R_{t-1}, $1 \leq t \leq m-1$, and consequently these orbits converge to 2-periodic points of the square $C_{0,0}$ (Figure 2).

In order to apply the process of mathematical induction, we start by studying the movement of squares $C_{m-j,m}$ and $C_{m,m-j}$, $j = 0, 1, \ldots, m$, as follows (for the proof, it suffices to evaluate the images of the vertices of the squares):

Lemma 4.11: *Let $m \geq 2$ be a positive integer. For any value $j \in \{0, 1, \ldots, m\}$:*

(a)

$$F(C_{m-j,m}) \subset \begin{cases} C_{m,-\frac{j}{2}} & \textit{if } j \textit{ is even,} \\ C_{m,\frac{-j-1}{2}} \cup C_{m,\frac{-j+1}{2}} & \textit{if } j \textit{ is odd.} \end{cases}$$

(b)

$$F(C_{m,j}) \subset \begin{cases} C_{j,\frac{m-j}{2}} & \textit{if } m-j \textit{ is even,} \\ C_{j,\frac{m-j-1}{2}} \cup C_{j,\frac{m-j+1}{2}} & \textit{if } m-j \textit{ is odd.} \end{cases}$$

Proof: (a) Notice that the vertices of $C_{m-j,m}$ are sent to:

$$F(2m-2j, 2m) = (2m, -j+1), \quad F(2m-2j, 2m+2) = (2m+2, -j),$$
$$F(2m-2j+2, 2m+2) = (2m+2, -j+1), \; F(2m-2j+2, 2m) = (2m, -j+2).$$

From here, it follows part (a). The proof of part (b) is similar and we omit it. ■

The following result, whose proof is direct, gives us the evolution of $C_{-j,m}$, $j = 0, 1, \ldots, m$.

Lemma 4.12: *Let $m \geq 2$ be a positive integer. For any value $j \in \{0, 1, \ldots, m\}$:*

$$F(C_{-j,m}) \subset \begin{cases} C_{m,\frac{-m-j}{2}} & \textit{if } m+j \textit{ is even,} \\ C_{m,\frac{-m-j-1}{2}} \cup C_{m,\frac{-m-j+1}{2}} & \textit{if } m+j \textit{ is odd.} \end{cases}$$

With respect to the iterates of $C_{m,-j}$, $0 \leq j \leq m$, we obtain the following lemma whose proof is direct.

Lemma 4.13: *Let $m \geq 2$ be a positive integer. For any value $j \in \{0, 1, \ldots, m\}$:*

$$F(C_{m,-j}) \subset \begin{cases} C_{-j,\frac{m-j+1}{2}} & \textit{if } m-j \textit{ is odd,} \\ C_{-j,\frac{m-j}{2}} \cup C_{-j,\frac{m-j+2}{2}} & \textit{if } m-j \textit{ is even.} \end{cases}$$

It remains to analyse the iterates of $C_{j,-m}$ and $C_{-m,j}$ for $-m \leq j \leq m$.

Lemma 4.14: *Let $m \geq 2$ be a positive integer. For any integer value $j \in [-m, m]$:*

$$F(C_{j,-m}) \subset \begin{cases} C_{-m,\frac{-m+j+1}{2}} & \textit{if } m-j \textit{ is odd,} \\ C_{-m,\frac{-m+j}{2}} \cup C_{-m,\frac{-m+j+2}{2}}, & \textit{if } m-j \textit{ is even.} \end{cases}$$

We stress that, for $j = -m$, we have $F(C_{-m,-m}) \subset C_{-m,-m} \cup C_{-m,-m+1}$. This means that this case has to be discussed carefully in order to obtain that its iterates enter to R_{m-1} at some moment. This will be clarified with the images $F(C_{m,-j})$ and with the tracing of the parallelograms obtained in successive steps.

Before, let us state the evolution of the images of $C_{-m,j}$ for integer values $j \in [-m, m]$. Again, we omit the proofs.

Lemma 4.15: *Let $m \geq 2$ be a positive integer. For any value $j \in \{0, 1, \ldots, m\}$:*

$$F(C_{-m,j}) \subset \begin{cases} C_{j,\frac{-m-j}{2}} & \text{if } m+j \text{ is even,} \\ C_{j,\frac{-m-j-1}{2}} \cup C_{j,\frac{-m-j+1}{2}}, & \text{if } m+j \text{ is odd.} \end{cases}$$

Lemma 4.16: *Let $m \geq 2$ be a positive integer. For any value $j \in \{0, 1, \ldots, m\}$:*

$$F(C_{-m,-j}) \subset \begin{cases} C_{-j,\frac{-m-j+1}{2}} & \text{if } m+j \text{ is odd,} \\ C_{-j,\frac{-m-j}{2}} \cup C_{-j,\frac{-m-j+2}{2}}, & \text{if } m+j \text{ is even.} \end{cases}$$

Finally, the promised result concerning the evolution of $C_{-m,-m}$:

Lemma 4.17: *Let $m \geq 2$ be an arbitrary integer. It holds $F(C_{-m,-m}) \subset C_{-m,-m} \cup C_{-m,-m+1}$. In particular:*

(a) *$F(C_{-m,-m})$ is the parallelogram with vertices $(-2m+2, -2m+3)$, $(-2m+2, -2m+2)$, $(-2m, -2m+1)$, $(-2m, -2m+2)$;*
(b) *$F^2(C_{-m,-m})$ is the parallelogram with vertices $(-2m+1, -2m+\frac{3}{2})$, $(-2m+2, -2m+2)$, $(-2m+3, -2m+\frac{7}{2})$, $(-2m+2, -2m+3)$;*
(c) *$F^2(C_{-m,-m}) \subset C_{-m,-m} \cup C_{-m,-m+1} \cup C_{-m+1,-m+1}$, and the part contained in $C_{-m,-m} \cup C_{-m,-m+1}$ is the triangle T_m with vertices $(-2m+1, -2m+\frac{3}{2})$, $(-2m+2, -2m+2)$, $(-2m+2, -2m+3)$;*
(d) *$F(T_m)$ is the triangle with vertices $(-2m+\frac{3}{2}, -2m+\frac{9}{4})$, $(-2m+2, -2m+3)$, $(-2m+3, -2m+\frac{7}{2})$. Therefore, $F(T_m) \subset C_{-m,-m+1} \cup R_{m-1}$.*

We collect the previous study in the following result.

Proposition 4.4: *Let*

$$(x, y) \in \bigcup_{-m \leq j \leq m} (C_{j,m} \cup C_{m,j} \cup C_{j,-m} \cup C_{-m,j}),$$

with $m \geq 1$. Then, there exists a positive integer N such that $F^N(x, y) \in R_{m-1}$.

Proof: The proof follows directly from Proposition 4.3 and Lemmas 4.11–4.17. ∎

As a consequence of all our study, we conclude:

Theorem 4.1: *Given the difference equation*

$$x_{n+1} = 1 - \frac{1}{2}|x_n| + \frac{1}{2}x_{n-1},$$

its dynamics is given by:

(a) *An equilibrium point,* $\overline{x} = 1$.
(b) *A continuum of 2-periodic sequences* $(\ldots, x, y, x, y, \ldots)$ *with* $0 \le x, y \le 2$, $x + y = 2$.
(c) *The rest of solutions converge to one of the 2-periodic solutions given in Part (b).*

Remark 4.1: It should be noticed that in [6], the authors detect the continuum of 2-periodic sequences for arbitrary a and b verifying the constraints $a > 0$ and either $a + b = 1$ or $b - a = 1$. In fact, they plot, fixing $b = 0.5$, a superposition of 20 attractors for the values $a = 0.1 \cdot n$, with $n = 1, \ldots, 20$. In particular, for $a = 0.5$, they show a segment line representing the attractor. In our work we have proved analytically this property of global attraction. It may be of some interest to study such problem for other values of a and b.

As a consequence of our study on the relationship between the Lozi map and suitable max-type equations (realize that $\delta = 1, \alpha + \beta + \gamma \neq 1$ and use Proposition 2.2), we deduce the following result.

Corollary 4.2: *Any element of the family of difference equations*

$$x_{n+1} = (x_n \cdot x_{n-1})^{\frac{1}{2}} \cdot \max\left\{\frac{1}{x_n}, A\right\},$$

defined for any arbitrary positive real initial conditions, with $0 < A < 1$, *presents the following dynamics: there exists a unique equilibrium point,* $\bar{x} = 1$; *there exists a continuum of 2-periodic solutions constructed from the initial conditions* $(\ldots, x, \frac{1}{x}, x, \frac{1}{x}, \ldots)$, *with* $x \in [A, \frac{1}{A}]$; *the rest of solutions converge to a 2-periodic solution.*

4.2. Case $a = -\frac{1}{2}$

We consider the parameter $a = -\frac{1}{2}$ for the Lozi map

$$x_{n+1} = 1 + \frac{1}{2}|x_n| - \frac{1}{2}x_{n-1}. \tag{27}$$

In this case, we are going to prove that the unique equilibrium point of Equation (27), namely $\bar{x} = 1$, is in fact a global attractor. The strategy is strongly similar to that developed in the case $a = \frac{1}{2}$, so we will only outline the proof.

The bidimensional map associated to the difference equation is given by $F_{-\frac{1}{2}}(x, y) = (y, 1 + \frac{1}{2}|y| - \frac{1}{2}x)$, and $DF_{-\frac{1}{2}}(1, 1) = \begin{pmatrix} 0 & 1 \\ -\frac{1}{2} & \frac{1}{2} \end{pmatrix}$; since the eigenvalues of this matrix are $\frac{1}{4} \pm \frac{i\sqrt{7}}{4}$, having a modulo less than 1, we deduce that, at least, the equilibrium point is locally asymptotically stable[9].

We maintain the notation employed in the analysis of the case $a = \frac{1}{2}$, and write $C_{m,n} = [2m, 2m + 2] \times [2n, 2n + 2]$. It is a simple matter to check (see Figure 3) that the square

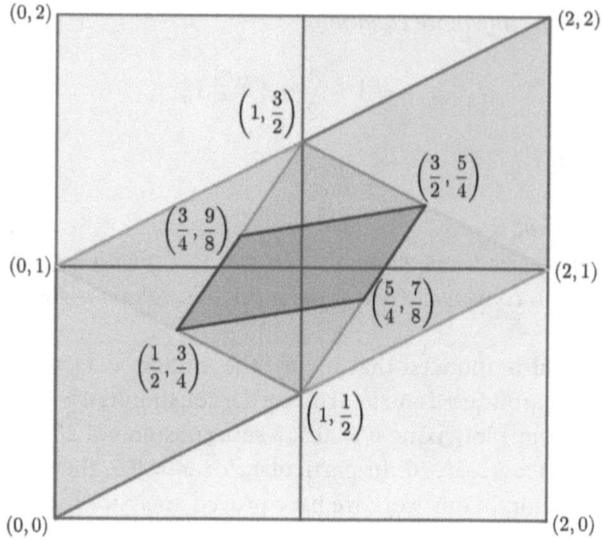

Figure 3. Dynamics of Equation (27) in the square [0, 2] × [0, 2].

$C_{0,0}$ is invariant by $F_{-\frac{1}{2}}$; this implies that we move in the upper half-plane H_u, and consequently the dynamics of the Lozi map $F_{-\frac{1}{2}}$ is governed by the linear difference equation $x_{n+1} = 1 + \frac{1}{2}x_n - \frac{1}{2}x_{n-1}$. Given two initial conditions $(x_{-1}, x_0) = (x, y) \in C_{0,0}$, its corresponding solution is

$$x_n = \left(\frac{1}{2}\right)^{\frac{n+1}{2}} \cdot \left((x-1)\cos((n+1)\theta) + \left[\frac{(y-1)\sqrt{2}}{\sin(\theta)} + (1-x)\cot(\theta)\right]\sin((n+1)\theta)\right) + 1,$$

for $n \geq -1$, as can be easily checked (here, $\theta = \arctan(\sqrt{7})$). Then,

$$\lim_{n\to+\infty} x_n = 1 = \bar{x}.$$

According to the last result, in order to prove that $\bar{x} = 1$ is a global attractor, it suffices to show that any square $C_{m,n}$ is eventually sent to $C_{0,0}$. For instance, $F_{-\frac{1}{2}}(C_{0,-1}) \subset C_{-1,0}$, and $F^3_{-\frac{1}{2}}(C_{-1,0}) \subset C_{0,0}$. The reasoning is completely analogous to that of case $a = \frac{1}{2}$, it is necessary to see that is always possible to descend from a level $R_m = \bigcup_{-m\leq i,j\leq m} C_{i,j}$ to the precedent level $R_{m-1} = \bigcup_{-m+1\leq i,j\leq m-1} C_{i,j}$ and check that, in fact, the iterates of points of $R_1 = \bigcup_{-1\leq i,j\leq 1} C_{i,j}$ eventually enter in $C_{0,0}$. We leave the details to the care of the reader.

5. Numerical simulations

As a consequence of the transformations developed in Section 2, the detection of a certain dynamical characteristic or property of a single max equation $x_{n+1} = \frac{\max\{x_n^k, A\}}{x_n^l x_{n-1}^m}$, for a concrete value $A > 1$ or $0 < A < 1$, will guarantee that all the elements of the family, with $A > 1$

or $0 < A < 1$, will share that property. In this sense, we have developed several numerical simulations in order to apply and illustrate the results proved in the previous sections. Moreover, we intend to progress in new open problems related to the topic.

In concrete, in this section, firstly we will deal with an equation studied by Abu-Saris and Allan in [2]. They show the existence of a *strange* attractor for a particular value of the parameter that is involved in the equation. Here, we will see that, in fact, the same behaviour appears for the whole uniparametric family of max-equations. It would be of interest to find other families of max-type equations exhibiting a strange attractor.

Next, as a first step of this search of complicated behaviour, we will focus on the evolution of the origin, $(x_{-1}, x_0) = (0, 0)$, under Lozi map in the particular case $a = b$, one of the cases that allow us to connect with max-equations. In this regard, we will present some numerical simulations related to the particular case $a = b$ of Lozi map. At first sight, it seems that no complicated behaviours are attained. It should be highlighted that, taking into account the study developed in the previous section, it will also be interesting to analysed the case $b = a + 2$.

5.1. A generic property

In [2], the authors show the figure of a *strange* attractor when they consider the difference equation

$$x_{n+1} = \frac{\max\{x_n^2, A\}}{x_n x_{n-1}}, \tag{28}$$

in the particular case $A = 2.3$. Notice that Equation (28) is topologically conjugate to

$$y_{n+1} = |y_n| - y_{n-1} - 1.$$

They also comment that *the solution in this case is also chaotic for certain values of A*. In fact, according to our study, and due to the fact that topological conjugation is a transitive relation, we deduce that all the elements of the family, $x_{n+1} = \frac{\max\{x_n^2, B\}}{x_n x_{n-1}}$, with $B > 1$, will present the *same* strange attractor, a homeomorphic copy of the attractor detected for $A = 2.3$. It is not difficult to reproduce the picture of Abu-Saris and Allan (see Figure 4), as well as new figures showing us the same dynamics, see Figure 5.

For $A < 1$, in order to do the transformation of a max-equation $x_{n+1} = \frac{\max\{x_n^2, A\}}{x_n x_{n-1}}$ into a generalized Lozi map, it must be $k - l - m - 1 = -\delta$, so in this case we conclude that it is topologically conjugate to $y_{n+1} = |y_n| - y_{n-1} + 1$. It seems that the same behaviour than the one exhibits when $A > 1$ takes place (see Figure 6).

In view of Figures 5 and 6, a natural question arises, to know if the systems are topologically conjugate.

5.2. Case $a = b$

Now, we will simulate the evolution of the origin, $(x_{-1}, x_0) = (0, 0)$, for different values of the parameters a and b in the particular case $a = b$, in order to deduce some consequences for the associated family of max-types equations. Moreover, we will focus on the unknown cases $a \notin [-\frac{1}{2}, \frac{1}{2}]$. In the sequel, we assume $(x_{-1}, x_0) = (0, 0)$.

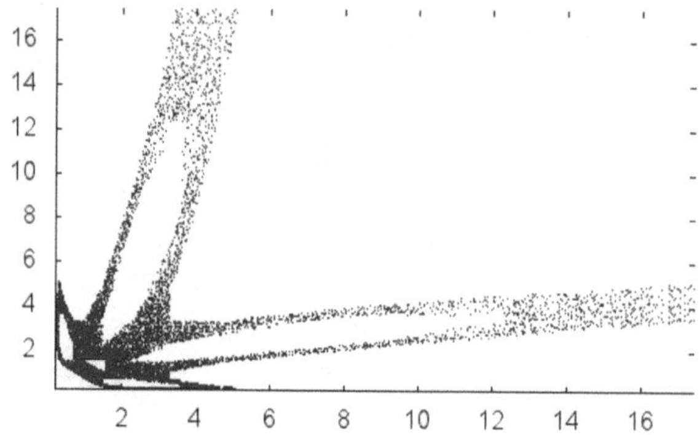

Figure 4. Reproduction of Abu-Saris Allan's Figure 4 in [2].

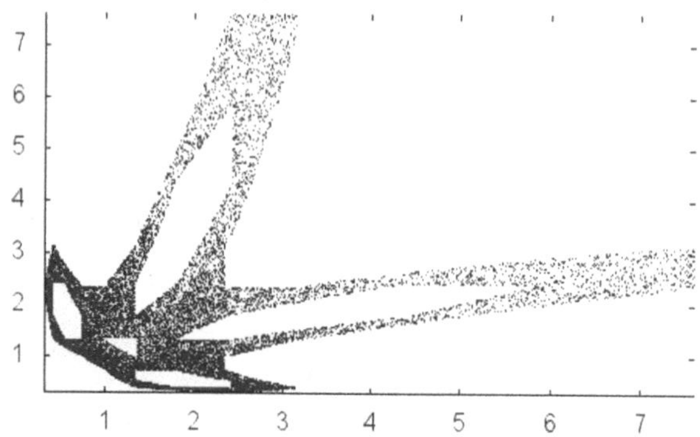

Figure 5. Simulation with $A = 1.8$ and initial conditions $x_{-1} = x_0 = 1.8^{0.8}$.

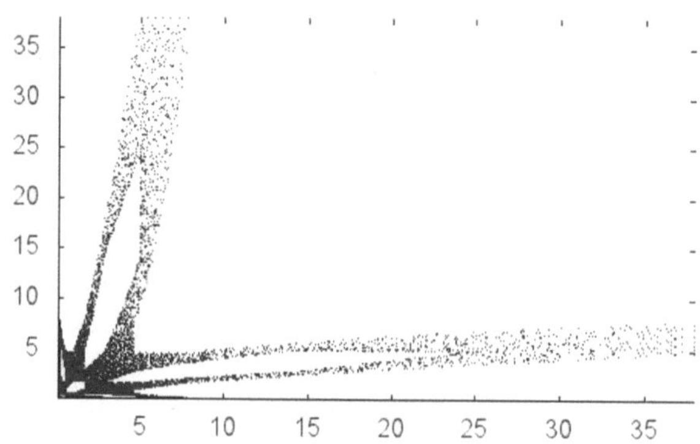

Figure 6. Simulation with $A = 0.35$ and two random initial conditions $x_{-1}, x_0 \in (0, 1)$.

Firstly, the case $a = b = -1$ yields to a 6-periodic orbit. When $a = b < -1$, it seems that the orbit of $(0, 0)$ tends to infinity in a spiral movement as Figure 7 shows. It would be interesting to prove if any initial conditions verify this class of dynamics.

Next, when $a \in (-1, -\frac{1}{2})$, it seem that the orbit of $(0, 0)$ is always trapped by an equilibrium point, which is different for distinct values of a. In this sense, it would be of interest

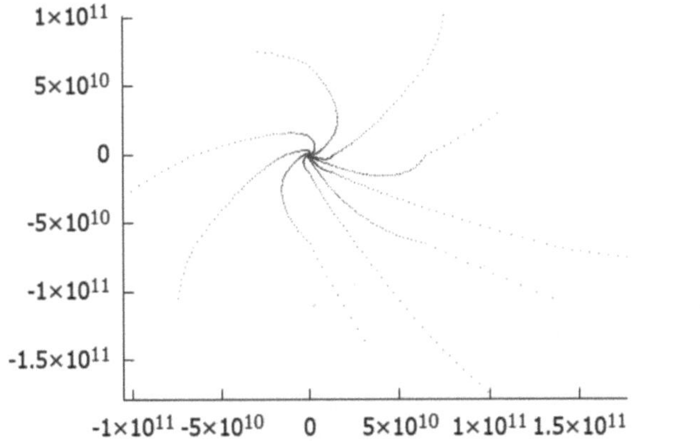

Figure 7. Case $a = b = -1.01$. A repulsive orbit.

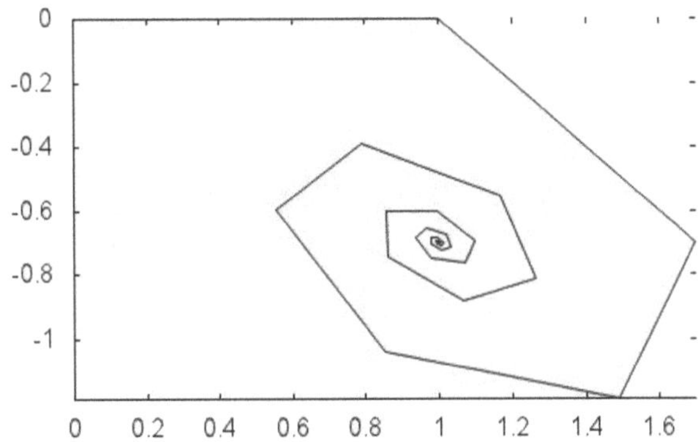

Figure 8. Case $a = b = -0.7$. The orbit of $(0, 0)$ goes to an equilibrium point.

to prove analytically whether or not the equilibrium point is a global attractor. Probably, a similar technique to that developed in Section 4 could be successful (Figure 8).

However, for values $a > \frac{1}{2}$, the situation changes drastically. Now, we have two unstable equilibrium points, namely $\bar{x} = 1$ and $\bar{x} = \frac{1}{1-2a}$, see Lemma 4.1. Furthermore, it seems that there exists a 2-periodic orbit which is a local attractor that attracts the origin. We illustrate in Figures 9 and 10 the cases $a = b = 0.99$ and $a = b = 0.9998$.

It is worth mentioning that, although it seems in Figure 10 that the orbit accumulates in 12 arcs, in fact it tends to a 2-periodic attractor. This is due to the fact that the eigenvalues associated to the linearization of F^2 at the periodic points are complex and their module is exactly $a = 0.998$, which is near 1. This yields to a spiral convergent movement that evolves very slowly to the 2-periodic attractor.

Finally, if $a = b = 1$, the orbit of $(0, 0)$ is 12-periodic. On the other hand, when $a = b > 1$, the orbit remains for some time near to the origin and finally goes to infinity by the third quadrant. Also, when we increase the value of a, the exit from the neighbourhood of $(0, 0)$ is more rapid (Figures 11 and 12).

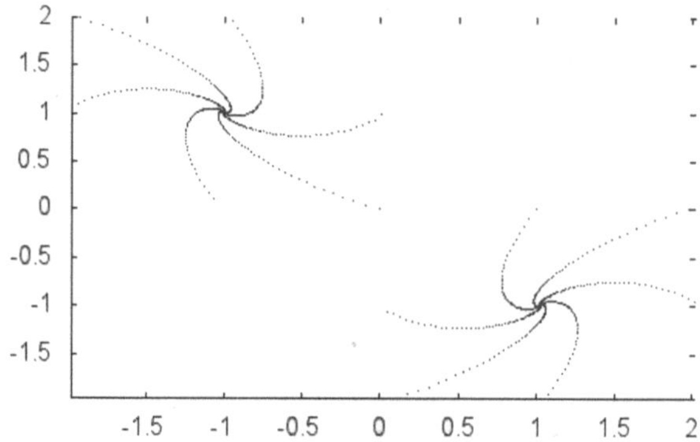

Figure 9. Case $a = b = 0.99$. The orbit of $(0, 0)$ goes to a 2-periodic orbit.

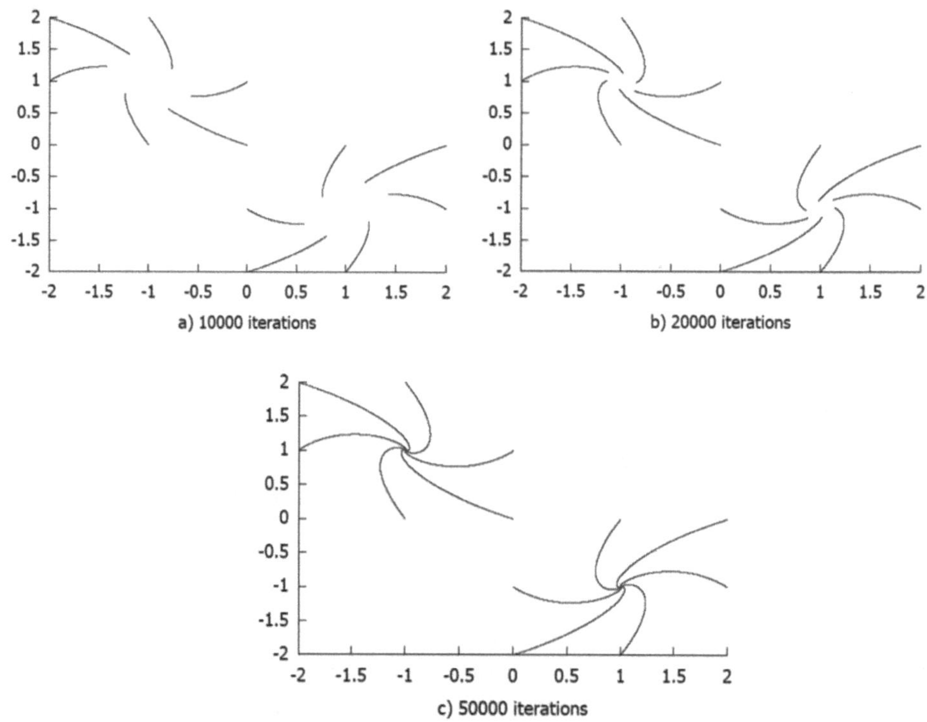

Figure 10. Case $a = b = 0.9998$.

6. Conclusions

The connection established between the Lozi map and difference equations of max-type allow us to extrapolate the dynamics of a single equation into a whole one-dimensional family constructed by the change of variables presented in Section 2. This broaden the scope of the treatment of dynamical aspects for max-type equations, usually restricted to properties on periodicity, boundedness,... In this sense, we can think in max-type equations as generators of complex dynamics (attractors, omega-limit sets,...), so their study will be interesting in the future. To this regard, it should be emphasized that the Lozi map

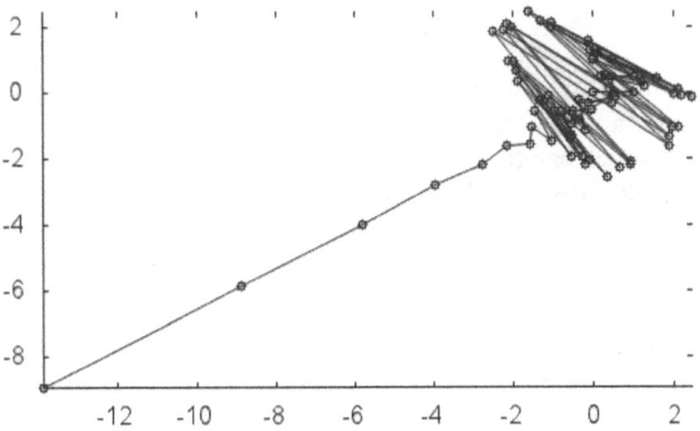

Figure 11. Case $a = b = 1.01$.

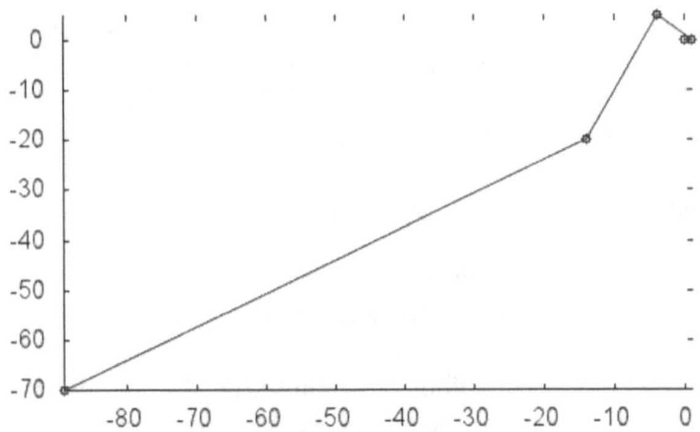

Figure 12. Case $a = b = 5$.

has applications in a variety of fields, such as control theory, game theory or synchronization theory, among others (see [10]), which can be translated automatically to max equations.

Moreover, we would like to stress that in order to consider the study of the dynamics of max-type equations we could take advantage of some techniques of differential equations or discrete dynamics, apart from the usual techniques employed in the literature that in most cases are strongly linked to arguments of real analysis. For more information related to max-type equations, see [14], a survey where the authors collect a large information on max-type difference equations and their known dynamics as well as the techniques used in that research.

Also, we would like to highlight some open problems related to the topic treated in this paper. On the one hand, to prove analytically the dynamics of the Lozi map in the case $a = b$ when $a \notin [\frac{1}{2}, -\frac{1}{2}]$, in order to see if the behaviour insinuated in the numerical simulations presented in Section 5 are true. On the other hand, to study in detail the other case deduced from the relation between the generalized Lozi map and max-type equations, namely, the case $b - a = 2$. Even, we could propose to deepen in the knowledge of the dynamics of generalized Lozi maps given by Equation 12 when $\delta = 0$ and $\alpha + \beta + \gamma = 1$.

Finally, it is worth mentioning that, nowadays, Lozi map is still a source of inspirations for research in different fields and, particularly, in the area of difference equations.

Notes

1. Based on [19], we will say that a system generates turbulence if its motion becomes very complicated, irregular and chaotic.
2. Roughly speaking, we will say that a fixed point is hyperbolic if the linearized equation at this point has no roots with absolute value equal to one. For more detail, consult [9].
3. A trapping region is a nonempty set that is mapped with its closure into its interior.
4. The basin of attraction corresponds to the set of points of the plane whose orbits tend to the strange attractor.
5. A fixed point is elliptic if his stability matrix has purely imaginary eigenvalues.
6. A difference equation is called globally periodic of period p when all the solutions are periodic of period (not necessarily minimal) p.
7. A discrete dynamical system is a pair (Y, φ), where Y is a topological space and $\varphi : Y \to Y$ is continuous. In the case that X is a topological space, we call associated dynamical system to the difference equation $x_{n+k} = f(x_{n+k-1}, \ldots, x_{n+1}, x_n)$ to (X^k, F), where the map $F : X^k \to X^k$ is given by $F(x_1, \ldots, x_k) = (x_2, \ldots, x_k, f(x_k, \ldots, x_2, x_1))$. In this sense, we say that the dynamical systems $\varphi : X \to X$ and $\psi : Y \to Y$, where X and Y are topological spaces, are topologically conjugate if there is an homeomorphism $\phi : X \to Y$, such that $\phi(\varphi(x)) = \psi(\phi(x))$ for all $x \in X$. We say that two difference equations are topologically conjugate when the associated dynamical systems so are. Notice that, in this case, the difference equations exhibit the same type of dynamics; for instance, they have the same number of equilibrium points or periodic orbits, or have chaotic attractors which are homeomorphic,...
8. Roughly speaking, we say that 2-periodic points are locally asymptotically stable if they are locally stable and for every set of initial conditions in a certain neighbourhood of the periodic points they converge to the periodic orbit. For a more accurate definition, consult for instance [9].
9. Roughly speaking, we say that an equilibrium point $\bar{x}$ is locally asymptotically stable if $\bar{x}$ is locally stable and if in addition is locally attractor. For a precise definition, consult [9].

Acknowledgments

We would like to thank the referees for their useful comments and suggestions.

Disclosure statement

No potential conflict of interest was reported by the author(s).

Funding

This work has been supported by the grant MTM2017-84079-P funded by (Spanish Ministry of Science and Innovation) MCIN/AEI/10.13039/501100011033 and by ERDF 'A way of making Europe', by the European Union.

References

[1] R. Abu-Saris, On the periodicity of the difference equation $x_{n+1} = \alpha|x_n| + \beta x_{n-1}$, J. Differ. Equ. Appl. 5 (1999), pp. 57–69.

[2] R. Abu-Saris and F. Allan, *Periodic and Nonperiodic Solutions of the Difference Equation* $x_{n+1} = \frac{\max\{x_n^2, A\}}{x}{}_n x_{n-1}$, Advances in Difference Equations (Veszprem, 1995), Gordon and Breach, Amsterdam, 1997, pp. 9–17.

[3] D. Baptista, R. Severino, and S. Vinagre, The basin of attraction of Lozi mappings, Int. J. Bifurcat. Chaos 19 (2009), pp. 1043–1049.

[4] A.F. Beardon, S.R. Bullet, and P.J. Rippon, Periodic orbits of difference equations, Proc. R. Soc. Edinburgh 125A (1995), pp. 657–674.
[5] M. Benedicks and L. Carleson, The dynamics of the Hénon map, Ann. Math. 133 (1991), pp. 73–169.
[6] V. Botella-Soler, J.M. Castelo, J.A. Oteo, and J. Ros, Bifurcations in the Lozi map, J. Phys. A: Math. Theor. 44 (2011), p. Art. ID 305101, 14 pages.
[7] M. Crampin, Piecewise linear recurrence relations, Math. Gazette 76 (1992), pp. 355–359.
[8] R.L. Devaney, A piecewise linear model for the zones of instability of an area-preserving map, Phys. 10D (1984), pp. 387–393.
[9] S. Elaydi, An Introduction to Difference Equations, Springer, New York, 2005.
[10] Z. Elhadj, *Lozi Mappings*, Theory and Applications, Taylor & Francis, CRC Press, Boca Raton, FL, 2014.
[11] J. Feuer, Periodic solutions of the Lyness max equation, J. Math. Anal. Appl 288 (2003), pp. 147–160.
[12] E.A. Grove and G. Ladas, Periodicities in Nonlinear Difference Equations, Chapman & Hall, CRC Press, Boca Raton, FL, 2005.
[13] M. Hénon, A two dimensional mapping with a strange attractor, Commun. Math. Phys. 50 (1976), pp. 69–77.
[14] A. Linero-Bas and D. Nieves-Roldán, *A survey on max-type equations*, in *Advances in Discrete Dynamical Systems*, S. Elaydi, S. Kalabušić, and M. Kulenović, eds., Difference Equations, and Applications, Proceedings of the International Conference on Difference Equations and Applications, ICDEA 2021, Springer, Sarajevo, 2021 (in press).
[15] E.N. Lorenz, Deterministic nonperiodic flow, J. Atmos. Sci. 20 (1963), pp. 130–141.
[16] R. Lozi, Un attracteur étrange du type attracteur de Hénon, J. Phys. 39 (1978), pp. 9–10.
[17] M. Misiurewicz, *Strange attractor for the Lozi mapping*, in *Non-linear Dynamics*, Annals of the New York Academy of Sciences, Vol. 357, 1980, pp. 348–358.
[18] E. Ott, Chaos in Dynamical Systems, Cambridge University Press, New York, 1993.
[19] D. Ruelle and F. Takens, On the nature of turbulence, Commun. Math. Phys. 20 (1971), pp. 167–192.

K-symbol fractional order discrete-time models of Lozi system

Rabha W. Ibrahim

ABSTRACT

In this investigation, we suggest different systems: the *k*-symbol fractional Lozi system (FLS), the *k*-symbol fractional Lozi system. We examine several of these systems' crucial dynamics. Moreover, we explore the sufficient and necessary conditions for the *k*-symbol fractional dynamic systems to be stable and asymptotically stable. Consequently, we provide the sufficient conditions on the *k*-symbol FLS to attain a steady and asymptotically steady zero resolutions. Additionally, we suggest in control laws designed to stabilize a combination of these systems. In this study, we assume 1D-controller and 2D-controller systems.

1. Introduction

It has been noted that the discrete fractional calculus (DFC) provides advantages over traditional calculus when simulating the dynamics of natural processes [14]. In 1974, Diaz and Olserin [7] made the initial proposal for the fundamental clarification of DFC, which is now regarded as one of the most important contemporary divisions of calculus [5]. In the past 10 years, a number of ideas and applications in this field have been developed with the goal of developing mathematical representations for a variety of natural events. Samples involve the use of DFC to describe ideal control systems [18] and to mimic the association of a bead on a wire [10]. In actuality, by discretizing their equivalent continuous operators, as in [7], a variety of differentiated discrete derivatives (DDDs), which are widely projected into the DFC field, could be engendered and demarcated. The Riemann–Liouville (RL) calculations [4], the Caputo derivative [1] and the Grunwald–Letnikov (GL) calculus [25] are just a few arbitrary-order derivatives that these operators are connected to. These three specific kinds of operators are known as fractional h-DDDs [21]. Applications of DFC can be found in the recent works [2,15,24].

The notion of differintegration heavily relies on the entire gamma function. The gamma function is a generalization of the factorial function. It provides a built-in capability to include memory effects in fractional derivatives. It is an important factor in the formula for the Mittag–Leffler function, sometimes known as the 'queen of special functions', which is used to solve practically all fractional differential and integral equations [31]. The idea of fractional calculus and its applications are thereby improved by any generalization, modification or extension of this function.

Among those models, the fractional chaotic discrete systems (FCDSs) are the greatest important [20]. The stage of chaotic performance and the category of attractors in fractional schemes have been established to have stretched dynamics in the realm of chaos. The fractional chaotic maps (FCMs) have been shown for the first time in [33]. In specific, Wu et al. [32] propose a fractional logistic map and statistically illustrate its chaotic performance. A fractional Lorenz map is proposed in [16], and on the basis of stability findings, its chaotic synchronization is examined. The new 2D chaotic logistic maps of different fractional order are presented by Farhan et al. [11] –[12] and the 3D-chaotic discrete system is studied in [30]. Limited by the standard discrete-time systems mentioned overhead have, as of yet been generalized to the arbitrary order difference situation [17]. We observe that various arbitrary order maps, though not in their distinct form, have been suggested in the nonfiction [19].

The concept of k-symbol calculus [8] is presented newly to generalize the well-known fractional calculus based on the k-gamma function

$$\Gamma_k(\eta) = \int_0^{\infty} \exp(-\vartheta^k/k)\vartheta^{\eta-1}\mathrm{d}\vartheta, \quad \Re(\eta) > 0, \quad k \in \mathbb{R}^+.$$

By using $\Gamma_k(\eta)$, the Riemann–Liouville fractional calculus is generalized in [13], Caputo fractional differential operator is modified in [3], the k-fractional Hilfer derivative is presented in [9], k-Appell 's integral function is considered in [29] and conformable fractional operator [28]. Special functions are also modified by using this types of calculus [25,26,34].

The k-symbol fractional Lozi system is suggested in this work. We investigate numerous strategic dynamics of these systems. In addition, we consider the satisfactory and compulsory requests for the solidity and asymptotic steadiness of the k-symbol fractional systems type Lozi frame. As a consequence, a generalization of the k-symbol FLS is formulated in a fractional difference delta operator based on Caputo difference operator. We also provide control principles to stabilize the combination of the suggested systems (k-symbol FLS). In a recent investigation, we deal with the one-dimensional and two-dimensional controller systems. Stability behaviour is theorized for general homogeneous and non-homogeneous systems. Based on this theory, we check the steady situation of k-symbol FLS and its control systems.

2. Methods

In this section, we deal with the following concepts.

2.1. K-symbol concept

This concept is formulated by Diaz and Osler [7,8], as follows:

Definition 2.1: The motivate gamma function, sometimes known as the k-symbol gamma function, is assumed by the formula that follows [8]:

$$\Gamma_k(\chi) = \lim_{n\to\infty} \frac{n!k^n(nk)^{\frac{\chi}{k}-1}}{(\chi)_{n,k}}, \tag{1}$$

where

$$(\chi)_{n,k} := \chi(\chi + k)(\chi + 2k) \cdots (\chi + (n-1)k)$$

and

$$(\chi)_{n,k} = \frac{\Gamma_k(\chi + nk)}{\Gamma_k(\chi)}.$$

Note that $\Gamma_k(\chi) \to \Gamma(\chi)$ when $k \to 1$, and

$$\Gamma_k(\chi + k) = \chi\Gamma_k(\chi), \quad \Gamma_k(k) = 1.$$

Definition 2.2: The k-symbol Riemann–Liouville fractional integral is given by the structure

$$J^{\sigma}_{a,k}\phi(\chi) = \frac{1}{k\Gamma_k(\sigma)} \int_a^{\tau} (\tau - \varsigma)^{\sigma/k-1}\phi(\varsigma)\mathrm{d}\varsigma.$$

Note that, when $\phi(\chi) = 1$, we have

$$J^{\sigma}_{a,k}\phi(\chi) = \frac{(\tau - a)^{\frac{\sigma}{k}}}{\Gamma_k(\sigma + k)}, \quad \sigma \in (0, 1].$$

As a special case, when $a = 0$, we have

$$J^{\sigma}_{k}\phi(\chi) = \frac{1}{k\Gamma_k(\sigma)} \int_0^{\tau} (\tau - \varsigma)^{\sigma/k-1}\phi(\varsigma)\mathrm{d}\varsigma.$$

The k-symbol Caputo derivative is formulated by

$${}^{C}_{k}D^{\sigma}\phi(\chi) = \begin{cases} k^{\nu}J_k^{\nu k-\sigma}\phi^{(\nu)}(\chi) = \frac{k^{\nu-1}}{\Gamma_k(\nu k-\sigma)}\int_0^{\tau}\frac{\varphi^{(\nu)}(\varsigma)}{(\tau-\varsigma)^{\sigma/k-\nu+1}}\mathrm{d}\varsigma, & (\nu-1)k < \sigma < \nu k \\ \frac{\mathrm{d}^{\nu}}{\mathrm{d}\tau^{\nu}}\phi(\tau), \quad \nu k = \sigma. \end{cases} \tag{2}$$

Note that when $k = 1$, we obtain the classical Caputo derivative.

Moreover, based on the k-symbol gamma function, Mittag–Leffler type function is defined as follows:

$$\Xi^{k}_{a,b}(\chi) = \sum_{n=0}^{\infty} \frac{\chi^n}{n!\Gamma_k(an + b)}.$$

Definition 2.3: The fractional difference operator in terms of Caputo calculus is defined as follows [1]:

$$\begin{aligned} {}^{C}\Delta_{\aleph}^{\sigma}\chi(\tau) &= \Delta_{\aleph}^{-(\nu-\sigma)}\Delta^{\nu}\chi(\tau) \\ &= \frac{1}{\Gamma(\nu-\sigma)}\sum_{\varsigma=\aleph}^{\tau-(\nu-\sigma)}(\tau-\varsigma-1)^{\nu-\sigma-1}\Delta^{\nu}\chi(\varsigma), \end{aligned} \tag{3}$$

where $\chi : \mathbb{N}_{\aleph} := \{\aleph, \aleph+1, \ldots\} \to \mathbb{R}, \nu = 1 + \lceil\sigma\rceil$, $\sigma \notin \mathbb{N}$ where $\tau \in \mathbb{N}_{\aleph+\nu-\sigma}$. Moreover, the factor $\tau^{(\sigma)}$ is formulated as follows:

$$\tau^{(\sigma)} = \frac{\Gamma(\tau+1)}{\Gamma(\tau+1-\sigma)}, \quad \sigma > 0.$$

It is possible to formulate the relevant discrete integral equation by

$$\chi(\tau) = \chi_0(\tau) + \frac{1}{\Gamma(\sigma)}\sum_{\varsigma=\aleph+\nu-\sigma}^{\tau-\sigma}(\tau-\varsigma-1)^{\sigma-1}\chi(\sigma+\varsigma-1). \tag{4}$$

Definition 2.4: The fractional difference operator in terms of k-symbol Caputo calculus is defined as follows:

$$\begin{aligned} {}^{C}_{k}\Delta_{\aleph}^{\sigma}\chi(\tau) &= \Delta_{\aleph}^{-(\nu-\frac{\sigma}{k})}\Delta^{\nu}\chi(\tau) \\ &= \frac{1}{\Gamma_k(\nu-\frac{\sigma}{k})}\sum_{\varsigma=\aleph}^{\tau-(\nu-\sigma)}(\tau-\varsigma-1)^{\nu-\frac{\sigma}{k}-1}\Delta^{\nu}\chi(\varsigma), \end{aligned} \tag{5}$$

for $\chi : \mathbb{N}_{\aleph} := \{\aleph, \aleph+1, \ldots\} \to \mathbb{R}$, $\tau \in \mathbb{N}_{\aleph+\nu-\frac{\sigma}{k}}$, $\frac{\sigma}{k} \notin \mathbb{N}$ and $\nu = 1 + \lceil\frac{\sigma}{k}\rceil$. Moreover, the term $\tau^{(\sigma)}$ is given by

$$\tau^{(\frac{\sigma}{k})} = \frac{\Gamma_k(\tau+1)}{\Gamma_k(\tau+1-\frac{\sigma}{k})}, \quad \sigma > 0.$$

This transformation results from the identification (the equivalent discrete integral equation)

$$\chi(\tau) = \chi_0(\tau) + \frac{1}{k\Gamma_k(\sigma)}\sum_{\varsigma=\aleph+\nu-\sigma}^{\tau-\sigma}(\tau-\varsigma-1)^{\frac{\sigma}{k}-1}\chi(\sigma+\varsigma-1). \tag{6}$$

2.2. K-symbol Lozi system

The Lozi chaotic map was developed by Rene Lozi and published in [22]

$$\begin{cases} \chi(\nu+1) = -\upsilon|\chi(\nu)| + \Upsilon(\nu) + 1 \\ \Upsilon(\nu+1) = \mu\chi(\nu), \end{cases} \tag{7}$$

where $\nu \in \mathbb{N}$, χ and Υ are the settings of the function and the constants υ and $\mu \in \mathbb{R}$. It is revealed that (7) has a chaotic attractive, for example, assumed the parameter values

Table 1. The behaviour of recurrence equation solution of System (7).

ν	Solution $\chi(n)$	Graph
1.7	$(-6/5)^n$	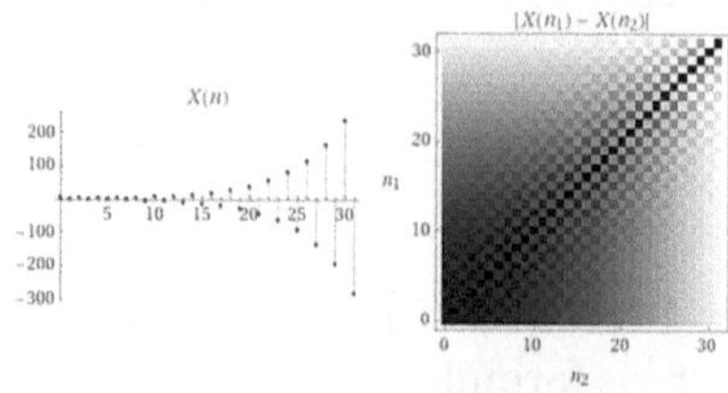
1.5	$(-1)^n$	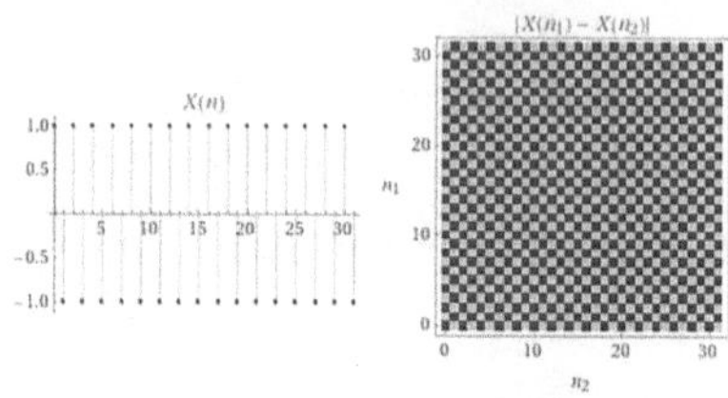
1.3	$(-\frac{4}{5})^n$	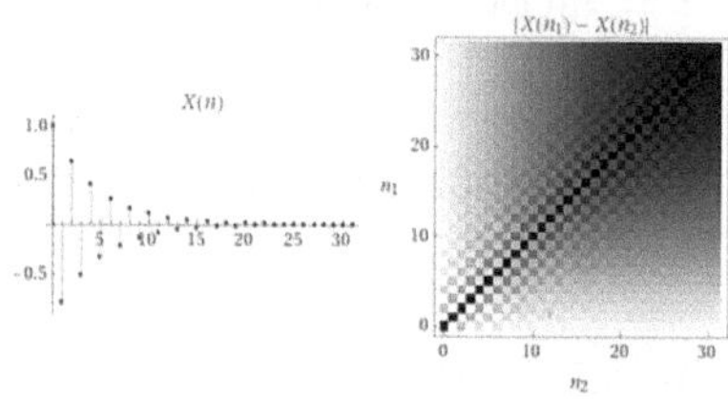
1.0	$(-\frac{1}{2})^n$	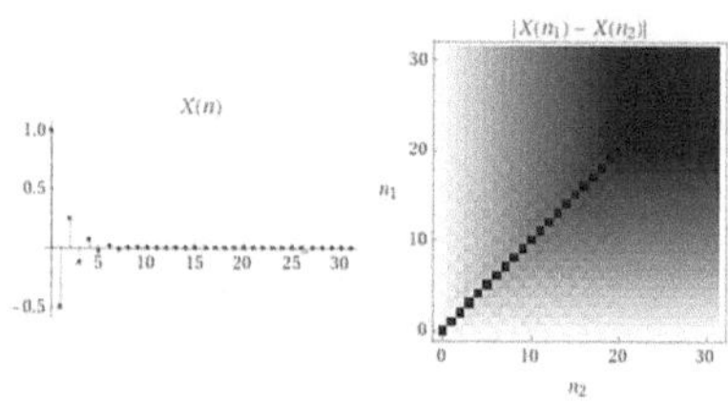

$(\nu, \mu) = (1.7, 0.5)$. The recurrent solution is given by (see Table 1)

$$\chi(\nu) = c_1 \left((\frac{1}{20}(-17 - \sqrt{489}))^\nu - (\frac{1}{20}(\sqrt{489} - 17))^\nu) + \frac{1}{11}(6(\frac{\sqrt{489}}{20} - \frac{17}{20})^\nu + 5 \right)$$

where c_1 is an arbitrary parameter

The first-order difference of (7) can be straightforwardly expressed as

$$\begin{cases} \Delta\chi(\nu) = -\nu|\chi(\nu)| + \Upsilon(\nu) + 1 - \chi(\nu) \\ \Delta\Upsilon(\nu) = \mu\chi(\nu) - \Upsilon(\nu). \end{cases} \tag{8}$$

The solution for System (8) is given by

$$\begin{cases} \chi(\nu) = 0.721803c_1 2^{-2\nu}5^{-\nu}((-42.7797)^{\nu} + 0.385419(-11.2203)^{\nu}) \\ \quad -0.633724c_2 2^{-2\nu}((-8.55595)^{\nu} - (-2.24405)^{\nu}) + 0.408163 \\ \Upsilon(\nu) = -0.316862c_1 2^{-2\nu}((-8.55595)^{\nu} - (-2.24405)^{\nu}) \\ \quad +0.278197c_2 2^{-2\nu}((-8.55595)^{\nu} + 2.59458(-2.24405)^{\nu}) + 0.102041 \end{cases}$$

and $(c_1, c_2) \in \mathbb{Z}^2$.

The FLS can be realized by utilizing the Caputo difference specified in Definition 2.3 with an Δ as the starting point [19]

$$\begin{cases} {}^{C}\Delta_{\aleph}^{\sigma}\chi(\tau) = -\nu|\chi(\tau-1+\sigma)| + \Upsilon(\tau-1+\sigma) + 1 - \chi(\tau-1+\sigma) \\ {}^{C}\Delta_{\aleph}^{\sigma}\Upsilon(\tau) = \mu\chi(\tau-1+\sigma) - \Upsilon(\tau-1+\sigma), \end{cases} \tag{9}$$

where $\sigma \in (0, 1]$ and $\tau \in \mathbb{N}_{\aleph+\nu-\sigma}$. By using the integral difference operator, System (8) becomes

$$\begin{cases} \chi(\tau) = \chi(\aleph) + \frac{1}{\Gamma(\sigma)} \sum_{\varsigma=\aleph+1}^{\tau-\sigma} (\tau-\varsigma-1)^{\sigma-1} \\ \quad (-\nu|\chi(\tau-1+\sigma)| + \Upsilon(\tau-1+\sigma) + 1 - \chi(\tau-1+\sigma)) \\ \Upsilon(\tau) = \Upsilon(\aleph) + \frac{1}{\Gamma(\sigma)} \sum_{\varsigma=\aleph+1}^{\tau-\sigma} (\tau-\varsigma-1)^{\sigma-1} \\ \quad (\mu\chi(\tau-1+\sigma) - \Upsilon(\tau-1+\sigma)), \end{cases} \tag{10}$$

where

$$\frac{(\tau-\varsigma-1)^{\sigma-1}}{\Gamma(\sigma)} = \frac{\Gamma(\tau-\varsigma)}{\Gamma(\sigma)\Gamma(\tau-\varsigma-\sigma+1)}$$

indicates the discrete kernel function.

The k-symbol FLS of fractional power can be realized by utilizing the k-symbol Caputo difference as follows:

$$\begin{cases} {}_{k}^{C}\Delta_{\aleph}^{\sigma}\chi(\tau) = -\nu|\chi(\tau-1+\sigma)| + \Upsilon(\tau-1+\sigma) + 1 - \chi(\tau-1+\sigma) \\ {}_{k}^{C}\Delta_{\aleph}^{\sigma}\Upsilon(\tau) = \mu\chi(\tau-1+\sigma) - \Upsilon(\tau-1+\sigma), \end{cases} \tag{11}$$

where $\sigma \in (0, 1]$ and $\tau \in \mathbb{N}_{\aleph+\nu-\sigma}$.

By using the definition of k-symbol gamma function and k-symbol fractional integration [27], we illustrate the following generalized k-symbol system:

$$\begin{cases} \chi(\tau) = \chi(\aleph) + \frac{1}{k\Gamma_k(\sigma)} \sum_{\varsigma=\aleph+1}^{\tau-\sigma} (\tau-\varsigma-1)^{\frac{\sigma}{k}-1} \\ \quad (-\nu|\chi(\tau-1+\sigma)| + \Upsilon(\tau-1+\sigma) + 1 - \chi(\tau-1+\sigma)) \\ \Upsilon(\tau) = \Upsilon(\aleph) + \frac{1}{k\Gamma_k(\sigma)} \sum_{\varsigma=\aleph+1}^{\tau-\sigma} (\tau-\varsigma-1)^{\frac{\sigma}{k}-1} \\ \quad (\mu\chi(\tau-1+\sigma) - \Upsilon(\tau-1+\sigma)), \end{cases} \tag{12}$$

where

$$\frac{(\tau-\varsigma-1)^{\frac{\sigma}{k}-1}}{k\Gamma_k(\sigma)} = \frac{\Gamma_k(\tau-\varsigma)}{k\Gamma_k(\sigma)\Gamma_k(\tau-\varsigma-\sigma+1)}$$

indicates the k-symbol discrete kernel function.

For the numerical case, when $\aleph = 0$, we have

$$\begin{cases} \chi(v) = \chi(0) + \frac{1}{k\Gamma_k(\sigma)} \sum_{j=1}^{v} \left(\frac{\Gamma_k(n-j+\frac{\sigma}{k})}{\Gamma_k(n-j+1)} \right) \\ \quad \left(-v|\chi(j-1)| + \Upsilon(j-1) + 1 - \chi(j-1) \right) \\ \Upsilon(v) = \Upsilon(0) + \frac{1}{k\Gamma_k(\sigma)} \sum_{j=1}^{v} \left(\frac{\Gamma_k(n-j+\frac{\sigma}{k})}{\Gamma_k(n-j+1)} \right) \left(\mu\chi(j-1) - \Upsilon(j-1) \right). \end{cases} \tag{13}$$

$$\left(v \in \mathbb{N}, \ \sigma \in (0,1], \ j \leq v \right)$$

Fractional calculus, as we just established, has the significant addition of an endless memory. In System (13), where the outcome $\chi(v)$, $v \in \mathbb{N}$ relies on all earlier estimates, this is plainly evident. Additionally, these answers will be iterated through the k-symbol procedure. The system's capacity is managed by this component. Figure 1 shows the behaviour of the k-symbol FLS, for various values of $\sigma \in (0,1]$ and $k \in \mathbb{N}$.

3. Stability

In this section, we investigate the global stability of the k-symbol FLS using the operator ${}_{k}^{C}D^{\sigma}\phi(\chi)$.

3.1. *Linear system*

We have the following description:

Definition 3.1: Assume that $\chi(\tau) = \chi(\tau; \tau_0, \chi_0)$ is a solution of the k-symbol FLS

$${}_{k}^{C}\Delta^{\sigma}\chi(\tau) = f(\tau, \chi), \quad \tau \in [\tau_0, \infty) \tag{14}$$

such that

- $\chi(\tau)$ is valid on the interval $[\tau_0, \infty)$;
- the point $(\tau, \chi(\tau)) \in S := \{(\tau, \chi(\tau)) : \tau \in (s_1, \infty), \|\chi\| < \tau_0, \tau > \tau_0\}$.

Then χ is called stable solution if there exists a positive number $\omega > 0$ for all solution $\chi(\tau) = \chi(\tau; \tau_0, \chi_0) \in S$ achieving

$$\|\chi_1 - \chi_0\| < \omega$$

and for available positive number $\varepsilon > 0$ there exists a positive number $0 < \varpi \leq \omega$ satisfying the next inequality

$$\begin{aligned} &\|\chi_1 - \chi_0\| < \varpi \Rightarrow \\ &\|\chi(\tau; \tau_0, \chi_0) - \chi(\tau; \tau_0, \chi_1)\| < \varepsilon, \quad \tau \in [\tau_0, \infty). \end{aligned}$$

Moreover, if

$$\lim_{\tau \to \infty} \|\chi(\tau; \tau_0, \chi_0) - \chi(\tau; \tau_0, \chi_1)\| = 0,$$

then the solution has reached its asymptotic stability.

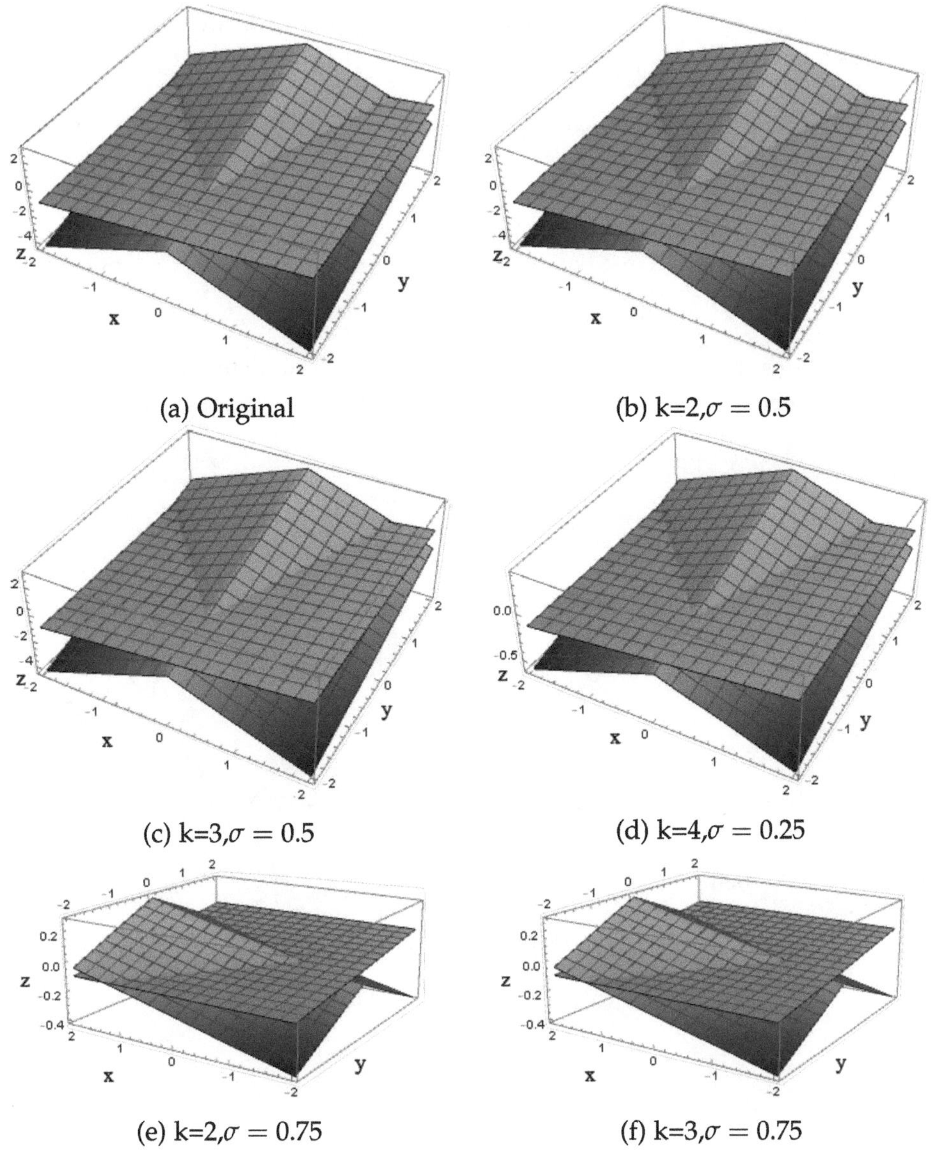

(a) Original (b) k=2,$\sigma = 0.5$

(c) k=3,$\sigma = 0.5$ (d) k=4,$\sigma = 0.25$

(e) k=2,$\sigma = 0.75$ (f) k=3,$\sigma = 0.75$

Figure 1. The 2D plots of k-symbol FLS with $(\nu, \mu) = (1.7, 0.5)$ for different values of $\sigma \in (0, 1]$, $k = 1, 2, \ldots$ and $\chi \in (-2, 2), \Upsilon \in (-2, 2)$: (a) original; (b) $k = 2$, $\sigma = 0.5$; (c) $k = 3, \sigma = 0.5$; (d) $k = 4, \sigma = 0.25$; (e) $k = 2, \sigma = 0.75$; (f) $k = 3, \sigma = 0.75$.

We have the following stability theories:

Theorem 3.2: *All solutions of the continuous system corresponding to (11)*

$$ {}_{k}^{C}\Delta^{\sigma} \begin{pmatrix} \chi(\tau) \\ \Upsilon(\tau) \end{pmatrix} = \Omega_{2\times 2}(\tau) \begin{pmatrix} \chi(\tau) \\ \Upsilon(\tau) \end{pmatrix} \tag{15} $$

are stable if and only if they are bounded. Furthermore, if the characteristic polynomial of Ω admits the stability behaviour, then the solutions become asymptotically stable.

Proof: Let's begin by developing a matrix-valued function with two variables that we'll name Σ as follows:

$$\Sigma(\tau, \tau_0) = \mathbb{I}_d + J_k^\sigma \Omega(\tau) + J_k^\sigma \Omega(\tau) J_k^\sigma \Omega(\delta_1) + \cdots + J_k^\sigma \Omega(\tau) J_k^\sigma \Omega(\delta_1) \cdots J_k^\sigma \Omega(\delta_{n-1}) + \cdots,$$

where $\mathbb{I}_d$ is the identity matrix. By the formula of J_k^σ, we obtain that

$$\Sigma(\tau_0, \tau_0) = \mathbb{I}_d.$$

Since all solutions of System (15) are bounded, then there occurs a fixed number $c > 0$ such that the fundamental matrix of (15) achieves $\|\Sigma\| < c$, where $\|.\|$ represents the max norm. This yields

$$\|\chi(\tau) - \chi_0(\tau)\| < \frac{\varepsilon}{2c}, \quad \|\Upsilon(\tau) - \Upsilon_0(\tau)\| < \frac{\varepsilon}{2c}, \quad \varepsilon > 0.$$

Consequently, we obtain

$$\|\chi(\tau; \tau_0, \chi_0) - \chi(\tau; \tau_0, \chi_1)\| = \|\Sigma(\tau, \tau_0)(\chi_0 - \chi_1)\| < \frac{c\varepsilon}{2c} = \frac{\varepsilon}{2}.$$

Similarly, we have

$$\|\Upsilon(\tau; \tau_0, \Upsilon_0) - \Upsilon(\tau; \tau_0, \chi_1)\| = \|\Sigma(\tau, \tau_0)(\Upsilon_0 - \Upsilon_1)\| < \frac{c\varepsilon}{2c} = \frac{\varepsilon}{2}.$$

Let $R = (\chi, \Upsilon)^\top$, then

$$\begin{aligned} \|R(\tau) - R_0(\tau)\| &\le \|\Sigma(\tau, \tau_0)\,(R(\tau) - R_0(\tau))\,\| \\ &\le c\|R(\tau) - R_0(\tau)\| \\ &< c\left(\frac{\varepsilon}{2c} + \frac{\varepsilon}{2c}\right) \\ &= \varepsilon. \end{aligned}$$

Hence, all solutions of System (3.1) are stable.

Conversely, by the definition of the stability, for $\varepsilon > 0$ there occurs a positive fixed number ℓ such that

$$\|R(\tau)\| < \ell \Rightarrow \|\Sigma(\tau)R(\tau)\| < \varepsilon.$$

In particular,

$$\|\chi(\tau)\| = \|\chi(\tau; \tau_0, \chi_0)\| < \varepsilon/2$$

and

$$\|\Upsilon(\tau)\| = \|\Upsilon(\tau; \tau_0, \Upsilon_0)\| < \varepsilon/2.$$

That is all resolutions are bounded.

Since the characteristic polynomial Ω admits a stability behaviour, this yields that the solutions

$$\begin{aligned}\|\chi(\tau;\tau_0,\chi_0)-\chi(\tau;\tau_0,\chi_1)\| &\le c\exp\left(\frac{\Omega(\tau-\tau_0)^{\sigma/k}}{(\sigma/k)}\right)\|\chi_1-\chi_0\|\\ &\le c\exp\left(-\varepsilon_1\frac{\tau^{\sigma/k}}{\sigma/k}\right),\quad 0<\varepsilon_1<\varepsilon\\ &=0,\quad \tau\to\infty.\end{aligned}$$

Similarly, for Υ, we have

$$\begin{aligned}\|\Upsilon(\tau;\tau_0,\Upsilon_0)-\Upsilon(\tau;\tau_0,\Upsilon_1)\| &\le c\exp\left(\frac{\Omega(\tau-\tau_0)^{\sigma/k}}{(\sigma/k)}\right)\|\Upsilon_1-\Upsilon_0\|\\ &\le c\exp\left(-\varepsilon_1\frac{\tau^{\sigma/k}}{\sigma/k}\right),\quad 0<\varepsilon_1<\varepsilon\\ &=0,\quad \tau\to\infty,\end{aligned}$$

which leads to asymptotically stable solutions. ∎

Corollary 3.3: *Suppose that $\|\Omega\| = \sup|\Omega| < 1$ such that its eigenvalues belong to the identity interval [0,1]. Then System (15) has asymptotically stable solutions if and only if they have the boundedness behaviour.*

Proof: Let the statement's presumptions hold. Then Ω admits invertible positive contraction property [23]. That is the characteristic polynomial $\Omega_{2\times 2}^{-1} - \mathbb{I}_d$ satisfies a positive semi-definite property, such that $Det(\Omega_{2\times 2}) > 0$ ([6]-Proposition 3.5). Consequently implies that the characteristic polynomial Ω is real stable. In view of Theorem 3.2, we have the asymptotically stable solutions. ∎

For non-homogeneous case, we have the following result:

Theorem 3.4: *The outcomes that fit (15)*

$$ {}^{C}_{k}D^{\sigma}\begin{pmatrix}\chi(\tau)\\ \Upsilon(\tau)\end{pmatrix} = \Omega_{2\times 2}(\tau)\begin{pmatrix}\chi(\tau)\\ \Upsilon(\tau)\end{pmatrix} + \begin{pmatrix}\psi_1(\tau)\\ \psi_2(\tau)\end{pmatrix} \tag{16}$$

are stable if and only if they are bounded and

$$\|\Psi\| < \kappa,\quad \kappa\in(0,\infty),\quad \Psi = (\psi_1,\psi_2)^{\top}.$$

Additionally, the solutions have the asymptotically stable behaviour if the characteristic polynomial is stable, with

$$c < \frac{e}{\kappa},\quad \kappa > 0,\quad \|\Sigma\| \le c.$$

Proof: Let $\|\Psi(\tau)\| < \kappa, \kappa > 0$. Then by the condition of the theorem, we have

$$\begin{aligned} \|\chi(\tau)\| &\leq c\exp\left(c\kappa\frac{\tau^{\sigma/k}}{\sigma/k}\right)\|\chi_0\| \\ &\leq c\exp\left((c\kappa - e)\frac{\tau^{\sigma/k}}{\sigma/k}\right)\|\chi_0\| \\ &= 0, \quad c\kappa - e < 0, \tau \to \infty. \end{aligned}$$

This proves the result. ■

A direct proof of Corollary 3.3 and application of Theorem 3.4, we have the following result.

Corollary 3.5: *Suppose that the sup norm $\|\Omega\| < 1$ and all its eigenvalues are in the interval [0, 1]. Also, consider $\|\Psi\| < \kappa, \kappa > 0$. Then System (16) has asymptotically stable solutions whenever*

$$c < \frac{e}{\kappa}, \quad \kappa > 0, \quad \|\Sigma\| \leq c.$$

3.2. Observation

Consider the k-symbol Lozi fractional system

$$\begin{cases} {}^C_kD^\sigma \chi(\tau) = -\nu|\chi(\tau)| + \Upsilon(\tau) + 1 - \chi(\tau) \\ {}^C_kD^\sigma \Upsilon(\tau) = \mu\chi(\tau) - \Upsilon(\tau), \quad \sigma \in (0, 1]. \end{cases} \tag{17}$$

System (17) can be divided into two systems

$$\begin{cases} {}^C_kD^\sigma \chi(\tau) = -\nu\chi(\tau) + \Upsilon(\tau) + 1 - \chi(\tau) \\ {}^C_kD^\sigma \Upsilon(\tau) = \mu\chi(\tau) - \Upsilon(\tau), \quad \sigma \in (0, 1] \end{cases} \tag{18}$$

and

$$\begin{cases} {}^C_kD^\sigma \chi(\tau) = \nu\chi(\tau) + \Upsilon(\tau) + 1 - \chi(\tau) \\ {}^C_kD^\sigma \Upsilon(\tau) = \mu\chi(\tau) - \Upsilon(\tau), \quad \sigma \in (0, 1]. \end{cases} \tag{19}$$

3.3. System (18)

The st of the eigenvalues is (see Figure 3)

$$\lambda_1 = \frac{1}{2}(-\sqrt{4\mu + \nu^2} - \nu - 1)$$

and

$$\lambda_2 = \frac{1}{2}(\sqrt{4\mu + \nu^2} - \nu - 1),$$

where the corresponding eigenvectors are

$$V_1 = \left(-\frac{\nu + \sqrt{4\mu + \nu^2}}{2\mu}, 1 \right)$$

and

$$V_2 = \left(-\frac{\nu - \sqrt{4\mu + \nu^2}}{2\mu}, 1 \right).$$

Based on the eigenvalues of the system and under the condition $\sqrt{4\mu + \nu^2} \geq \nu + 1$, we have two domains for the solutions,

$$[\Delta_1]_{\lambda_1} = \{(\chi, \Upsilon) \in \mathbb{R}^2 : \lambda < 0\}$$
$$[\Delta_2]_{\lambda_2} = \{(\chi, \Upsilon) \in \mathbb{R}^2 : \lambda \geq 0\}.$$

The equilibrium point of System (18) is

$$(\chi_0, \Upsilon_0) = \left(\frac{1}{\nu - \mu + 1}, \frac{-\mu}{-\nu + \mu - 1} \right).$$

Additionally, the fixed point is

$$(\chi_{fix}, \Upsilon_{fix}) = \left(\frac{2}{2\nu - \mu + 4}, \frac{-\mu}{-2\nu + \mu - 4} \right).$$

If the inequality

$$\frac{4}{3}\mu \leq \nu + 3$$

holds, then the eigenvalues λ_1 and λ_2 are in the interval [0,1] and hence, by Corollary 3.5, the system is asymptotically stable. Denote by $Q = \sqrt{4\mu + \nu^2}$ then the eigenvalues are in [0,1] when Q satisfies the following inequalities (see Figure 2)

$$-Q - 3 < \nu < -Q - 1, \quad Q \in \mathbb{R}^+$$

and

$$Q - 3 < \nu < Q - 1, \quad Q \in \mathbb{R}^+.$$

Moreover, a computation implies that $\|\Psi\| = 1$ then by taking sufficient value of $\kappa > 1$, Corollary 3.5 yields that all solutions are asymptotically stable. Hence, solutions that begin sufficiently close to the equilibrium not only stay sufficiently close but also gradually converge to it.

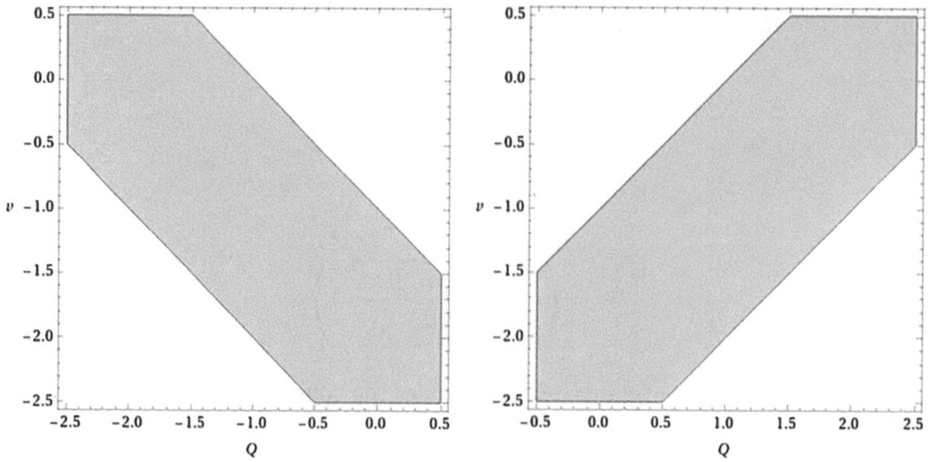

Figure 2. Inequality plot for $-Q-3<\nu<-Q-1$ and $Q-3<\nu<Q-1$ respectively.

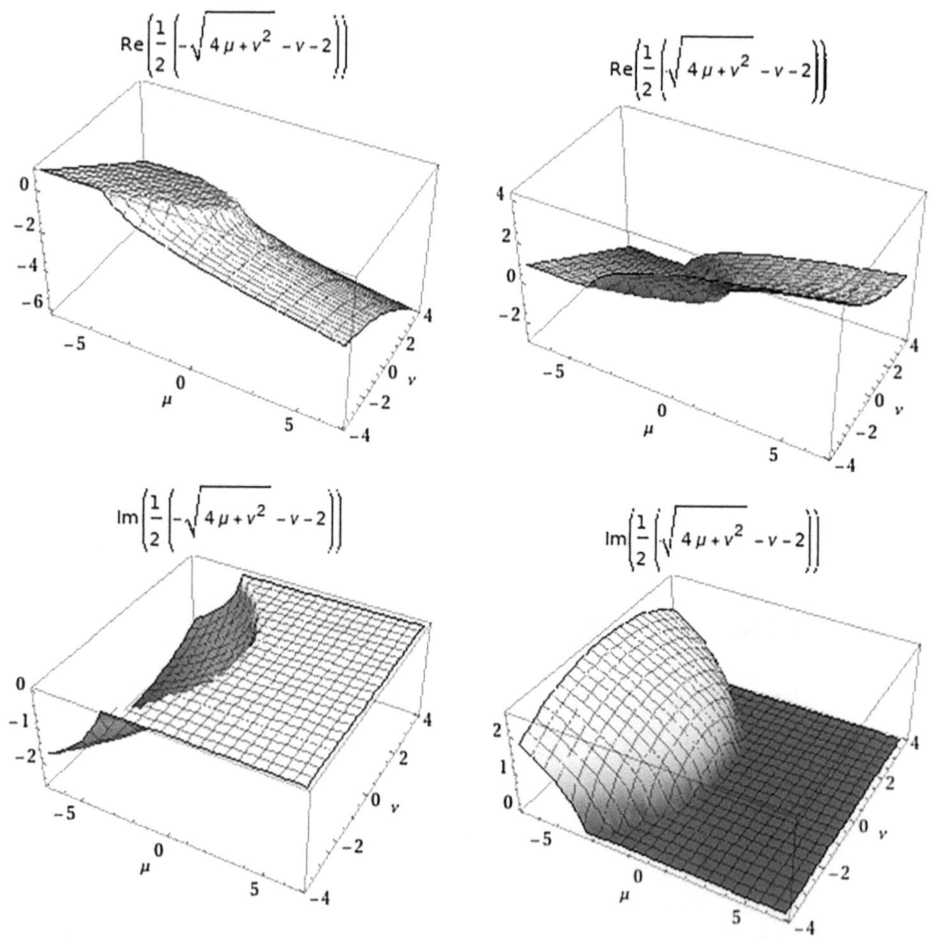

Figure 3. The set of eigenvalues of System (18).

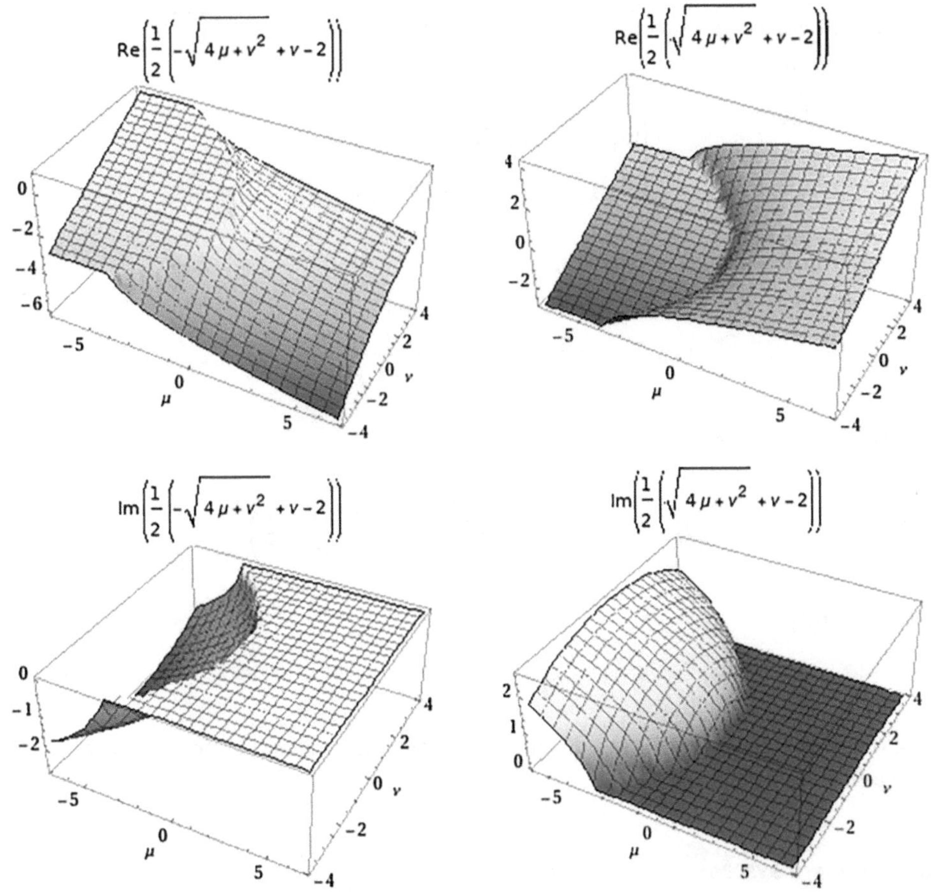

Figure 4. The set of eigenvalues of System (19).

3.4. System (19)

The st of the eigenvalues is (see Figure 4)

$$\lambda_1 = \frac{1}{2}(-\sqrt{4\mu + \nu^2} + \nu - 2)$$

and

$$\lambda_2 = \frac{1}{2}(\sqrt{4\mu + \nu^2} + \nu - 2),$$

where the corresponding eigenvectors are

$$V_1 = \left(-\frac{-\nu + \sqrt{4\mu + \nu^2}}{2\mu}, 1 \right)$$

and

$$V_2 = \left(-\frac{-\nu - \sqrt{4\mu + \nu^2}}{2\mu}, 1 \right).$$

Based on the eigenvalues of the system and under the condition $\sqrt{4\mu + \nu^2} \geq 2 - \nu$, we have two domains for the solutions,

$$[\Theta_1]_{\lambda_1} = \{(\chi, \Upsilon) \in \mathbb{R}^2 : \lambda < 0\}$$
$$[\Theta_2]_{\lambda_2} = \{(\chi, \Upsilon) \in \mathbb{R}^2 : \lambda \geq 0\}.$$

The equilibrium point of System (19) is

$$(\chi_0, \Upsilon_0) = \left(\frac{1}{-\nu - \mu + 1}, \frac{-\mu}{\nu + \mu - 1}\right).$$

Also, the fixed point is

$$(\chi_{fix}, \Upsilon_{fix}) = \left(\frac{-2}{2\nu + \mu - 4}, \frac{-\mu}{2\nu + \mu - 4}\right).$$

If the inequality

$$\mu \leq 2(2 - \nu)$$

holds, then the eigenvalues λ_1 and λ_2 are in the interval [0,1] and hence, by Corollary 3.5, the system is asymptotically stable.

3.5. Nonlinear system

Here, we suppose the (2×2) k-symbol FLS (or $(n \times n)$ k-symbol system)

$$ {}^C_k\Delta^\sigma \begin{pmatrix} \chi(\tau) \\ \Upsilon(\tau) \end{pmatrix} = \Omega_{2\times 2}(\tau) \begin{pmatrix} \chi(\tau) \\ \Upsilon(\tau) \end{pmatrix} + \begin{pmatrix} f_1(\tau, \chi, \Upsilon) \\ f_2(\tau, \chi, \Upsilon) \end{pmatrix}, \tag{20}$$

where the continuous function $f(\tau, \chi, \Upsilon) = (f_1(\tau, \chi, \Upsilon), f_2(\tau, \chi, \Upsilon))^\top$ satisfies the following property:

$$\|f\| \leq m\|R\| := \|(\chi, \Upsilon)\|, \quad m > 0,$$

uniformly with respect to τ.

Theorem 3.6: *Let the characteristic polynomial of Ω is stable. If*

$$\frac{m\tau^{\sigma/k}}{\Gamma_k(\sigma + 1)} < 1, \quad m > 0, \tau \in [0, \infty)$$

then all the solutions of System (20) admit the stability. Moreover, if $cm < e$, then the zero solution of System (20) is asymptotically stable.

Proof: By using the k-symbol fractional integral formula, we have

$$\begin{aligned}\|R(\tau)\| &\leq \|R_0(\tau)\| + \frac{1}{k\Gamma_k(\sigma)}\int_0^{\tau} \|(\tau-\varsigma)^{\sigma/k-1}f(\varsigma)\|d\varsigma \\ &\leq c + \frac{m\|R(\tau)\|}{\Gamma_k(\sigma)}\frac{\tau^{\sigma/k}}{\sigma},\end{aligned}$$

which implies that

$$\|R\| \leq \frac{c}{1-\frac{m\tau^{\sigma/k}}{\Gamma_k(\sigma+1)}}.$$

This results in the conclusion that the solutions are bounded, which implies that they are stable given the proof of Theorem 3.2. Also, we have

$$\begin{aligned}\|\chi(\tau)\| &\leq c\exp\left(cm\frac{\tau^{\sigma/k}}{\sigma/k}\right)\|\chi_0\| \\ &\leq c\exp\left((cm-e)\frac{\tau^{\sigma/k}}{\sigma/k}\right)\|\chi_0\| \\ &= 0, \quad cm-e<0.\end{aligned}$$

Similarly, for the variable Υ

$$\begin{aligned}\|\Upsilon(\tau)\| &\leq c\exp\left(cm\frac{\tau^{\sigma/k}}{\sigma/k}\right)\|\Upsilon_0\| \\ &\leq c\exp\left((cm-e)\frac{\tau^{\sigma/k}}{\sigma/k}\right)\|\Upsilon_0\| \\ &= 0, \quad cm-e<0.\end{aligned}$$

Hence, the asymptotically stable is valid. ■

4. Stabilizing

The design of control techniques to attain stability is the important key elements in the investigation of chaotic systems, whether in discrete time or continuously. To stabilize the aforementioned arbitrary order discrete-time systems, we shall offer different possible nonlinear control laws in this section. Joying a newfangled time-alter parameter, $\mho(\tau)$, to the specific system's states and figuring out an adaptive closed-form technique for these parameters to drive the system states to zero in a timely manner constitute stabilization.

Theorem 4.1: *System (11) is controlled by the 1D-control equality*

$$\mho_\chi(\tau) = \nu|\chi(\tau-1+\sigma)| - \Upsilon(\tau-1+\sigma) - 1.$$

Proof: The time-altering control parameter $\mho_\chi(\tau)$ is used in the controlled k-symbol FLS, which is denoted by

$$\begin{cases} {}^C_k\Delta^\sigma \chi(\tau) = -\nu|\chi(\tau-1+\sigma)| + \Upsilon(\tau-1+\sigma) + 1 - \chi(\tau-1+\sigma) + \mho_\chi(\tau) \\ {}^C_k\Delta^\sigma \Upsilon(\tau) = \mu\chi(\tau-1+\sigma) - \Upsilon(\tau-1+\sigma). \end{cases} \tag{21}$$

The considered control equality $\mho\chi(\tau)$ is replaced into Equation (21) to have

$$\begin{cases} {}^C_k\Delta^\sigma \chi(\tau) = -\chi(\tau-1+\sigma) \\ {}^C_k\Delta^\sigma \Upsilon(\tau) = \mu\chi(\tau-1+\sigma) - \Upsilon(\tau-1+\sigma). \end{cases} \tag{22}$$

Therefore, the set of eigenvalues admits the boundedness property with$\lambda_{1,2} = -1$ conforming the eigenvectors $V = (\frac{1}{\mu}, 0), \mu \neq 0$. Thus, by Theorem 3.2, the solution admits the asymptotically stable behaviour. ∎

Theorem 4.2: *System (11) is controlled by the 2D-control equality*

$$U_\chi(\tau) = \nu|\chi(\tau-1+\sigma)| - \Upsilon(\tau-1+\sigma) - 1 + \left(1 - \frac{\nu}{4}\right)\chi(\tau-1+\sigma)$$

$$V_\Upsilon(\tau) = -\mu\chi(\tau-1+\sigma) + \frac{3}{4}\Upsilon(\tau-1+\sigma), \quad |\nu| < 3.$$

Proof: Consider the item $(U_\chi(\tau), V_\Upsilon(\tau))$, where is used in the 2D-controlled k-symbol FLS

$$\begin{cases} {}^C_k\Delta^\sigma \chi(\tau) = -\nu|\chi(\tau-1+\sigma)| + \Upsilon(\tau-1+\sigma) + 1 - \chi(\tau-1+\sigma) + U_\chi(\tau) \\ {}^C_k\Delta^\sigma \Upsilon(\tau) = \mu\chi(\tau-1+\sigma) - \Upsilon(\tau-1+\sigma) + V_\Upsilon(\tau). \end{cases} \tag{23}$$

The proposed 2D control law $(U_\chi(\tau), V_\Upsilon(\tau))$ is substituted into (23) to get the simplified dynamics

$$\begin{cases} {}^C_k\Delta^\sigma \chi(\tau) = -\frac{\nu}{4}\chi(\tau-1+\sigma) \\ {}^C_k\Delta^\sigma \Upsilon(\tau) = -\frac{1}{4}\Upsilon(\tau-1+\sigma). \end{cases} \tag{24}$$

Thus the set of eigenvalues is bounded, with $\|\Omega\| < 1$, where

$$\lambda_1 = -1/4, \quad \lambda_2 = -\nu/4, \quad |\nu| < 3.$$

The asymptotically stability is valid by Corollary 3.3. ∎

The following result indicates the stabilizing two-dimensional parametric 2D control law FLS, which are ν and μ.

Theorem 4.3: *System (11) is controlled by the 2D control equality*

$$P_\chi(\tau) = \nu|\chi(\tau-1+\sigma)| - \Upsilon(\tau-1+\sigma) - 1 + \left(1 - \frac{\nu}{4}\right)\chi(\tau-1+\sigma)$$

$$Q_\Upsilon(\tau) = \frac{3}{4}\Upsilon(\tau-1+\sigma), \quad |\nu| < 3.$$

Proof: Consider the term $(P_\chi(\tau), Q_\Upsilon(\tau))$, which is used in the 2D-controlled k-symbol FLS

$$\begin{cases} {}^C_k\Delta^\sigma \chi(\tau) = -\nu|\chi(\tau-1+\sigma)| + \Upsilon(\tau-1+\sigma) + 1 - \chi(\tau-1+\sigma) + P_\chi(\tau) \\ {}^C_k\Delta^\sigma \Upsilon(\tau) = \mu\chi(\tau-1+\sigma) - \Upsilon(\tau-1+\sigma) + Q_\Upsilon(\tau). \end{cases} \tag{25}$$

The suggested 2D control equation is replaced to obtain the dynamics 2D control law $(P_\chi(\tau), Q_\Upsilon(\tau))$ into (23)

$$\begin{cases} {}^C_k\Delta^\sigma \chi(\tau) = -\frac{\nu}{4}\chi(\tau-1+\sigma) \\ {}^C_k\Delta^\sigma \Upsilon(\tau) = \mu\chi(\tau-1+\sigma) - \frac{1}{4}\Upsilon(\tau-1+\sigma). \end{cases} \tag{26}$$

Thus the set of eigenvalues is bounded, with $\|\Omega\| < 1$, where

$$\lambda_1 = -\frac{1}{4}, \quad \lambda_2 = -\frac{\nu}{4}, \quad |\nu| < 3, \quad \mu \neq 0,$$

conforming the eigenvectors $W = (-\frac{\nu-1}{4\mu}, 1)$. Then by Corollary 3.3, the solution is asymptotically stable. ■

5. Conclusion

In this study, we looked at different k-symbol discrete arbitrary-order chaotic maps. The suggested maps are viewed as the well-known Lozi map's fractional k-symbol generalizations. It was decided to examine the dynamics of the suggested systems using DFC. It goes without saying that the DFC has advantages over traditional calculus, involving the limitless memory impact, which results in a higher degree of dynamics. We also suggested a method to make the suggested maps more stable. The stabilization techniques are one-dimensional and two-dimensional, therefore, to ensure that entirely system states converge to zero, we only need to adaptability manage one of the system's states. Additionally, we suggest a mixed stabilization system in which the k-symbol fractional Lozi map attends under different controller laws. We suggested one- and two-dimensional controller systems of one- and two-dimensional parametric laws. For future works, one can suggest to use the q-quantum calculus or (q, k)-calculus to generalize the fractional Lozi system and amps.

Disclosure statement

No potential conflict of interest was reported by the author(s).

ORCID

Rabha W. Ibrahim http://orcid.org/0000-0001-9341-025X

References

[1] T. Abdeljawad, On Riemann and Caputo fractional differences, Comput. Math. Appl. 62(3) (2011), pp. 1602–1611.

[2] I. Aldawish and R.W. Ibrahim, A new mathematical model of multi-faced COVID-19 formulated by fractional derivative chains, Adv. Contin. Discrete Models 2022(1) (2022), pp. 1–10.
[3] M.H. AlSheikh, N.M. Al-Saidi, and R.W. Ibrahim, Dental X-ray identification system based on association rules extracted by k-Symbol fractional haar functions, Fractal Fract. 6(11) (2022), pp. 669.
[4] F.M. Atici and P.W. Eloe, A transform method in discrete fractional calculus, Int. J. Differ. Equ. 2(2) (2007), pp. 165–176.
[5] D. Baleanu, A. Jajarmi, S.S. Sajjadi, and J.H. Asad, The fractional features of a harmonic oscillator with position-dependent mass, Commun. Theor. Phys. 72(5) (2020), pp. 055002.
[6] J Borcea, P. Branden, and T Liggett, Negative dependence and the geometry of polynomials, J. Am. Math. Soc. 22(2) (2009), pp. 521–567.
[7] J.B. Diaz and T.J. Osler, Differences of fractional order, Math. Comput. 28(125) (1974), pp. 185–202.
[8] R. Diaz and E. Pariguan, On hypergeometric functions and pochhammer K-symbol, Divulgaciones Mat. 15 (2007), pp. 179–192.
[9] G.A. Dorrego and R.A. Cerutti, The k-fractional Hilfer derivative, Int. J. Math. Anal 7(11) (2013), pp. 543–550.
[10] V.P. Dubey, R. Kumar, D. Kumar, I. Khan, and J. Singh, An efficient computational scheme for nonlinear time fractional systems of partial differential equations arising in physical sciences, Adv. Differ. Equ. 2020(1) (2020), pp. 1–27.
[11] M.S. Fadhil, A.K. Farhan, M.N. Fadhil, and N.M.G. Al-Saidi, A new lightweight AES using a combination of chaotic systems, In 2020 1st. Information Technology to Enhance e-learning and Other Application IT-ELA, IEEE, 2020, pp. 82–88.
[12] A.K. Farhan, A. Tariq, R.S. Ali, and N.M.G. Alsaidi, A new pseudo random bits generator via 2D chaotic system, diffusion, and permeation, 2019.
[13] D. Foukrach and B Meftah, Some new generalized result of Gronwall–Bellman–Bihari type inequality with some singularity, Filomat 34(10) (2020), pp. 3299–3310.
[14] C. Goodrich and A.C. PetersonDiscrete Fractional Calculus, Vol.10, Springer, Cham, 2015.
[15] S.B. Hadid and R.W. Ibrahim, Fractional dynamic system simulating the growth of microbe, Adv. Differ. Equ. 2021(1) (2021), pp. 1–15.
[16] L.L. Huang, D. Baleanu, G.C. Wu, and S.D. Zeng, A new application of the fractional logistic map, 2016.
[17] R.W. Ibrahim and D. Baleanu, Global stability of local fractional Henon–Lozi map using fixed point theory, AIMS Math. 7(6) (2022), pp. 11399–11416.
[18] A. Jajarmi, M. Hajipour, E. Mohammadzadeh, and D. Baleanu, A new approach for the nonlinear fractional optimal control problems with external persistent disturbances, J. Franklin Inst. 355(9) (2018), pp. 3938–3967.
[19] A.A. Khennaoui, A. Ouannas, S. Bendoukha, G. Grassi, R.P. Lozi, and V.T Pham, On fractional-order discrete-time systems: chaos, stabilization and synchronization, Chaos Soliton. Fractal. 119 (2019), pp. 150–162.
[20] A.A Khennaoui, A. Ouannas, S. Bendoukha, X. Wang, and V.T. Pham, On chaos in the fractional-order discrete-time unified system and its control synchronization, Entropy 20(7) (2018), pp. 530.
[21] X. Liu, F. Du, D. Anderson, and B. Jia, Monotonicity results for Nabla fractional h-difference operators, Math. Methods. Appl. Sci. 44(2) (2021), pp. 1207–1218.
[22] R Lozi, A strange attractor of the Henon attractor type, J. Phys. Colloquia 39(C5) (1978), pp. C5–C9.
[23] R Lyons, A note on tail triviality for determinantal point processes, Electron. Commun. Probab. 23 (2018), pp. 1–3.
[24] C. Meshram, R.W. Ibrahim, A.J. Obaid, S.G. Meshram, A. Meshram, and A.M. Abd El-Latif, Fractional chaotic maps based short signature scheme under human-centered IoT environments, J. Adv. Res. 32 (2021), pp. 139–148.

[25] D. Mozyrska and E. Girejko, Overview of fractional h-difference operators, In Advances in harmonic analysis and operator theory, Basel, Birkhauser, 2013, pp. 253–268.

[26] S. Mubeen and G.M Habibullah, k-Fractional integrals and application, Int. J. Contemp. Math. Sci.7(2) (2012), pp. 89–94.

[27] N.R. Ojeda and M.V Diaz, On fractional Laplace equation in plane polar and spherical coordinates, J. Fract. Calc. Appl. 13(2) (2022), pp. 116–125.

[28] G. Rahman, K.S. Nisar, A. Ghaffar, and F. Qi, Some inequalities of the Gruss type for conformable k-fractional integral operators, Rev. La Real Acad. Ciencias Exact. Fisic. Natur. Ser. A. Mat. 114(1) (2020), pp. 1–9.

[29] G. Rahman, K.S. Nisar, and S. Mubeen, On generalized k-fractional derivative operator. Preprints 2017, 2017120101 (doi:10.20944/preprints201712.0101.v1).

[30] S.H. Salih and N.M. Al-Saidi, 3D-Chaotic discrete system of vector borne diseases using environment factor with deep analysis, AIMS Math. 7(3) (2022), pp. 3972–3987.

[31] M.P. Velasco, D. Usero, S. Jimenez, L. Vazquez, J.L. Vazquez-Poletti, and M. Mortazavi, About some possible implementations of the fractional calculus, Mathematics 8(6) (2020), pp. 893.

[32] G.C. Wu, D. Baleanu, H.P. Xie, and F.L. Chen, Chaos synchronization of fractional chaotic maps based on the stability condition, Phys. A: Stat. Mech. Appl. 460 (2016), pp. 374–383.

[33] G.C. Wu and D. Baleanu, Discrete fractional logistic map and its chaos, Nonlinear. Dyn. 75(1) (2014), pp. 283–287.

[34] S. Zhao, S.I. Butt, W. Nazeer, J. Nasir, M. Umar, and Y Liu, Some Hermite–Jensen–Mercer type inequalities for k-Caputo-fractional derivatives and related results, Adv. Differ. Equ. 2020(1) (2020), pp. 1–17.

Border collision bifurcations in a piecewise linear duopoly model

Laura Gardini and Davide Radi

ABSTRACT
We consider a duopoly model characterized by a two-dimensional non-invertible continuous map *T* given by piecewise linear functions, with several partitions, defining a duopoly game. The structure of the game is such that it has separate second iterate so that its dynamics can be studied via a one-dimensional composite function, that is piecewise linear with multiple partitions in which the definition of the map changes. The number of partitions may change from 2 to 5, depending on the parameters. The dynamics are characterized by degenerate bifurcations and border collision bifurcations, which are typical in maps having kink points. Here the peculiarity is the multiplicity of the partitions, which leads to bifurcations different from those occurring in maps with only one kink point. We show several bifurcations, coexistence of cycles, attracting and superstable, as well chaotic attractors and chaotic repellors, related to the outcome of particular border collision bifurcations.

1. Introduction

In this work, we consider a duopoly model that is described by a continuous two-dimensional map characterized by piecewise linear reaction functions having several different definitions. The duopoly model, say T, is of Cournot type, i.e. $x_{n+1} = r_1(y_n)$ and $y_{n+1} = r_2(x_n)$ so that its second iterate is characterized by a system with separate variables: $x_{n+2} = r_1 \circ r_2(x_n) = F(x)$ and $y_{n+2} = r_2 \circ r_1(y) = G(y_n)$. This kind of maps appears in many works in the economic literature (see the seminal papers [14,31]). In [14] it has been described how the two functions F and G are related, so that it is enough to consider only the dynamics of one of these one-dimensional maps. In [10] it is described how to derive the properties of the two-dimensional map from those of the one-dimensional map F (or equivalently from map G). A characteristic of this kind of systems is the multistability, and several coexisting different cycles of the duopoly model T are related to just one cycle of map F, and also the local and global properties of T can be obtained from those of map F.

An extension to a three-dimensional triopoly model of Cournot type, in which the third iterate has separate variables has also been considered (see e.g. [1,2,7]).

Clearly, here we shall make use of the same technique. The economic formulation and meaning of the duopoly is quite new, as evidenced in [33], and the main novelty in terms of the dynamic behaviours of the one-dimensional map F is related to its *piecewise linear* (*PWL* for short) formulation, with several kink points, that are the points where the definition of the function F changes. So, depending on the parameters of the model, it may result in a *PWL* map with j kink points, where $j \in \{1, 2, 3, 4\}$.

The properties of the dynamics of one-dimensional *PWL* maps have been mainly studied for maps with one kink point, and the bifurcations occurring in the skew tent map that is the *one-dimensional border collision normal form* (*BCNF* for short) have great relevance in the study of generic piecewise smooth maps with one kink point. This has been evidenced in several works in the recent literature. We refer to [27,28] which introduced the term *border collision bifurcation* (*BCB*, for short), and several other works ([6,16,20,21], to cite a few). Since then, the term *BCB* is used to describe this particular kind of bifurcations, which occur when an invariant set (typically the periodic point of a cycle) merges (or collides), when some parameter is varied, with a kink point, a border of definition, so that the map has a different definition on the right and left sides of the kink point. The result of such a border collision has been investigated in the last decades in piecewise smooth and *PWL* maps. Moreover, notice that the local bifurcations occurring in one-dimensional *PWL* maps are different from those occurring in smooth maps. In fact, the appearance/disappearance of cycles can occur via a *fold border collision bifurcation*, and differently from what occurs in smooth maps, the two cycles associated with this bifurcation may also be both repelling. Also the usual bifurcation with eigenvalue -1, which in a smooth system is related to a cycle of period 2, in a *PWL* map is quite different. The result of this *degenerate flip bifurcation*, may be an attracting cycle of double period, but also an attracting set consisting in cyclic chaotic intervals (besides a repelling cycle of double period) or a chaotic repellor (as we shall see also in this work), we refer to [5,34,36] for details.

As remarked above, the map of interest in this work has in general several kink points, and this leads to other peculiarities, although, as we shall see, in several cases the investigation may use the results of the *BCNF*. However, also new dynamic behaviours are observed, due to the multiplicity of kink points in our map. New kind of bifurcations (of the 2-cycles as well as of other cycles and chaotic intervals) exist when the map is bimodal in an absorbing interval. The investigation of a bimodal map with two kink points is also not new, see [29,30]. In those papers, the authors have investigated regimes in which the slopes in the partitions are smaller than 1 in absolute value, or when one slope is larger than 1 in modulus, but without flat definitions for the map. In the current analysis, besides the existence of flat branches of the map that play a relevant role, many kink points are involved and in two partitions of the map the functions may be steep.

We remark that our system is characterized by several parameters, and the possible different dynamics occurring in map F are really many. Here, we have mainly focused our attention on one particular case, depending on two parameters, that is representative of the possible dynamics in this kind of maps. This is clearly only a preliminary result on the system, but it shows which are the relevant facts to be considered, and the kind of bifurcations that may occur, so that also considering different parameter settings, an analysis similar to the one here performed can be used. Thus, the present work is not conclusive for the analysis of our duopoly game, but shows the road for its analysis.

Regarding the application, our work enriches the literature on nonlinear dynamics in duopoly games, which is made of continuous piecewise linear maps, see, e.g. [9], discontinuous piecewise-linear maps, see, e.g. [22], smooth quadratic maps, see, e.g. [8], and discontinuous maps, see, e.g. [23]. For an overview on nonlinear dynamics in oligopoly games, see also [3,11]. With respect to the previous literature on nonlinear duopoly games, the main novelty of the current work is the presence of a map with multiple kink points.

After this introduction, the rest of the paper is as follows. In Section 2 we introduce the duopoly model T, deriving the one-dimensional composite map F, and giving some preliminary properties to be used in the other sections. Moreover, it is shown the relation between the dynamics of the one-dimensional map F and those of the two-dimensional map T, by using known results for this kind of maps (having separate second iterate) that are also recalled in the Appendix, for the sake of completeness. Sections 3 and 4 include our results. We completely describe the *BCBs* occurring to fixed points and 2-cycles of the map, which belong to several partitions, as well as the structure of the basins of attraction of the existing attracting sets. Besides some general properties of our system, we consider a specific case in which we fix all the parameters except two, in order to better investigate and describe the possible dynamics. In several subsections we consider the detailed properties and bifurcations, evidencing atypical behaviours and some particular bifurcations related to the multiple kink points of the map as well as bifurcations related to the flat branches of map F in some partitions. Moreover, the coexistence of an attracting fixed point and superstable cycles is also a peculiarity of our model. Section 5 concludes.

2. The model

In this section we recall the system that, as described in [33], represents a Cournot-duopoly model with worst-case maximizers. In [32] we have considered a symmetric duopoly game combining ambiguity aversion and constant expectations. Ambiguity, also known as Knightian uncertainty, is the lack of trust in a unique probability description of a random event. The aversion to this type of uncertainty influences the decision process of economic agents. The maximin preferences axiomatized in [19] offers a theoretical framework used to model ambiguity aversion. When the lack of trust involves all possible probability distributions over a finite set of possible realizations the ambiguity aversion degenerates in the worst-case approach. Employing this kind of approach, [12,13] show that when firms are uncertain about the values of the parameters of the inverse demand functions, the Nash equilibria of a duopoly game are modified. Mixing the worst-case approach to this uncertainty with constant expectations about the next-period level of production of the competitor, we have shown for the first time in [32] that complex dynamics may occur (when the Nash equilibria are repelling) also in a simple and stylized duopoly with linear demand functions and constant marginal costs of production. In [33] the system proposed in [32] has been improved introducing asymmetry in firms' uncertainty sets, showing that also in that more generic case, the system is characterized by multistability and chaotic dynamics. Our goal here is to investigate the bifurcations, mainly due to border collisions, as well as other global bifurcations (as homoclinic bifurcations and contact bifurcations) leading to relevant changes in the dynamics.

As in [32], the Cournot-tâtonnement process is described by a *PWL* two-dimensional system T with separate variables in the second iterate, that can be studied via a one-dimensional *PWL* map F. Now, differently from [32], the one-dimensional map F that describes the global dynamics of the model is not a bimodal map of type 'decreasing-increasing-decreasing'. The new model consists in a different number of branches, from two to five, and the shape of the map in the absorbing interval is of type 'increasing-decreasing-increasing' so that it may have up to five Cournot–Nash equilibria.

The results of the possible dynamics evidenced in this work confirm the results in [32] and show the robustness of the route to randomness obtained by mixing ambiguity aversion and non-perfect foresight expectations.

The model is as follows. Let (x, y) be the vector of the quantities produced in the duopoly by firm X and firm Y. Products of the two firms are sold in the market at the prices

$$P_X\left(x, y\right) = \max\left\{a_1 - b_1 x - d_1 b_1 y;\ 0\right\} \tag{1}$$

and

$$P_Y\left(x, y\right) = \max\left\{a_2 - d_2 b_2 x - b_2 y;\ 0\right\} \tag{2}$$

The choke prices $a_1 > 0$ and $a_2 > 0$ are assumed to be known by the firms and constant over time. Instead, the values of the slope of the inverse-demand functions $b_1 \geq 0$ and $b_2 \geq 0$ as well as the degrees of product substitutability $d_1 \in [0, 1]$ and $d_2 \in [0, 1]$ are random variables. This modelling assumption is justified by consumers and consumers' tastes that change over time. Moreover, firms are uncertain about the probability description of the realizations of these parameters. The uncertainty regards also the support set of all possible realizations of these parameters, which leads to ambiguity about the composition of the game. Indeed, observing only the past prices and the level of production of an opponent, and with the values of b_i and γ_i that are random realizations, it is not possible to exclude the existence of more than one competitor. Therefore, firms do not know if they are playing a duopoly game (as it is) or an oligopoly game with N players. To easily account for this uncertainty, the inverse-demand functions can be rewritten as follows:

$$P_X\left(x, y; b_1, \gamma_1\right) = \max\left\{a_1 - b_1 x - \gamma_1 y;\ 0\right\} \tag{3}$$

and

$$P_Y\left(x, y; b_2, \gamma_2\right) = \max\left\{a_2 - \gamma_2 x - b_2 y;\ 0\right\} \tag{4}$$

where $\gamma_1 = d_1(N-1)b_1$ and $\gamma_2 = d_2(N-1)b_2$. When it holds $\gamma_1 > b_1 > 0$ (resp. $\gamma_2 > b_2 > 0$) then the economic meaning is as follows: Considering $\gamma_1 > Nb_1$ (resp. $\gamma_2 > Nb_2$), with $N \geq 1$ means that firm X (resp. firm Y) considers the possibility that the oligopoly is populated by at least N competitors.

To close the model, constant marginal costs of production are assumed for both firms:

$$C_X\left(x\right) = c_1 x + \delta_1 \quad \text{and} \quad C_Y\left(y\right) = c_2 y + \delta_2 \tag{5}$$

where the marginal costs c_1 and c_2, as well as the fixed costs δ_1 and δ_2, are assumed to be non-negative values known to the firms.

Given this, we have the profit functions of the firms:

$$\Pi_X(x, y; b_1, \gamma_1) = P_X(x, y; b_1, \gamma_1) - C_X(x) \tag{6}$$

and

$$\Pi_Y(x, y; b_2, \gamma_2) = P_Y(x, y; b_2, \gamma_2) - C_Y(y) \tag{7}$$

A firm determines its level of production by maximizing its profit function in (6)–(7). The ambiguity regarding parameters b_1, b_2, γ_1 and γ_2 is characterized by the lack of trust in any probability distribution over a sample space of possible realizations given by $U^* = U_1^* \times U_2^*$ where $(b_1, \gamma_1) \in U_1^*$ and $(b_2, \gamma_2) \in U_2^*$. Let us rename firm X and Y as firm 1 and 2, respectively. Noting that U_i^* is the only uncertainty set relevant for the decision process of firm i, with $i = 1, 2$, we assume that firm i has a belief about U_i^* which is denoted hereafter by U_i. As in [32], we further assume that U_i is a consistent and conservative belief on U_i^*. It is a consistent belief if a firm cannot exclude this uncertainty set based on historical observations, while it is a conservative belief when the maximum-guaranteed payoff computed according to this uncertainty set is not higher than the one computed according to the real set of possible realizations for whatever quantities produced by the firms. To accommodate this type of uncertainty, firm i is a worst-case maximizer with respect to U_i. According to these maximin preferences, the only relevant elements of U_i are those realizations that can be worst case for some level of production of firm i and of its competitors. Therefore, we define U_i by focusing on the worst-case realizations only (neglecting to specify those possible realizations that are never worst case). Specifically, we assume that firm 1 relies on an uncertainty set made of two (possible) worst-case realizations, $U_1 = \{(\overline{b}_1, \underline{\gamma}_1), (\underline{b}_1, \overline{\gamma}_1)\}$ with $\overline{b}_1 > \underline{b}_1$ and $\overline{\gamma}_1 > \underline{\gamma}_1$, while firm 2 relies on an uncertainty set made of a unique worst-case realization, $U_2 = \{(b_2, \gamma_2)\}$. Then, the best-reply function of firm 1 is given by

$$R_1(y) = \arg\max_{q_1 \geq 0} \left[\arg\min_{(b_1, \gamma_1) \in U_1} \Pi_X(x, y; b_1, \gamma_1)\right] \tag{8}$$

while the best-reply function of firm 2 is given by

$$R_2(x) = \arg\max_{q_2 \geq 0} \Pi_Y(x, y; b_2, \gamma_2) \tag{9}$$

Moreover, to simplify the analysis we assume $a = a_1 - c_1 = a_2 - c_2$ and hereafter we assume the following constraints on the parameters:

$$\overline{b}_1 > \underline{b}_1 \geq 0, \quad b_2 > 0, \quad \gamma_2 \geq 0, \quad \overline{\gamma}_1 > \underline{\gamma}_1 \geq 0, \quad a > 0 \tag{10}$$

Solving problems (8) and (9) under the assumptions in (10) we have

$$R_1(y) = \begin{cases} f_1(y) := \mu_L - s_L y & \text{if } 0 \leq y \leq y_L \\ f_2(y) := ry & \text{if } y_L \leq y \leq y_U \\ f_3(y) := \mu_U - s_U y & \text{if } y_U \leq y \leq y_M \\ f_4(y) := 0 & \text{if } y \geq y_M \end{cases} \tag{11}$$

with

$$\mu_L = \frac{a}{2\overline{b}_1}; \quad s_L = \frac{\underline{\gamma}_1}{2\overline{b}_1}; \quad r = \frac{\overline{\gamma}_1 - \underline{\gamma}_1}{\overline{b}_1 - \underline{b}_1}; \quad \mu_U = \frac{a}{2\underline{b}_1}; \quad s_U = \frac{\overline{\gamma}_1}{2\underline{b}_1} \tag{12}$$

$$\begin{aligned} y_L &= \frac{\mu_L}{s_L + r} = \frac{a}{\underline{\gamma}_1 + 2\overline{b}_1 r} \\ y_U &= \frac{\mu_U}{s_U + r} = \frac{a}{2\underline{b}_1 r + \overline{\gamma}_1} \\ y_M &= \frac{a}{\overline{\gamma}_1} \end{aligned} \tag{13}$$

where it is always $y_L < y_U < y_M$ and

$$R_2(x) = \begin{cases} g_1(x) := \alpha - s_x x & \text{if } 0 \leq x \leq x_M \\ g_2(x) := 0 & \text{if } x \geq x_M \end{cases} \tag{14}$$

with

$$\alpha = \frac{a}{2b_2}; \quad s_x = \frac{\gamma_2}{2b_2}; \quad x_M = \frac{a}{\gamma_2} \tag{15}$$

The game is played at any time $n \in N$ and each firm determines its next-period of production according to $x_{n+1} = R_1(y^e_{n+1})$ and $y_{n+1} = R_2(x^e_{n+1})$. Assuming a constant expectation scheme, that is $x^e_{n+1} = x_n$ and $y^e_{n+1} = y_n$, we obtain the following map:

$$\left(x_{n+1}, y_{n+1}\right) = T\left(x_n, y_n\right) \quad \text{with } T\left(x_n, y_n\right) = \left(R_1\left(y_n\right), R_2\left(x_n\right)\right) \tag{16}$$

according to which the output dynamics of the duopoly game with ambiguity aversion is determined. A trajectory generated by map T represents a so-called Cournot-tâtonnement process, see, e.g. [11]. The possible asymptotic behaviours of our Cournot-tâtonnement process are analysed in the following sections.

2.1. Deduction of map F and some preliminaries

The dynamics of the two-dimensional map $T(x, y) = (R_1(y), R_2(x))$ defined in $(x, y) \in \mathcal{R}^2_+$ can be studied via the one-dimensional composite map F:

$$F(x) = R_1 \circ R_2(x), \quad x \geq 0 \tag{17}$$

or equivalently via the one-dimensional composite map $G(y) = R_2 \circ R_1(y)$ for $y \geq 0$ where F and G are called conjugate, since the dynamics of G can be obtained from those of F and *viceversa*. The relevant property is that

$$T^{2k}(x, y) = \left(F^k(x), G^k(y)\right), \quad \forall k \geq 1 \tag{18}$$

and in the Appendix we recall some useful properties of this kind of maps, that are used below to describe the dynamics of map T from those of F.

Concerning the analytical expression of F, from standard computations we obtain a map which may have up to four kink points ξ_i (and thus five different definitions), that is, assuming the four kink points all positive:

$$F(x) = \begin{cases} F_0(x) := 0 & \text{if } 0 \le x \le \xi_1 \\ F_1(x) := \mu_U - s_U(\alpha - s_x x) & \text{if } \xi_1 \le x \le \xi_2 \\ F_2(x) := r(\alpha - s_x x) & \text{if } \xi_2 \le x \le \xi_3 \\ F_3(x) := \mu_L - s_L(\alpha - s_x x) & \text{if } \xi_3 \le x \le \xi_4 \\ F_4(x) := \mu_L & \text{if } x \ge \xi_4 \end{cases} \tag{19}$$

where

$$\begin{aligned} &\xi_1 := \frac{\alpha - y_M}{s_x} = \frac{a}{\gamma_2}\left(1 - \frac{2b_2}{\overline{\gamma}_1}\right), \quad \xi_2 := \frac{\alpha - y_U}{s_x} = \frac{a}{\gamma_2}\left(1 - \frac{2b_2}{\overline{\gamma}_1 + 2\underline{b}_1 r}\right) \\ &\xi_3 := \frac{\alpha - y_L}{s_x} = \frac{a}{\gamma_2}\left(1 - \frac{2b_2}{\underline{\gamma}_1 + 2\overline{b}_1 r}\right), \quad \xi_4 := x_M = \frac{a}{\gamma_2} \end{aligned} \tag{20}$$

considering the parameters as in (12), (13), (15) and the conditions in (10) satisfied.

Let us consider some preliminary properties of the F. First of all, note that in map T, and thus in map F, parameter a is a scaling factor. Indeed, operating the change of variables $\mathcal{X} := x/a$ and $\mathcal{Y} := y/a$, we obtain a map that does not depend on a. Therefore, hereafter we shall consider $a = 1$.

Moreover, since the map is defined only for $x \ge 0$, we have the following properties. For parameters leading to:

- (C5) $\xi_1 \ge 0$, then all the other kink points are positive and map F consists in five branches F_i, $i = 0, 1, 2, 3, 4$;
- (C4) $\xi_1 < 0$ and the other three kink points positive, then map F consists in four branches F_i, $i = 1, 2, 3, 4$, with $F_1(x)$ defined in $[0, \xi_2]$;
- (C3) $\xi_1 < 0$, $\xi_2 < 0$ and the other two kink points positive, then map F consists in three branches F_i, $i = 2, 3, 4$, with $F_2(x)$ defined in $[0, \xi_3]$;
- (C2) $\xi_1 < 0$, $\xi_2 < 0$ and $\xi_3 < 0$, then map F consists in two branches F_i, $i = 3, 4$ with $F_3(x)$ defined in $[0, \xi_4]$ (since it is $\xi_4 > 0$).

In order to investigate the dynamics of map F, we can state that it will be relevant the slopes of the three intermediate branches (since the first one and the last one are flat branches), and we can immediately see that these three branches are increasing/decreasing/increasing. In fact, we have:

$$\begin{aligned} F_1' &= s_U s_x = \frac{\overline{\gamma}_1}{2\underline{b}_1}\frac{\gamma_2}{2b_2} > 0 \\ F_2' &= -r s_x = -r\frac{\gamma_2}{2b_2} < 0 \\ F_3' &= s_L s_x = \frac{\underline{\gamma}_1}{2\overline{b}_1}\frac{\gamma_2}{2b_2} > 0 \end{aligned} \tag{21}$$

The kink points and their images/preimages are relevant tools in the study of the dynamics of *PWL* maps. In the notation used in [25,26], the kink points are critical points of rank -1 and their images and preimages are critical points of the related rank, so that $F(\xi_i)$ are critical points of rank 0, $F^2(\xi_i)$ are critical points of rank 1, and so on. Moreover, when $\xi_i > 0$, then the kink point ξ_2 is always a point of local maximum and ξ_3 always a point of local minimum. Thus, the values of the critical points $F(\xi_i)$ are relevant in what follows, and are given by:

$$
\begin{aligned}
F_1(\xi_2) &= F_2(\xi_2) = r(\alpha - s_x\xi_2) = ry_U = r\frac{\mu_U}{s_U + r} \\
F_2(\xi_3) &= F_3(\xi_3) = r(\alpha - s_x\xi_3) = ry_L = r\frac{\mu_L}{s_L + r} \\
F_3(\xi_4) &= F_4(\xi_4) = \mu_L
\end{aligned} \tag{22}
$$

as well as the possible fixed points related to each partition. Clearly, in the first partition it is $x_0^* = 0$ (superstable) for $\xi_1 > 0$, then let x_i^* be the fixed point of $F_i(x)$, we have

$$
\begin{aligned}
x_1^* &= \frac{2b_2 - \overline{\gamma}_1}{4\underline{b}_1 b_2 - \overline{\gamma}_1\gamma_2}, \quad \text{existing for } \max\{0, \xi_1\} \le x_1^* \le \xi_2 \\
x_2^* &= \frac{r}{2b_2 + r\gamma_2}, \quad \text{existing for } \max\{0, \xi_2\} \le x_2^* \le \xi_3 \\
x_3^* &= \frac{2b_2 - \underline{\gamma}_1}{4\overline{b}_1 b_2 - \underline{\gamma}_1\gamma_2}, \quad \text{existing for } \max\{0, \xi_3\} \le x_3^* \le \xi_4 \\
x_4^* &= \frac{1}{2\overline{b}_1}\ (= \mu_L), \quad \text{existing superstable for } x_4^* \ge \xi_4, \quad \text{that is, for } \gamma_2 \ge 2\overline{b}_1
\end{aligned} \tag{23}
$$

Notice that for $\xi_1 > 0$ (case (C5)), since the branch F_1 is increasing from ξ_1, the fixed point $x_1^* > 0$, when existing, is necessarily repelling (see Figure 1(a)). Differently, x_2^* and x_3^* (when existing) may be attracting or repelling (depending on $F_2' = -r\frac{\gamma_2}{2b_2} \gtrless -1$ and $F_3' = \frac{\underline{\gamma}_1}{2\overline{b}_1}\frac{\gamma_2}{2b_2} \lessgtr 1$), some examples are shown in Figure 2.

As it is used in piecewise smooth systems, it is convenient to represent a trajectory by symbols related to the partition to which a point belongs to. We use the symbol σ_i to denote that the point x of a trajectory belongs to the domain of the function F_i (that

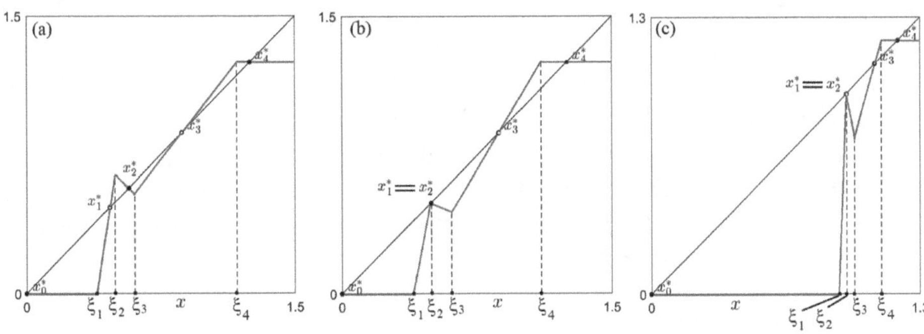

Figure 1. Examples of map $F(x)$ in the case (C5).

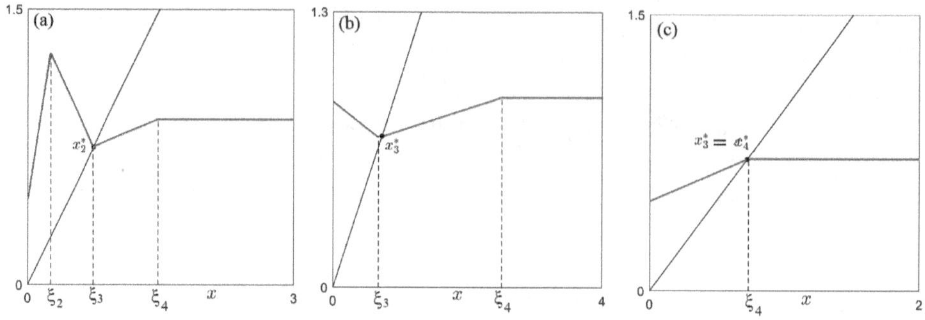

Figure 2. Examples of map $F(x)$ in the cases (C4), (C3) and (C2).

is, F_i is the function to be applied to x). This notation is particularly useful to represent a cycle. If F has an n-period cycle $\{x_1, x_2, \ldots, x_n\}$, we say that the cycle has symbolic sequence $\sigma_{i_1}\sigma_{i_2}, \ldots, \sigma_{i_n}$ where each symbol σ_{i_j} (with $i_j \in \{1, 2, 3, 4, 5\}, j \leq n$) is related to the partition, or branch, to which the periodic point belongs.

Moreover, x_2^* (when existing) is the only fixed point of F that may undergo a degenerate flip bifurcation. The result of this bifurcation may be determined by using the skew tent map as a border collision normal form (*BCNF*), see [5,36]. In fact, when the fixed point x_2^* is closer to ξ_2 then at the degenerate flip bifurcation ($F_2' = -1$) the invariant interval $[\xi_2, F(\xi_2)]$ is filled with cycles of period 2 and the endpoints undergo a border collision, so that after the bifurcation, for $F_2' < -1$, a 2-cycle with symbolic sequence $\sigma_1\sigma_2$ exists (attracting or repelling). Differently, when the fixed point x_2^* is closer to ξ_3 then at the degenerate flip bifurcation the invariant interval $[F(\xi_3), \xi_3]$ is filled with 2-cycles and the endpoints undergo a border collision, so that a 2-cycle with symbolic sequence $\sigma_2\sigma_3$ exists (attracting or repelling) after the bifurcation. Clearly, since map F is defined in many partitions, and considering map F as a function of the parameters which are varied, we may also have a 2-cycle with symbolic sequence $\sigma_1\sigma_3$, or $\sigma_2\sigma_4$, that are related to other different bifurcations (in the next sections we shall see several examples).

When $F_1' < 1$, and the parameters satisfy

$$BC_0: \quad \xi_1 = 0 \tag{24}$$

for map F we have that the transition through $\xi_1 = 0$ (i.e from $\xi_1 < 0$ to $\xi_1 > 0$ or *vice versa*) corresponds to the appearance/disappearance of the superstable fixed point $x_0^* = 0$.

When $F_1' > 1$ the condition $\xi_1 = 0$ ($F_1(0) = 0$) corresponds to a border collision, $x_0^* = x_1^*$ at which two fixed points x_0^* and x_1^* merge and appear/disappear ($\xi_1 > 0/\xi_1 < 0$) (superstable x_0^* and repelling x_1^*) and also it corresponds to the qualitative change of map F consisting in 5 branches for $\xi_1 > 0$ and 4 branches for $\xi_1 < 0$.

A similar transition, but only related to a qualitative change of map F, occurs at parameters leading to $\xi_2 = 0$ and $\xi_3 = 0$ at which we have the transition for map F consisting in 3/4 branches (3 for $\xi_2 < 0$ / 4 for $\xi_2 > 0$) and the transition in 2/3 branches (2 for $\xi_3 < 0$ / 3 for $\xi_3 > 0$), respectively. Explicitly we have:

$$\xi_1 = 0 \text{ for } \overline{\gamma}_1 - 2b_2 = 0$$

$$\xi_2 = 0 \text{ for } \overline{\gamma}_1 \left(\overline{b}_1 + \underline{b}_1\right) - 2\left(\underline{b}_1\underline{\gamma}_1 + b_2\left(\overline{b}_1 - \underline{b}_1\right)\right) = 0$$

$$\xi_3 = 0 \text{ for } \underline{\gamma}_1 \left(3\overline{b}_1 - \underline{b}_1\right) - 2\left(\overline{b}_1\overline{\gamma}_1 + b_2\left(\overline{b}_1 - \underline{b}_1\right)\right) = 0 \tag{25}$$

For the analysis of the dynamics of map F, besides the border collision of the fixed point x_0^* commented above, other relevant border collisions, related to the merging of two fixed points, may occur as a function of the parameters, when the parameters satisfy

$$BC_i: \quad F(\xi_{i+1}) = \xi_{i+1} \quad \text{for } i = 0, 1, 2, 3 \tag{26}$$

and the following conditions hold:

$$BC_3: \quad x_3^* = x_4^* = \xi_4, \quad \text{that is, } \mu_L = \frac{1}{\gamma_2}, \quad \text{for } \gamma_2 = 2\overline{b}_1 \tag{27}$$

$$BC_2: \quad x_2^* = x_3^* = \xi_3 > 0, \quad \text{that is, for } r\frac{\mu_L}{s_L + r} = \xi_3 \tag{28}$$

$$BC_1: \quad x_1^* = x_2^* = \xi_2 > 0, \quad \text{that is, for } r\frac{\mu_U}{s_U + r} = \xi_2 \tag{29}$$

$$BC_0: \quad x_0^* = x_1^* = 0, \quad \text{that is, } F_1(0) = 0 \quad \text{for } \overline{\gamma}_1 = 2b_2 \tag{30}$$

These border collisions (due to the merging of two fixed points and a kink point) may be related to different kinds of bifurcations (when we consider the dynamics of map F varying some parameters).

A *fold BCB* is always related to the *appearance/disappearance of a pair of fixed points* which are merging with a kink point at the bifurcation. Differently from what occurs in smooth systems, in *PW* smooth maps we can have a fold *BCB* related to the appearance/disappearance of a pair of repelling fixed points, as in the example shown in Figure 1(c) for $x_1^* = x_2^*$, or to the appearance/disappearance of a pair of fixed points one attracting and one repelling (as in smooth systems) as in the example shown in Figure 1(b) for $x_1^* = x_2^*$. Clearly, the occurrence of one or the other kind of fold *BCB* depends on the two slopes at the kink point merging with the fixed points. Two repelling fixed points occur when both the derivatives of the two branches involved at the kink point are of different sign and in modulus larger than 1, while the case attracting/repelling occurs when one derivative is larger than 1, and the other smaller than 1, in absolute value.

The border collision (*BC*, for short) related to the merging of two fixed points may also be related to the *persistence of a fixed point*, when an existing fixed point changes the branch of map F to which it belongs. In such a transition we may have a change in the stability of the fixed point, as in case (C4) of Figure 2(a) for $x_2^* = x_3^*$, or persistence of attracting fixed points, as in case (C3) of Figure 2(b) for $x_2^* = x_3^*$ or in case (C2) of Figure 2(c) for $x_3^* = x_4^*$, and also persistence of repelling fixed points.

Before the next section, where we comment some bifurcations occurring in our duopoly model, let us describe some relationship between the one-dimensional map F and the two-dimensional map T.

Let us consider the case (C5) shown in Figure 1(a) where map F has five fixed points x_i^*, $i = 0, 1, \ldots, 4$ whose stability is alternating attracting/repelling. The two repelling fixed points x_1^* and x_3^* separate the intervals which are the basins of attraction of the other stable fixed points. In this case the Cartesian product $\{x_0^*, \ldots, x_4^*\} \times \{R_2(x_0^*), \ldots, R_2(x_4^*)\}$ consists of 25 points, all periodic points of map T, and where the fixed points related to

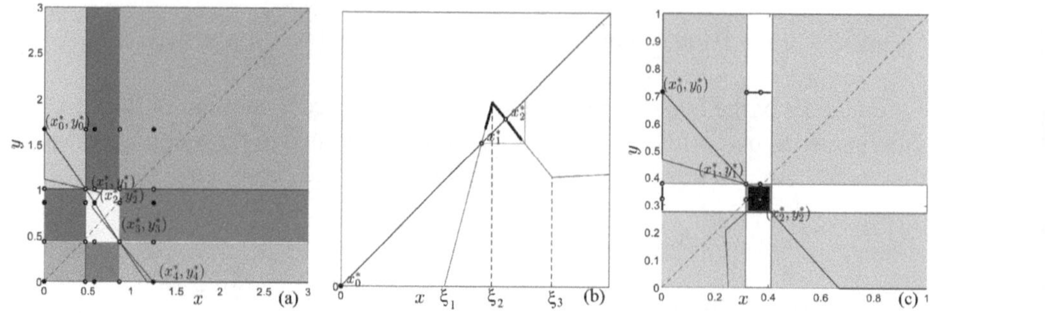

Figure 3. In (a) it is shown the phase plane (x, y) of map T in the case reported in Figure 1(a), at the following values of the parameters: $\overline{b}_1 = 0.4, \underline{b}_1 = 0.1, b_2 = 0.3, \gamma_2 = 0.85, \overline{\gamma}_1 = 0.9, \underline{\gamma}_1 = 0.7, a = 1$. The reaction curves $R_1(y)$ and $R_2(x)$ are shown, whose intersection points give the fixed points (x_i^*, y_i^*), $i = 0, 1, 2, 3, 4$. In (b) an example of map $F(x)$ in the case (C5) when the attracting sets are the origin and a chaotic segment, at the following values of the parameters: $\overline{b}_1 = 0.5, \underline{b}_1 = 0.1, b_2 = 0.3, \gamma_2 = 0.6, \overline{\gamma}_1 = 0.9, \underline{\gamma}_1 = 0.45, a = 1$. In (c) it is shown the phase plane (x, y) of map T in the case reported in (b). The reaction curves $R_1(y)$ and $R_2(x)$ are also shown.

those of map F are given by the points $(x_i^*, y_i^*) = (x_i^*, R_2(x_i^*))$, $i = 0, 1, \ldots, 4$. Here map T has 3 attracting fixed points $(x_i^*, R_2(x_i^*))$, for $i = 0, 2, 4$, and two repelling fixed points $(x_1^*, R_2(x_1^*))$ and $(x_3^*, R_2(x_3^*))$, then three attracting 2-cycles $\{(x_0^*, R_2(x_4^*)), (x_4^*, R_2(x_0^*))\}$, $\{(x_0^*, R_2(x_2^*)), (x_2^*, R_2(x_0^*))\}$ and $\{(x_2^*, R_2(x_4^*)), (x_4^*, R_2(x_2^*))\}$, five 2-cycles saddles and other two 2-cycles that are repelling nodes. All the unstable cycles belong to the basin boundary of the attracting sets, as it can be seen in Figure 3(a).

Another example of case (C5) is shown in Figure 3(b), the last branch of map F is not visible in the figure, but all the points larger than ξ_3 converge to the fixed point x_0^*. Besides this attractor, a second attracting set consists in the chaotic interval $[F(\xi_2), \xi_2]$ having basin of attraction (x_1^*, x_1^{*-1}) where $x_1^{*-1} = F_2^{-1}(x_1^*)$ is the rank -1 preimage of the repelling fixed point x_1^*. Figure 3(c) shows the attracting sets and related basins of attraction for map T. Besides the attracting fixed point $(x_0^*, R_2(x_0^*))$ and the chaotic attractor $[F(\xi_2), \xi_2] \times [R_2(F(\xi_2)), R_2(\xi_2)]$ map T has an attractor consisting in two cyclical chaotic intervals given by the segment on the y-axis with endpoints $(x_0^*, R_2(F(\xi_2)))$ and $(x_0^*, R_2(\xi_2))$, and its image on the line $y = R_2(x_0^*)$ with endpoints $(F(\xi_2), R_2(x_0^*))$ and $(\xi_2, R_2(x_0^*))$, whose basin of attraction is the white region in Figure 3(c).

3. Bifurcations occurring in map *F*

In this section we consider some of the relevant cases that can be observed in the dynamics of F (and thus having a correspondence in the dynamics of map T).

It can be immediately stated that the cases (C2) and (C3) are the simplest ones, since, when the parameters lead to these cases, map F has a unique globally attracting fixed point, so that also map T has a unique globally attracting fixed point.

Proposition 3.1: *Let the parameters satisfy cases* (C2) *or* (C3), *then map F (and thus map T) has a unique globally attracting fixed point.*

Proof of Proposition 3.1: Let us consider case (C2), then if the fixed point exists in the branch of F_3 the slope is smaller than 1, if it exists in the last branch then x_4^* is superstable. So, depending on the parameters, for $F(\xi_4) < \xi_4$ the fixed point x_3^* is globally attracting, while for $F(\xi_4) > \xi_4$ the fixed point x_4^* is globally attracting. For $F(\xi_4) = \xi_4$ a persistence border collision of attracting fixed points occurs, and the fixed point is globally attracting (an example is shown in Figure 2(c)). Consider now the parameters satisfying case (C3) and let us consider two subcases: $F(\xi_3) > \xi_3$ and $F(\xi_3) < \xi_3$. For $F(\xi_3) > \xi_3$ the dynamics can only be as in the previous case (C2), that is, map F can only have one fixed point, either x_3^* or x_4^*, and it is globally attracting (this is because the points of the interval $[0, \xi_3)$ are mapped in the absorbing interval that has lower boundary in the critical point $F(\xi_3)$). For $F(\xi_3) < \xi_3$, instead, it can be seen that the slope of the branch F_2 in this case must satisfy $F_2' = -rs_x > -1$, and that it must be $F(\xi_4) < \xi_4$. It follows that map F can only have the fixed point x_2^*, globally attracting. The condition $F(\xi_3) = \xi_3$ leads to a persistence border collision of attracting fixed points (an example is shown in Figure 2(b)), the fixed point crosses the kink point, on the diagonal, changing the branch of the map to which it belongs. ■

The dynamic behaviour of map F may become more complex in the other two cases, and in particular when the slope of the map in the partition of the branch F_1 is larger than 1 and that of F_2 smaller than -1. In order to see how the dynamics may change, we fix the following parameters:

$$\overline{b}_1 = 2, \quad \underline{b}_1 = 0.05, \quad b_2 = 0.15, \quad \underline{\gamma}_1 = 0.2 \tag{31}$$

and consider the bifurcations occurring as the parameters vary in the plane $(\overline{\gamma}_1, \gamma_2)$ with $\overline{\gamma}_1 > 0.2$ and $\gamma_2 > 0$.

A peculiarity of the chosen set of parameters is that the slope of the third branch is smaller than 1 in the range of interest, since it is

$$F_3' = s_L s_x = \frac{\underline{\gamma}_1}{2\overline{b}_1}\frac{\gamma_2}{2b_2} = \frac{\gamma_2}{6} \tag{32}$$

and the two-dimensional bifurcation diagram in Figure 4(a) shows the region of interest (in which it is $\gamma_2 < 6$, and thus $F_3' < 1$). We say that this is the region of interest because for $\gamma_2 > 4$ the parameter region is above the bifurcation curve BC_3, and we shall see that for map F this means that either x_4^* is globally attracting, or x_0^* and x_4^* are attracting, with the repelling fixed point x_1^* that separates the two basins of attraction. Thus, the region shown in Figure 4 is the relevant one for the chosen example. It is worth mentioning that other different sets of parameters may result in qualitatively different sequences for the dynamics of the map. However, the elements that we are evidencing here can be used in other cases, performing similar investigations.

The region in yellow in Figure 4(a) corresponds to values of the parameters at which map T has an attracting fixed point or two coexisting attracting fixed points, which are evidenced in Figure 4(b). Attracting fixed points are marked with a circle. The white region denotes the existence of chaotic dynamics, while regions in different colours denote the existence of attracting cycles of different periods.

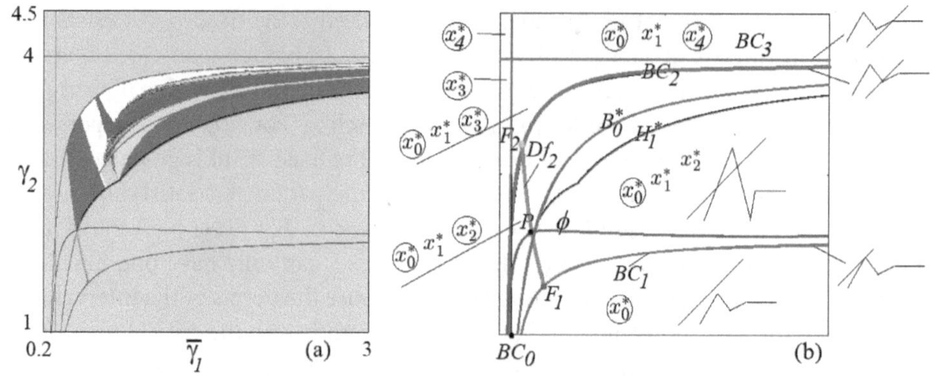

Figure 4. Two-dimensional bifurcation diagram in the parameter plane $(\overline{\gamma}_1, \gamma_2)$. The other parameters as in (31).

In Figure 4(b) we have reported (in light blue) the curves BC_i, for $i = 1, 2, 3$, corresponding to the *BCBs* associated with the merging of a pair of fixed points with a kink point, related to each *BC* we have also shown the shape of the map at such a transition.

Note that the value at which BC_0 takes place is $\overline{\gamma}_1 = 2b_2 = 0.3$ and it corresponds to the bifurcation leading to $\xi_1 > 0$, and when this holds then it is also $\xi_2 > 0$ and $\xi_3 > 0$. It follows that for $\overline{\gamma}_1 > 2b_2$ the one-dimensional map F has five branches in the positive quadrant, that is, it is of case (C5), and this implies that the attracting fixed point x_0^* always exists for parameters on the right side of the vertical line in Figure 4, denoted BC_0. For parameters satisfying $\overline{\gamma}_1 < 2b_2$ map F has only globally attracting fixed points (from Proposition 3.1), so that in the following we shall consider $\overline{\gamma}_1 > 2b_2$.

3.1. Basin of attraction of x_0^*

In the simplest cases the basin of attraction of x_0^* is an interval. However, when, besides the fixed point x_0^*, another attracting set exists, we may have disconnected basins of attraction.

Proposition 3.2: *Let the parameters of map F satisfy* (31). *If*

(i) $F(\xi_2) < \xi_2$, *then* x_0^* *is globally attracting;*
(ii) $F(\xi_2) > \xi_2$ *and* $F(\xi_3) > x_1^*$, *then the basin of attraction of the origin is* $B(x_0^*) = [0, x_1^*)$;
(iii) $F(\xi_2) > \xi_2$ *and* $F(\xi_3) < x_1^*$, *then the basin of attraction of the origin is* $B(x_0^*) = [0, x_1^*) \cup_{n \geq 0} F^{-n}[F(\xi_3), x_1^*)$, *and* x_1^* *is homoclinic for* $F^2(\xi_2) < x_1^*$.

Proof of Proposition 3.2: Let us prove the three cases:

(i) In this case the fixed point x_1^* does not exist, then the branch related to F_2 is decreasing, the one related to F_3 is increasing with slope smaller than 1, and the last branch is flat, so that the result follows.
(ii) Since the immediate basin of x_0^* is the interval $[0, x_1^*)$ and ξ_3 is a point of local minimum, it follows that when the parameters satisfy $F(\xi_3) > x_1^*$ then this immediate

basin does not have a preimage, so that it is also the total basin, $\mathcal{B}(x_0^*)$, and the points $x > x_1^*$ have a different asymptotic behaviour.

(iii) Differently, for $F(\xi_3) < x_1^*$ the basin of attraction of x_0^* includes other preimages of the interval $(0, x_1^*)$ which are all the preimages of any rank of the interval $[F(\xi_3), x_1^*)$. Thus, $\mathcal{B}(x_0^*)$ may be disconnected with a finite or infinite number of components, or connected but with a chaotic repellor on its boundary (an example is given below). These two cases depend on the existence of homoclinic orbits for the repelling fixed point x_1^*, and the homoclinic orbits of x_1^* depend on the second branch F_2, so that when besides the first inequalities in *(iii)* it is also $F^2(\xi_2) < x_1^*$, then x_1^* is homoclinic and the same occurs for the repelling cycles in the interval $[x_1^*, x_1^{*-1}]$ where $x_1^{*-1} = F_2^{-1}(x_1^*)$ is the rank -1 preimage of the repelling fixed point x_1^*, so that a chaotic repellor exists (see [15,17,18]).

■

Case *(i)* of Proposition 3.2 occurs in the parameter region in Figure 4(b) below the border collision curve BC_1.

Moreover, the curve of equation $F(\xi_3) = x_1^*$ (related to the immediate basin $\mathcal{B}(x_0^*)$), that is

$$B_0^* : \quad \frac{(\overline{\gamma}_1 - \underline{\gamma}_1)}{\underline{\gamma}_1(\overline{b}_1 - \underline{b}_1) + 2\overline{b}_1(\overline{\gamma}_1 - \underline{\gamma}_1)} = \frac{2b_2 - \overline{\gamma}_1}{4\underline{b}_1 b_2 - \overline{\gamma}_1\gamma_2} \tag{33}$$

is shown in brown colour in Figure 4(b). Below the curve B_0^* and above the border collision curve BC_1 (where $F(\xi_2) > \xi_2$ holds), case *(iii)* occurs, while above the curve B_0^* we have case *(ii)*.

The curve of equation $F^2(\xi_2) = x_1^*$ (related to the existence of homoclinic orbits of x_1^*), that is

$$H_1^* : \quad r\left(\alpha - s_x r y_U\right) = \frac{2b_2 - \overline{\gamma}_1}{4\underline{b}_1 b_2 - \overline{\gamma}_1\gamma_2} \tag{34}$$

is shown in gray in Figure 4(b). On the left side of the curve H_1^* it is $F^2(\xi_2) > x_1^*$ and the fixed point x_1^* is not homoclinic, a chaotic repellor does not exist, and in case *(iii)* the basin $\mathcal{B}(x_0^*)$ is disconnected, while on the right side of the curve H_1^* it is $F^2(\xi_2) < x_1^*$, the fixed point x_1^* is homoclinic, and a chaotic repellor exists, even if almost all the points have a trajectory convergent to x_0^*.

3.2. Border collision bifurcations of fixed points

We have already commented the transition occurring at the bifurcation BC_0, which leads to an attracting fixed point x_0^* existing for any $\overline{\gamma}_1 > 2b_2$. In order to comment the other border collisions involving the fixed points, let us consider a point Z_1 belonging to the set BC_1 at which it holds $F_2' = -1$ (red point in Figure 4(b)). It separates the points of the set BC_1 in two parts: below the point Z_1 it is $0 > F_2' > -1$, while above the point Z_1 it is $F_2' < -1$. Similarly, let us consider a point Z_2 belonging to the set BC_2 at which it holds $F_2' = -1$ (green point in Figure 4(b)). It separates the points of the set BC_2 in two parts:

below the point Z_2 it is $0 > F_2' > -1$, while above the point Z_2 it is $F_2' < -1$. We have the following results.

Proposition 3.3: *Let the parameters of map F satisfy* (31). *Then the BC occurring*

(j) *at BC_3 is a persistence BC of attracting fixed points;*
(jj) *at BC_2 is a persistence BC of attracting fixed points when it is crossed below the point Z_2 while it is a change of stability when it is crossed above the point Z_2;*
(jjj) *at BC_1 is a fold BCB, crossing BC_1 below the point Z_1, increasing γ_2, two fixed points exist, x_1^* repelling and x_2^* attracting, while crossing BC_1 above Z_1, increasing γ_2, two repelling fixed points, x_1^* and x_2^*, exist, and the trajectory of almost all the points converges to x_0^*.*

Proof of Proposition 3.3: Let us prove the three cases:

(j) The *BC* occurring at BC_3 has been already commented (in Proposition 3.1), is a persistence *BC* in which crossing the curve, increasing γ_2, the attracting fixed point x_3^* moves to the flat branch becoming x_4^* superstable.
(jj) The bifurcation occurring at BC_2 above the green point Z_2 is a change of stability (crossing the curve, increasing γ_2, the repelling fixed point x_2^* moves to the branch F_3 becoming the attracting fixed point x_3^*). Differently, the bifurcation occurring at BC_2 below the green point Z_2 is a persistence *BC*: crossing the curve increasing γ_2 the attracting fixed point x_2^* moves to the branch F_3 becoming the attracting fixed point x_3^*.
(jjj) The bifurcation occurring at BC_1 is a fold *BCB* whose effect differs if the curve is crossed (increasing γ_2) on the left side of the red point Z_1 evidenced on the curve or on its right side. In fact, crossing BC_1 below the point Z_1 (entering the region in which the slope of the branch F_2 satisfies $F_2' > -1$) the *BC* fold bifurcation leads to the appearance of two more fixed points, x_1^* (repelling) and x_2^* (attracting, whose immediate basin of attraction is the interval (x_1^*, x_1^{*-1})). Differently, crossing BC_1 above the point Z_1, the two slopes F_1' and F_2' are both in modulus larger than 1, so that the *BC* fold bifurcation leads to the appearance of two fixed points, x_1^* and x_2^*, both repelling, and in our case the following inequality holds:

$$\left(F_1' - 1\right)\left(F_2' + 1\right) > 1 \tag{35}$$

which implies, by using the skew tent map as the *BCNF* (see [5,36]), that a bounded absorbing interval is not created after the bifurcation. In fact, this part of the bifurcation curve BC_1 is below the curve B_0^* (so that case (iii) of Proposition 3.2 occurs for the basin of attraction of the origin) and it is on the right side of the curve H_1^*, so that it is $F^2(\xi_2) < x_1^*$, and the fixed point x_1^* appearing at the fold *BCB* is homoclinic, a chaotic repellor exists inside the interval $[x_1^*, x_1^{*-1}]$ and the trajectory of almost all the points converges to x_0^*.

■

3.3. Degenerate flip bifurcation of x_2^* (left boundary of the stability region of the 2-cycle C_{23})

In the region between the two curves BC_1 and BC_2, where besides x_0^* two more fixed points exist, x_1^* and x_2^*, we know that x_1^* is always repelling, while x_2^* may be attracting or repelling, depending on the slope of the branch F_2. The degenerate flip bifurcation at which x_2^* becomes unstable occurs when the parameters satisfy $F_2' = -1$, that is $r\frac{\gamma_2}{2b_2} = 1$, leading to

$$Df_2: \quad \overline{\gamma}_1 = \underline{\gamma}_1 + 2\frac{b_2}{\gamma_2}\left(\overline{b}_1 - \underline{b}_1\right) \tag{36}$$

and the curve is evidenced by the green arc in Figure 4(b). Clearly, this arc intersects the BC bifurcation curves BC_1 and BC_2 in the points defined above, Z_1 and Z_2, respectively (evidenced by the red point on BC_1 and the green point on BC_2 in Figure 4(b)).

The result of such a bifurcation when the flip curve is crossed from the left side to the right one, may be an attracting 2-cycle or a chaotic attractor, the separator between these two cases is a point P evidenced by a black point of the degenerate flip bifurcation curve at which the condition $F(\xi_2) = \xi_3$ holds.

The curve of equation $F(\xi_2) = \xi_3$, say ϕ, that is

$$\phi: \quad \frac{\left(\overline{\gamma}_1 - \underline{\gamma}_1\right)}{\underline{\gamma}_1(\overline{b}_1 - \underline{b}_1) + 2\overline{b}_1\left(\overline{\gamma}_1 - \underline{\gamma}_1\right)} = \frac{2b_2 - \overline{\gamma}_1}{4\underline{b}_1 b_2 - \overline{\gamma}_1\gamma_2} \tag{37}$$

is shown in red in Figure 4(b). Above (resp. below) the curve ϕ it is $F(\xi_2) > \xi_3$ (resp. $F(\xi_2) < \xi_3$). The point P is the intersection of the flip curve Df_2 with the curve ϕ.

We have the following results.

Proposition 3.4: *Let the parameters satisfying* (31) *belong to the region between the bifurcation curves* BC_1 *and* BC_2. *A degenerate flip bifurcation of* x_2^* *occurs when it holds* $F_2' = -1$. *For* $F_2' = -1$ *and* $F(\xi_2) > \xi_3$ *the degenerate flip bifurcation of* x_2^* *leads to an attracting 2-cycle with symbolic sequence* $\sigma_2\sigma_3$ *(say* C_{23}*). For* $F_2' = -1$ *and* $F(\xi_2) < \xi_3$ *the degenerate flip bifurcation of* x_2^* *leads to a repelling 2-cycle with symbolic sequence* $\sigma_1\sigma_2$ *and to attracting chaotic intervals.*

Proof of Proposition 3.4: Let P be the point at which the set satisfying $F(\xi_2) = \xi_3$ intersects the set $F_2' = -1$ (that is, the intersection point between the curves Df_2 and ϕ). Then, crossing the flip bifurcation curve **above point** P it is $F(\xi_2) > \xi_3$, which implies $F(\xi_3) < F^2(\xi_2)$, and at the flip bifurcation value the fixed point x_2^* is closer to the kink point ξ_3 (than to ξ_2), so that the two slopes involved at the bifurcation are F_2' and F_3' whose product is in modulus smaller than 1. At the bifurcation value the segment $[F(\xi_3), \xi_3]$ is filled with cycles of period two, and the 2-cycle at the end points of this interval, with periodic points $\{F(\xi_3), \xi_3\}$ will become the attracting 2-cycle after the flip bifurcation, with one periodic point at the other side of the kink point (and thus on the branch F_3) and the other periodic point on the branch F_2. So we have that when the fixed point x_2^* becomes unstable, an attracting 2-cycle appears, C_{23}, with periodic points on the two branches F_2

and F_3, having symbolic sequence $\sigma_2\sigma_3$ (pink region at the right side of the flip bifurcation curve in Figure 4(a)), and after the flip bifurcation the attracting 2-cycle belongs to an absorbing interval bounded by the local extrema, that is $[F(\xi_3), F(\xi_2)]$. Moreover, since this portion of the flip bifurcation curve is above the curve in brown colour B_0^*, it follows (from Proposition 3.2*(ii)*) that the unstable fixed point x_1^* separates the two basins of attraction of the attracting fixed point x_0^* and the attracting 2-cycle. That is, $\mathcal{B}(x_0^*) = [0, x_1^*)$ and $\mathcal{B}(\mathcal{C}_{23}) = (x_1^*, +\infty)$.

Differently, when the flip curve is crossed **below point** P, it is $F(\xi_2) < \xi_3$, then at the bifurcation value the fixed point x_2^* is closer to the kink point ξ_2, so that the two slopes involved at the bifurcation are F_1' and F_2' whose product is in modulus larger than 1. At the bifurcation value the segment $[\xi_2, F(\xi_2)]$ is filled with cycles of period two, and the 2-cycle at the end points of this interval, with periodic points $\{\xi_2, F(\xi_2)\}$ after the bifurcation will become the repelling 2-cycle, with one periodic point at the other side of the kink point (and thus on the branch F_1) and the other periodic point on the branch F_2, that is, with symbolic sequence $\sigma_1\sigma_2$. Moreover, since the flip curve is on the left side of the homoclinic curve H_1^*, the fixed point x_1^* is not homoclinic when the flip bifurcation takes place, so that an absorbing interval exists bounded by the images of the critical point ξ_2. That is, chaos occurs in $[F^2(\xi_2)F^4(\xi_2)] \cup [F^3(\xi_2), F(\xi_2)]$, when x_2^* is not homoclinic, and the basin of attraction is the interval $(x_1^*, x_1^{*-1})\backslash\{x_2^*\}$. The homoclinic bifurcation of x_2^* occurs at $F^3(\xi_2) = x_2^*$. When x_2^* is homoclinic then $[F^2(\xi_2), F(\xi_2)]$ is an attracting chaotic interval, and its basin of attraction is the interval (x_1^*, x_1^{*-1}). ■

It is worth noticing that in this second case the degenerate flip bifurcation leads to a one-piece chaotic interval, but only in a very small region of the parameters (see the white strip at the right side of the flip bifurcation curve in Figure 4(a), below the black point). This is due to the steep slope of the branch F_1 which is such that the homoclinic bifurcation of the fixed point x_1^* (that is H_1^* $(F^2(\xi_2) = x_1^*)$) occurs very close to the flip bifurcation of x_2^*. This can be better appreciated in Figure 5.

In Figure 5(a), before the flip bifurcation, the fixed point x_2^* is attracting with immediate basin of attraction given by the interval (x_1^*, x_1^{*-1}). At the flip bifurcation the segment $[\xi_2, F(\xi_2)]$ is filled with cycles of period 2, after the bifurcation map F becomes chaotic in this segment $[F^2(\xi_2), F(\xi_2)]$ with immediate basin of attraction (x_1^*, x_1^{*-1}) (see Figure 5(b)).

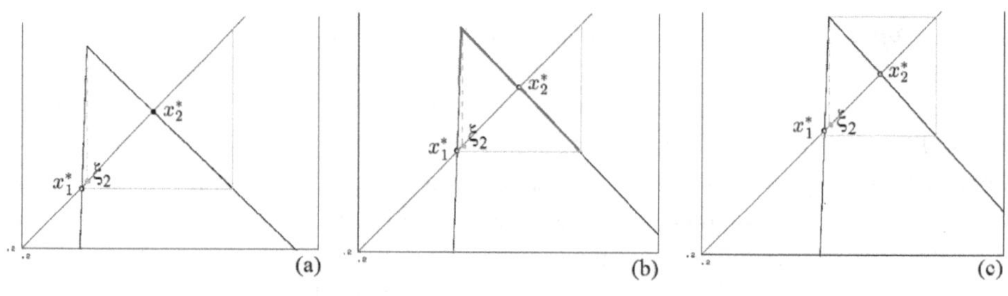

Figure 5. Map $F(x)$ at the parameters as in (31) and $\gamma_2 = 1.8$. In (a) $\overline{\gamma}_1 = 0.5$, the fixed point x_2^* is attracting. In (b) $\overline{\gamma}_1 = 0.53$, x_2^* is repelling and an attracting chaotic interval exists. In (c) $\overline{\gamma}_1 = 0.55$, it is $F^2(\xi_2) < x_1^*$ and almost all the trajectories converge to x_0^*.

When the parameters satisfy $x_1^* = F^2(\xi_2)$ (increasing $\overline{\gamma}_1$ or γ_2), the first homoclinic bifurcation of x_1^* occurs, after which it is $F^2(\xi_2) < x_1^*$ (see Figure 5(c)) and almost all the points of the interval $[x_1^*, x_1^{*-1}]$ converge to the origin x_0^* although a chaotic repellor exists and persists in this interval, a chaotic Cantor set of points that includes all the repelling cycles of the former chaotic interval and related preimages (see [15,17,18]).

In Figure 4(b) we have evidenced the curves defined above, B_0^* $(F(\xi_3) = x_1^*)$, H_1^* $(F^2(\xi_2) = x_1^*)$, Df_2 $(F_2' = -1)$ and ϕ $(F(\xi_2) = \xi_3)$, and all seem crossing in the same point, but this is not the case. Clearly, three of them are crossing in one point, because if $F(\xi_2) = \xi_3$ then it is also $F^2(\xi_2) = F(\xi_3)$ so that for $F(\xi_3) = x_1^*$ we also have $F^2(\xi_2) = x_1^*$, which means that the three curves B_0^*, H_1^* and ϕ must intersect in one point, but this is not related to the degenerate flip bifurcation Df_2. Let us show in Figure 6(a) an enlargement evidencing the behaviours of these curves.

In Figure 6(a) we have marked four points (A, B, C and D) and in Figure 7 it is shown the result of the dynamics of map F at the first three of these points. In A (before the flip bifurcation) the fixed point x_2^* is attracting, in B (after the flip bifurcation) the fixed point x_2^* is repelling and two cyclical attracting chaotic intervals exist. In these two cases the basin

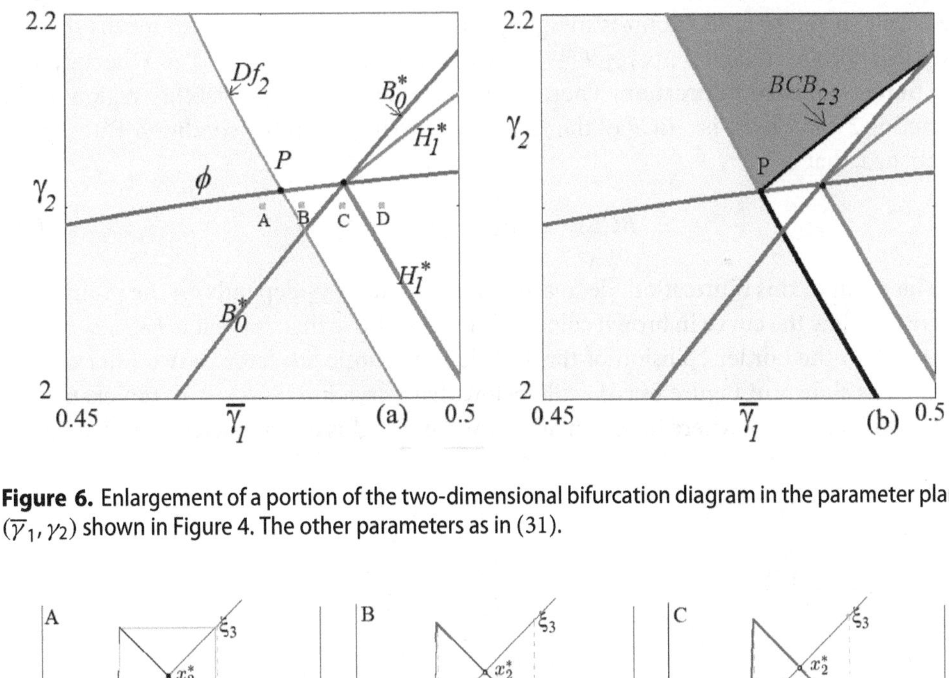

Figure 6. Enlargement of a portion of the two-dimensional bifurcation diagram in the parameter plane $(\overline{\gamma}_1, \gamma_2)$ shown in Figure 4. The other parameters as in (31).

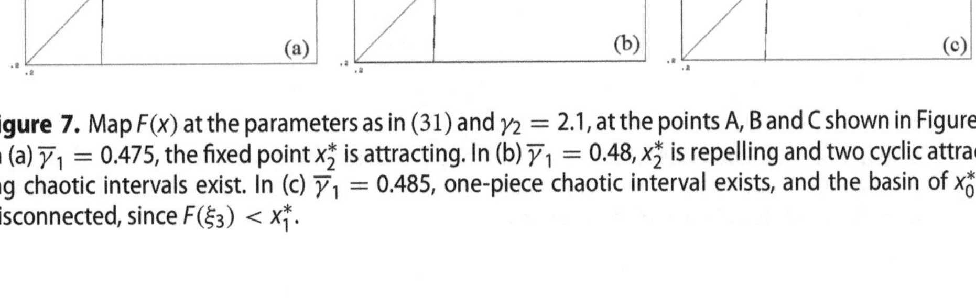

Figure 7. Map $F(x)$ at the parameters as in (31) and $\gamma_2 = 2.1$, at the points A, B and C shown in Figure 6. In (a) $\overline{\gamma}_1 = 0.475$, the fixed point x_2^* is attracting. In (b) $\overline{\gamma}_1 = 0.48$, x_2^* is repelling and two cyclic attracting chaotic intervals exist. In (c) $\overline{\gamma}_1 = 0.485$, one-piece chaotic interval exists, and the basin of x_0^* is disconnected, since $F(\xi_3) < x_1^*$.

of attraction of the origin is $[0, x_1^*)$ while that of the other attractor is the interval $(x_1^*, +\infty)$, because we are above the curve B_0^* ($F(\xi_3) = x_1^*$) so that it is $F(\xi_3) > x_1^*$. In C the attracting chaotic set is one interval (it is after the homoclinic bifurcation of x_2^*), moreover, now the basin of attraction of the origin is disconnected (since below B_0^* it is $F(\xi_3) < x_1^*$). In point D it is $F^2(\xi_2) < x_1^*$ so that almost all the points converge to x_0^* even if a chaotic repellor (which includes the homoclinic fixed point x_1^*) exists in the interval $[x_1^*, x_1^{*-1}]$.

3.4. Lower boundary of the stability region of the 2-cycle C_{23} (border collision)

In order to evidence in Figure 6(a) the boundary of the pink region associated with the attracting 2-cycle appearing after the degenerate flip bifurcation of x_2^* we need to draw one more curve, that is evidenced in black in Figure 6(b), that corresponds to the equation $F^2(\xi_2) = \xi_2$. It is drawn in black also in Figure 4(b) and it is very close to the gray curve H_1^* ($F^2(\xi_2) = x_1^*$) so they cannot be well distinguished, but in the enlarged part of Figure 6(b) this is clearly visible.

Considering a parameter point inside the stability region of C_{23}, when the parameter point approaches the lower boundary, the periodic point of the 2-cycle belonging to the branch F_2 approaches the kink point ξ_2 so that at the lower boundary of the stability region we have that the periodic point is merging with the kink point ξ_2, which means that the 2-cycle is with periodic points $\{\xi_2, F(\xi_2)\}$ (or, equivalently, $F^2(\xi_2) = \xi_2$) and the 2-cycle is at a border collision bifurcation. Thus, the lower boundary of the stability region of the attracting 2-cycle $\sigma_2\sigma_3$ is a *BCB* of the 2-cycle, occurring when the condition $F^2(\xi_2) = \xi_2$ is satisfied, that is

$$BCB_{23}: \quad F_3 \circ F_2 (\xi_2) = \xi_2 \tag{38}$$

The result of this bifurcation, decreasing the parameter γ_2 depends on the point being above or below the curve in brown colour B_0^*. In fact, above that curve it is $F(\xi_3) > x_1^*$ and the result of the border collision of the 2-cycle is a chaotic attractor in two intervals (an example is shown in Figure 8(a)), while below that curve it is $F(\xi_3) < x_1^*$ (an example is shown in Figure 8(b), where the basin of the fixed point x_0^* is disconnected), the fixed point

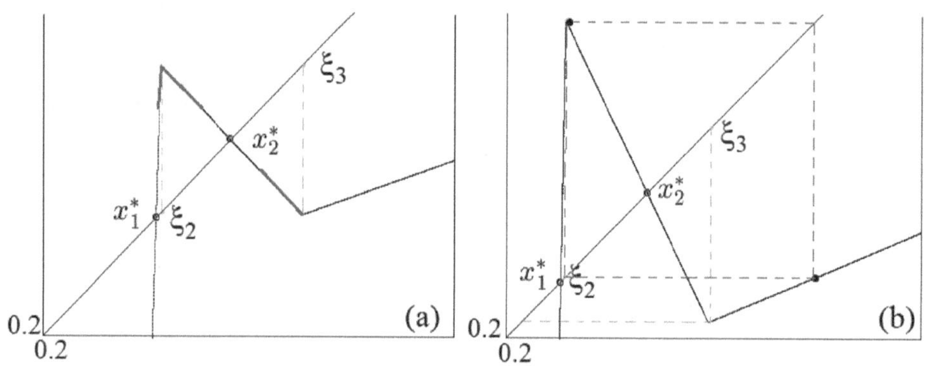

Figure 8. Map $F(x)$ at the parameters as in (31). In (a) $\overline{\gamma}_1 = 0.48$ and $\gamma_2 = 2.11$, the fixed point x_2^* is repelling and two cyclic attracting chaotic intervals exist. In (b) $\overline{\gamma}_1 = 0.675$ and $\gamma_2 = 2.5$, x_2^* is repelling and a 2-cycle is almost at its *BCB* with the kink point ξ_2.

x_1^* is homoclinic, and the result of the border collision of the 2-cycle is convergence to x_0^* for almost all the points, although a chaotic repellor exists in the interval $[x_1^*, x_1^{*-1}]$.

We have so proved the following proposition.

Proposition 3.5: *Let the parameters satisfying* (31) *belong to the region between the bifurcation curves BC_1 and BC_2. The lower boundary of the existence region of the attracting cycle C_{23} is a border collision bifurcation curve at which $F_3 \circ F_2(\xi_2) = \xi_2$, denoted BCB_{23}. Moreover, let Q be the intersection point between the curves B_0^* and BCB_{23}, then the result of this border collision bifurcation is a chaotic attractor if the crossing occurs below Q. If the crossing occurs above Q the result of the border collision bifurcation is a chaotic repellor, and the origin attracts almost all the points in* $(0, +\infty)$.

The curve of equation $F^2(\xi_2) = \xi_2$ is also the lower boundary of the existence region of a superstable cycle of symbolic sequence $\sigma_2\sigma_3$ but with a different meaning, as it will be commented below.

3.5. Right boundary of the stability region of the 2-cycle C_{23} (degenerate flip)

The borders of the region of the attracting 2-cycle with symbolic sequence $\sigma_2\sigma_3$ consist, besides of the curves already commented on the left side (degenerate flip bifurcation), the upper side (the BC bifurcation BC_2) and the lower boundary (the border collision bifurcation BCB_{23}), of a boundary on the right side, that we comment now. These borders can be seen in Figure 4 and also better observable in the enlargement of Figure 9.

The bifurcation curve on the right side of the stability region of the 2-cycle $\sigma_2\sigma_3$ is related to its degenerate flip bifurcation, which occurs when the parameters satisfy $F_2'F_3' = -1$, that is $r(\frac{\gamma_2}{2b_2})^2\frac{\underline{\gamma}_1}{2\overline{b}_1} = 1$, leading to

$$Df_{23}: \quad \overline{\gamma}_1 = \underline{\gamma}_1 + 8\frac{\overline{b}_1\,(b_2)^2}{\underline{\gamma}_1\,(\gamma_2)^2}\left(\overline{b}_1 - \underline{b}_1\right) \tag{39}$$

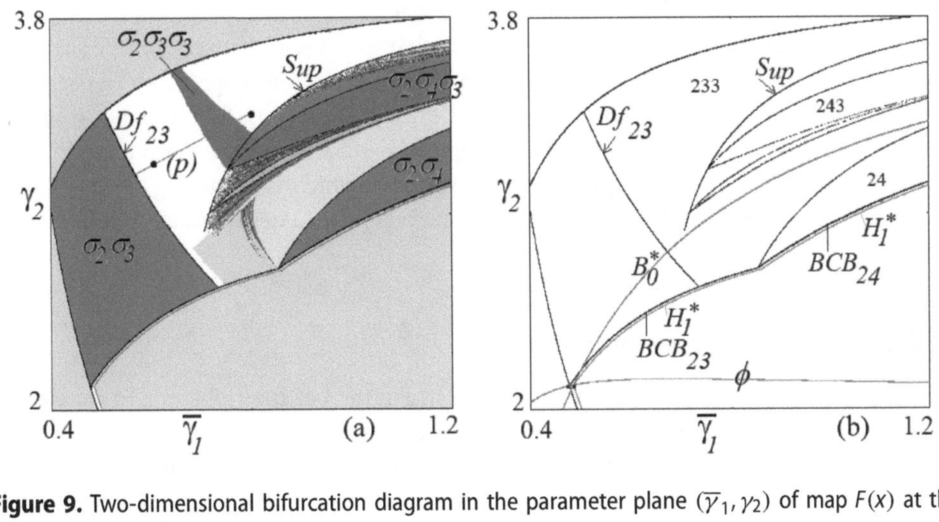

Figure 9. Two-dimensional bifurcation diagram in the parameter plane $(\overline{\gamma}_1, \gamma_2)$ of map $F(x)$ at the parameters as in (31).

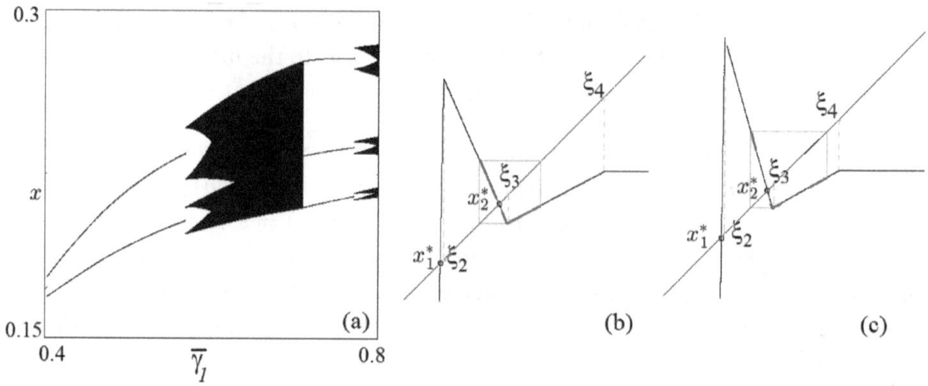

Figure 10. In (a), one-dimensional bifurcation diagram along the path (p) of equation $\gamma_2 = 1.2\overline{\gamma}_1 + 2.4$ shown in Figure 9(a). In (b), map $F(x)$ at $(\overline{\gamma}_1, \gamma_2) = (0.61, 3.132)$ where there are 2-cyclic chaotic segments. In (c), map $F(x)$ at $(\overline{\gamma}_1, \gamma_2) = (0.8, 3.36)$ where there are 3-cyclic chaotic segments.

This bifurcation necessarily results in chaotic intervals, and 2^n intervals may occur for $n > 1$. The exact number can be detected by using the two slopes F_2' and F_3' and the skew tent map as the *BCNF* (see [5,36]).

In our example we have observed the transition to 4-cyclical chaotic intervals.

In Figure 10(a) it is reported the one-dimensional bifurcation diagram along the path (p) of equation $\gamma_2 = 1.2\overline{\gamma}_1 + 2.4$ shown in Figure 9(a) as a segment in azure colour. In Figure 10(b,c) it is also shown the shape of the map at the two points that are highlighted on the path (p): at $(\overline{\gamma}_1, \gamma_2) = (0.61, 3.132)$ where there are two chaotic intervals, and at $(\overline{\gamma}_1, \gamma_2) = (0.8, 3.36)$ where there are three chaotic intervals, which occur after the crossing of the stability region of an attracting 3-cycle with symbolic sequence $\sigma_2\sigma_3\sigma_3$.

For the parameters in this region of the $(\overline{\gamma}_1, \gamma_2)$ plane, although the local maximum via the kink point ξ_2 ends up in the flat blanch (as we can see in Figure 10(b,c)), the local minimum with the kink point ξ_3 leads to an absorbing interval that does not include the branch F_4, and inside which map F is unimodal, defined via the only branches F_2 and F_3. It follows that we can indeed expect a dynamic scenario similar to the one occurring in the skew tent map. This can be observed in the one-dimensional bifurcation diagram in Figure 10(a). That is, the degenerate flip bifurcation of the 2-cycle $\sigma_2\sigma_3$ is followed by 4-cyclical chaotic intervals, which become 2 chaotic intervals and then a one-piece chaotic interval. Then a fold *BCB* leads to a pair of 3-cycles, one of which is attracting, with symbolic sequence $\sigma_2\sigma_3\sigma_3$ that also leads to 6 chaotic intervals (after a degenerate flip bifurcation) and so on.

4. Two different parameter regions

For parameters belonging to the region shown in Figure 4 at the right side of the degenerate flip bifurcation curve of x_2^* and between the two *BCB* curves BC_1 and BC_2, among the fixed points the only attracting one is the origin x_0^*. However, in that region we have other attracting sets (cycles or chaotic intervals), whose existence is related to two different dynamic occurrences:

- On one side cycles and chaotic sets, that are due to the dynamic behaviour of the map inside an absorbing interval after the degenerate flip bifurcation of x_2^* when it leads to an attracting 2-cycle with symbolic sequence $\sigma_2\sigma_3$, and an absorbing interval $[F(\xi_3), F^2(\xi_3)]$ exists, and as long as it involves only the branches of F_2 and F_3 we can take advantage of the knowledge of the skew tent map, and the transition to chaotic intervals is to be expected, as it is shown in Figure 10;
- On the other side, stable cycles may exist when the absorbing interval of map F includes a part of the branch F_4, since in that case superstable cycles may appear. They are called superstable because the related eigenvalue is zero.

This is indeed what we can observe and describe.

4.1. Upper boundary of the superstable cycles

The two kink points ξ_2 and ξ_3 associated with the local maximum and local minimum, respectively, are the relevant points to determine the existence of the smallest absorbing interval, inside which we can restrict the analysis, to detect the bifurcations occurring in map F.

Notice that when the second iterate of the local minimum $F^2(\xi_3)$ is smaller than the kink point ξ_4 the map has the absorbing interval not involving the flat branch F_4, and in this case, as commented above, map F behaves as a skew tent map in the interval $[F(\xi_3), F^2(\xi_3)]$. The occurrence of the flat branch inside the absorbing interval of the map can take place when $F^2(\xi_3) > \xi_4$. It follows that a parameter set separating the two different kinds of dynamics is obtained for parameters that satisfy the equation $F^2(\xi_3) = \xi_4$, that is:

$$S_{up}: \quad 2\overline{b}_1 r\left(\alpha - rs_x\frac{\mu_L}{s_L + r}\right) = 1 \tag{40}$$

We know that when the parameters are below the bifurcation BC_3 the fixed point x_4^* is not involved in the dynamics. However, the flat branch inside the absorbing interval of map F, occurring after the bifurcation labelled S_{up}, may lead to the existence of superstable cycles. This can be better observed in the enlarged part of the two-dimensional parameter plane $(\overline{\gamma}_1, \gamma_2)$ shown in Figure 9, where the curve labelled S_{up} is evidenced, and below it we can see that periodicity regions of superstable cycles exist.

4.2. Lower boundary of the superstable cycles

The curve S_{up} here defined can be considered an upper boundary of the region with superstable cycles (and it is located above the curve B_0^*, so that the basin of the origin is only the interval $[0, x_1^*)$). However, a lower boundary also exists, related to a different bifurcation. In fact, let us consider the parameters in the region in which the origin x_0^* attracts almost all the points (below the curve B_0^*), an example is shown in Figure 11(a), where we can see that $F(\xi_4)$ is smaller than the repelling fixed point x_1^* but very close to it, and increasing γ_2 we have the bifurcation occurring when it holds $F(\xi_4) = x_1^*$. This bifurcation creates a new attracting set (it is also the BCB of a cycle). In fact, for $F(\xi_4) > x_1^*$ an attracting 2-cycle of symbolic sequence $\sigma_1\sigma_4$ exists (an example is shown in Figure 11(b)). Notice that the basin of the origin is still disconnected (since $F(\xi_3) < x_1^*$), and $F(\xi_4)$ is smaller but very close to

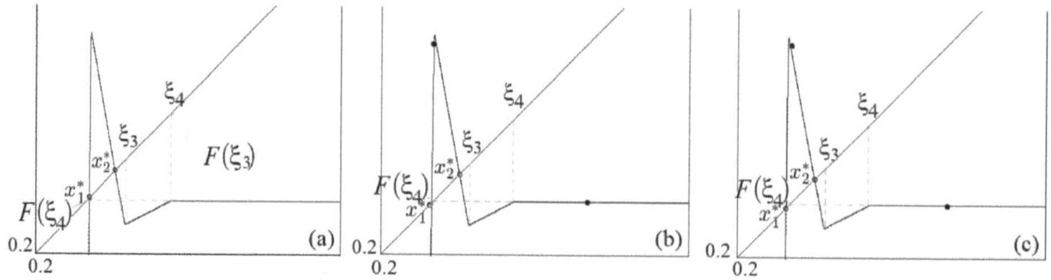

Figure 11. Map $F(x)$ at the parameters as in (31) and $\overline{\gamma}_1 = 1.2$. In (a) $\gamma_2 = 3$, almost all the trajectories converge to x_0^*. In (b) $\gamma_2 = 3.04$, an attracting 2-cycle $\sigma_1\sigma_4$, the basin of x_0^* is disconnected. In (c) $\gamma_2 = 3.06$, an attracting 2-cycle $\sigma_2\sigma_4$, the basin of x_0^* is disconnected.

the kink point ξ_2. When the parameters satisfy $F(\xi_4) = \xi_2$ the 2-cycle of symbolic sequence $\sigma_1\sigma_4$ undergoes a border collision bifurcation, which is of persistence type, leading to an attracting 2-cycle of symbolic sequence $\sigma_2\sigma_4$ (an example is shown in Figure 11(c)).

Noticing that in our case the local maximum ends in the flat branch of map F, since it is $F(\xi_2) > \xi_4$, we have that the bifurcation occurring for $F(\xi_4) = x_1^*$ also corresponds to the bifurcation already defined as H_1^* $(F^2(\xi_2) = x_1^*)$ (since $F^2(\xi_2) = F(\xi_4)$) and that the border collision occurring at $F(\xi_4) = \xi_2$ also corresponds to the bifurcation already defined as $F^2(\xi_2) = \xi_2$ but now it is defined with different branches of map F since it is

$$BCB_{24}: \quad F_4 \circ F_2 (\xi_2) = \xi_2 \tag{41}$$

These are the two curves, very close to each other, in gray colour (H_1^*) and in black (BCB_{24}) already shown in Figure 4(b) and better visible in the enlargement of Figure 9. Moreover, increasing γ_2 the flat branch of map F is still involved, so that this parameter region determines a region of existence of superstable cycles.

4.3. *Superstable cycles*

In the previous subsection we have commented the lower boundary of the parameter region that is characterized by superstable cycles, leading to attracting 2-cycles, at first with symbolic sequence $\sigma_1\sigma_4$ and soon after with symbolic sequence $\sigma_2\sigma_4$. Recall that this kind of cycles, having eigenvalue zero, cannot undergo flip bifurcations, so their changes are due to different bifurcations. Clearly a cycle may undergo a border collision, when one of the periodic point of a cycle merges with one kink point of the map.

As an example, in Figure 12 we show the 2-cycle $\sigma_2\sigma_4$ that undergoes a BC with the kink point ξ_4 leading to a superstable 4-cycle with symbolic sequence $\sigma_2\sigma_4\sigma_2\sigma_3$ that, on its turn, undergoes a BC with the kink point ξ_3 (of persistence type) leading to a superstable 4-cycle with symbolic sequence $\sigma_2\sigma_4\sigma_2\sigma_2$. But then we have a change into a superstable cycle of period 8, without any border collision of the periodic points of the 4-cycle with any one of the kink points of map F. Indeed, this is a different mechanism leading to a change in the period of superstable cycles. As we can see from the graph of map F, the value of the map in the kink point ξ_4, that is $F(\xi_4)$, is close to the repelling fixed point x_2^*. This means that an infinite number of preimages of this point in the branch F_2, given by $F_2^{-n}(F(\xi_4))$ for any $n \geq 1$ exist. It follows that when the periodic point of a superstable cycle, that is the closest

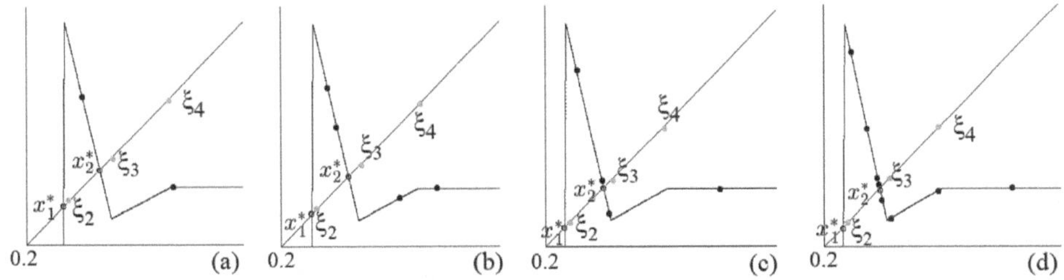

Figure 12. Map $F(x)$ at the parameters as in (31) and $\overline{\gamma}_1 = 1$. In (a) $\gamma_2 = 3.035$, a superstable 2-cycle $\sigma_2\sigma_4$ is close to its *BCB*. In (b) $\gamma_2 = 3.1$, a superstable 4-cycle $\sigma_2\sigma_4\sigma_2\sigma_3$ exists. In (c) $\gamma_2 = 3.25$, a superstable 4-cycle $\sigma_2\sigma_4\sigma_2\sigma_2$ exists. In (d) $\gamma_2 = 3.259$, a superstable 8-cycle exists.

to x_2^*, merges with a preimage of $F(\xi_4)$ in the neighbourhood of x_2^*, then the period of the cycle increases when the merging occurs below x_2^*, while it decreases when the merging occurs above x_2^* (this mechanism has been explained in [34]).

This kind of bifurcations leads to a particular one-dimensional bifurcation diagram, characteristic for superstable cycles, as shown in Figure 13(a) and its enlargement in Figure 13(b). Notice that for this kind of maps, all the points that are mapped in the flat branch of map F have the same ω-limit set, that of the critical point $F(\xi_4)$.

Moreover, we also have some particular attractors, that occur when the value $F(\xi_4)$ merges with a periodic point of some repelling cycle. In our case there exists a repelling 2-cycle with symbolic sequence $\sigma_2\sigma_3$ (due to the skew tent map shape close to the kink point ξ_3) and also the repelling fixed point x_2^*. It is plain that when it occurs the merging of $F(\xi_4)$ with one repelling cycle, at such a particular value that cycle becomes a Milnor attractor (see [24]), since it is locally repelling but attracts points as a superstable cycle, that is, its stable set becomes of positive measure, including the flat branch and its preimages. This leads to the particular cobweb structure observed in Figure 13 occurring in continuous or discontinuous systems having a flat branch (see, e.g. [4,35,37,38] to cite a few).

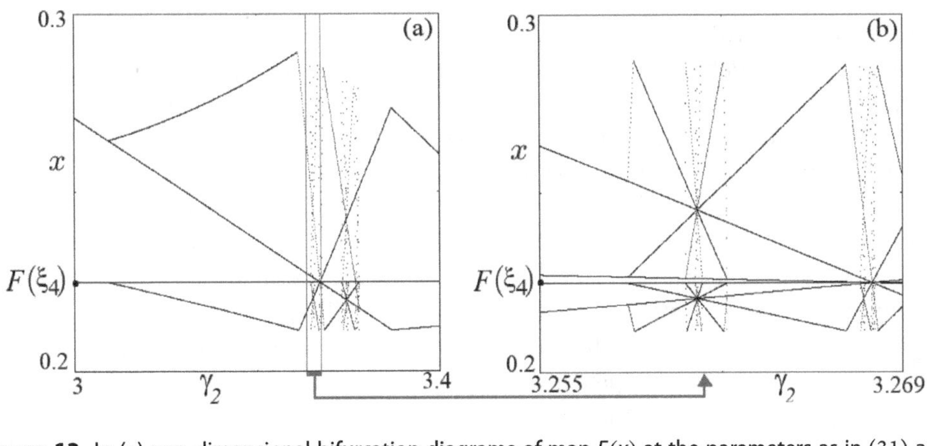

Figure 13. In (a) one-dimensional bifurcation diagrams of map $F(x)$ at the parameters as in (31) and $\overline{\gamma}_1 = 1$, showing superstable cycles. In (a) $\gamma_2 \in (3, .3.4)$. In (b) an enlargement for $\gamma_2 \in (3.255, .3.269)$.

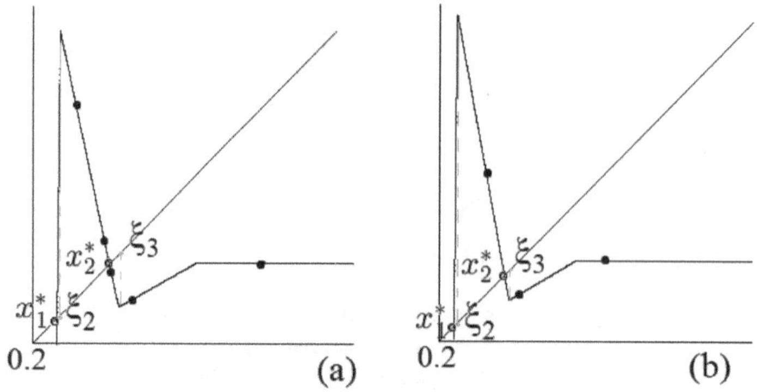

Figure 14. Map $F(x)$ at the parameters as in (31). In (a) $\overline{\gamma}_1 = 1$ and $\gamma_2 = 3.28$, a superstable 5-cycle exists. In (b) $\overline{\gamma}_1 = 1.1$ and $\gamma_2 = 3.5$, a superstable 3-cycle exists.

Notice that the *BCB* of the 2-cycle shown in Figure 12(a) takes place when the parameters satisfy

$$F^2(\xi_4) = \xi_4 \tag{42}$$

see the black curve bounding the upper part of the pink region related to the 2-cycle $\sigma_2\sigma_4$ in Figure 9.

In a similar way we can detect other border collision bifurcation curves for the appearance/disappearance of superstable cycles. Two examples are shown in Figure 9, having the equations

$$F^3(\xi_4) = \xi_4\,,\ F^5(\xi_4) = \xi_4 \tag{43}$$

bounding the existence region of a superstable cycle of period 3 and 5 (two examples are shown in Figure 14).

5. Conclusions

In this work we have considered a duopoly model characterized by a continuous two-dimensional map defined by linear functions having the second iterate with separate variables. This implies that the dynamic behaviours of the two-dimensional system can be obtained from the dynamics of a one-dimensional composite function, that results in a continuous piecewise linear map $F(x)$ defined for $x \geq 0$, and with a number of kink points that depends on the values of the parameters, and can be defined in two partitions or more, up to five partitions. We have described some generic properties of this type of systems, and recalled some properties in the Appendix. To describe the rich dynamic behaviour that the system can have, in Section 3 we have considered a specific case, fixing all the parameters except two, so to better describe the bifurcations and the peculiarities of the system by using a two-dimensional parameter plane. This has allowed a quite complete description of the different dynamics, and the characterization of the border collision bifurcations that may occur, due to the many kink points and the different fixed points that may exist, as well as of the homoclinic bifurcations of the fixed points belonging to the branches F_1 and F_2. In

Section 3 we have described what occurs in the four boundaries of the existence region of an attracting 2-cycle with symbolic sequence $\sigma_2\sigma_3$. In Section 4 we have described how the resulting dynamics depends on the existence of an absorbing interval that includes a flat branch of the map. This leads to particular dynamics, associated with superstable cycles and repelling cycles that are Milnor attractors, and we have shown the existence of two different kinds of bifurcations that change the periods of the superstable cycles.

Clearly, the possible outcomes in our duopoly model are really many, and we have considered a particular case. However, also considering different parameter settings, the possible results in the dynamics of the map are similar to those evidenced in our example, and the analytical tools to use for the analysis of the dynamics are the same. Of course the bifurcation sequence may be different, allowing for different results. For example in our system we can have bistability but only with the attracting fixed point in the origin and another attracting set (regular or chaotic). It is possible to have parameters of the system that also lead to bistability among two different attracting sets far from the origin, due to the bimodal shape of the map inside the two flat branches. This is also quite interesting for the applied duopoly model, and it is left as future research work.

Acknowledgments

The work of both authors has been done within the activities of the research project on 'Dynamic models in economics and finance' of the Department DESP of the University of Urbino.

Disclosure statement

No potential conflict of interest was reported by the author(s).

Funding

Davide Radi acknowledges the financial support from the Grantová Agentura České Republiky (Czech Scientific Foundation) under project 23-06282S and from the VŠB–TU Ostrava under the SGS research project SP2023/19. This work was supported by Grantová Agentura České Republiky.

References

[1] A. Agliari, *Coexisting cycles in a class of 3-D discrete maps*, in *ESAIM: Proceedings ECIT 2010*, Vol. 17, EDP Sciences, Les Ulis Cedex A, 2012. pp. 170–179

[2] A. Agliari, D. Fournier-Prunaret, and A.K. Taha, *Periodic orbits and their bifurcations in 3D maps with a separate third iterate*, in *Global Analysis of Dynamic Models in Economics and Finance*, G.-I. Bischi, C. Chiarella and I. Sushko, eds., Springer, Berlin Heidelberg, 2013, pp. 397–427

[3] M. Anufriev, D. Radi, and F. Tramontana, *Some reflections on past and future of nonlinear dynamics in economics and finance*, Decis. Econ. Finance 41 (2018), pp. 91–118.

[4] V. Avrutin, B. Futter, L. Gardini, and M. Schanz, *Unstable orbits and Milnor attractors in the discontinuous flat top tent map*, Eur. Ser. Appl. Ind. Math. (ESAIM) 36 (2012), pp. 126–158.

[5] V. Avrutin, L. Gardini, I. Sushko, and F. Tramontana, *Continuous and Discontinuous Piecewise Smooth One-Dimensional Maps: Invariant Sets and Bifurcation Structures*, World Scientific, Singapore, 2019.

[6] S. Banerjee, M.S. Karthik, G. Yuan, and J.A. Yorke, *Bifurcations in one-dimensional piecewise smooth maps – theory and applications in switching circuits*, IEEE Trans. Circ. Syst. I 47 (2000), pp. 389–394.

[7] F. Bignami and A. Agliari, *Chaotic dynamics in a three-dimensional map with separate third iterate: The case of Cournot duopoly with delayed expectations*, Chaos Solitons Fractals 110 (2018), pp. 216–225.

[8] G.I. Bischi and M. Kopel, *Equilibrium selection in a nonlinear duopoly game with adaptive expectations*, J. Econ. Behav. Organ. 46(1) (2001), pp. 73–100.

[9] G.I. Bischi and F. Lamantia, *Routes to complexity induced by constraints in Cournot oligopoly games with linear reaction functions*, Stud. Nonlinear Dyn. Econ. 16 (2012), pp. 1–28.

[10] G.I. Bischi, C. Mammana, and L. Gardini, *Multistability and cyclic attractors in duopoly games*, Chaos Solitons Fractals 11 (2000), pp. 543–564.

[11] G.I. Bischi, C. Chiarella, M. Kopel, and F. Szidarovszky, *Nonlinear Oligopolies: Stability and Bifurcations*, Springer-Verlag, Berlin, 2010.

[12] G.P. Crespi, D. Radi, and M. Rocca, *Robust games: Theory and application to a Cournot duopoly model*, Decis. Econ. Finance 40(1–2) (2017), pp. 177–198.

[13] G.P. Crespi, D. Radi, and M. Rocca, *Insights on the Theory of Robust Games*, arXiv preprint, arXiv:2002.00225. (2020)

[14] R.A. Dana and L. Montrucchio, *Dynamic complexity in duopoly games*, J. Econ. Theory 40 (1986), pp. 40–56.

[15] R. Devaney, *An Introduction to Chaotic Dynamical Systems*, Menlo Park, California: The Benjamin, Cummings Publishing Corporation, 1986.

[16] M. di Bernardo, C.J. Budd, A.R. Champneys, and P. Kowalczyk, *Piecewise-smooth Dynamical Systems: Theory and Applications*, Applied Mathematical Sciences; 163, Springer, London, 2008.

[17] L. Gardini, *Homoclinic bifurcations in n-dimensional endomorphisms, due to expanding periodic points*, Nonlinear Anal. Theory Methods Appl. 23(8) (1994), pp. 1039–1089.

[18] L. Gardini, I. Sushko, V. Avrutin, and M. Schanz, *Critical homoclinic orbits lead to snap-back repellers*, Chaos Solitons Fractals 44 (2011), pp. 433–449.

[19] I. Gilboa and D. Schmeidler, *Maxmin expected utility with non-unique prior*, J. Math. Econ. 18 (1989), pp. 141–153.

[20] S. Ito, S. Tanaka, and H. Nakada, *On unimodal transformations and chaos II*, Tokyo J. Math. 2 (1979), pp. 241–259.

[21] K. Ichimura and M. Ito, *Dynamics of skew tent maps*, RIMS Kokyuroku 1042 (1998), pp. 92–98.

[22] F. Lamantia, M. Pezzino, and F. Tramontana, *Dynamic analysis of discontinuous best response with innovation*, J. Econ. Dyn. Control 91 (2018), pp. 120–133.

[23] F. Lamantia, M. Pezzino, B. Scardamaglia, and F. Tramontana, *A discontinuous model of duopoly with isoelastic demand and innovation costs*, Chaos Solitons Fractals 158 (2022), Article ID 112015.

[24] J. Milnor, *On the concept of attractor*, Comm. Math. Phys. 99 (1985), pp. 177–195.

[25] C. Mira, *Chaotic Dynamics. From the One-Dimensional Endomorphism to the Two-Dimensional Diffeomorphism*, World Scientific, Singapore, 1987.

[26] C. Mira, L. Gardini, A. Barugola, and J.C. Cathala, *Chaotic Dynamics in Two-Dimensional Noninvertible Maps*, World Scientific Series on Nonlinear Science; Vol. 20, World Scientific, New Jersey, 1996.

[27] H.E. Nusse and J.A. Yorke, *Border-Collision bifurcations including 'period two to period three' bifurcation for piecewise smooth systems*, Phys. D: Nonlinear Phenom. 57(1–2) (1992), pp. 39–57.

[28] H.E. Nusse and J.A. Yorke, *Border-collision bifurcations for piecewise smooth one dimensional maps*, Int. J. Bifurcat. Chaos 5 (1995), pp. 189–207.

[29] A. Panchuk, I. Sushko, B. Schenke, and V. Avrutin, *Bifurcation structures in a bimodal piecewise linear map: Regular dynamics*, Int. J. Bifurcat. Chaos 23 (2013), Article ID 1330040.

[30] A. Panchuk, I. Sushko, and V. Avrutin, *Bifurcation structures in a bimodal piecewise linear map: Chaotic dynamics*, Int. J. Bifurcat. Chaos 25(3) (2015), Article ID 1530006.

[31] D. Rand, *Exotic phenomena in games and duopoly models*, J. Math. Econ. 5 (1978), pp. 173–184.

[32] D. Radi and L. Gardini, *Ambiguity aversion as a route to randomness in a duopoly game*, Working paper (2022)

[33] D. Radi, L. Gardini, and D. Goldbaum, *A robust route to randomness in a simple Cournot duopoly game where ambiguity aversion meets constant expectations*, Working Paper (2022)
[34] I. Sushko and L. Gardini, *Degenerate bifurcations and border collisions in piecewise smooth 1D and 2D maps*, Int. J. Bifurcat. Chaos 20 (2010), pp. 2045–2070.
[35] I. Sushko, L. Gardini, and K. Matsuyama, *Superstable credit cycles and U-sequence*, Chaos Solitons Fractals 59 (2014), pp. 13–27.
[36] I. Sushko, V. Avrutin, and L. Gardini, *Bifurcation structure in the skew tent map and its application as a border collision normal form*, J. Differ. Equ. Appl. 22(8) (2016), pp. 1040–1087.
[37] F. Tramontana, L. Gardini, and T. Puu, *Global bifurcations in a piecewise-smooth Cournot duopoly game*, Chaos Solitons Fractals 43 (2010), pp. 15–24.
[38] F. Tramontana, L. Gardini, and T. Puu, *Mathematical properties of a combined Cournot-Stackelberg model*, Chaos Solitons Fractals 44 (2011), pp. 58–70.

Appendix

In this Appendix we recap some results from [10] related to a map $T(x, y) = (R_1(y), R_2(x))$. Consider the functions

$$F(x) = R_1 \circ R_2(x) \quad \text{and} \quad G(y) = R_2 \circ R_1(y) \tag{A1}$$

Then, the following properties hold for each integer $k \geq 1$ and for each integer $n \geq 1$:

(1) $T^{2k}(x, y) = (F^k(x), G^k(y))$;
(2) $R_1 \circ G^k(y) = F^k \circ R_1(y)$;
(3) $R_2 \circ F^k(x) = G^k \circ R_2(x)$;
(4) If $\{x_1, \ldots, x_n\}$ is an n-cycle of F then $\{y_1, \ldots, y_n\} = \{R_2(x_1), \ldots, R_2(x_n)\}$ is the conjugate n-cycle of G. The two conjugate cycles have the same stability property;
(5) If $\{y_1, \ldots, y_n\}$ is an n-cycle of G then $\{x_1, \ldots, x_n\} = \{R_1(y_1), \ldots, R_1(y_n)\}$ is the conjugate n-cycle of F. The two conjugate cycles have the same stability property;
(6) The cycles of the two-dimensional map T are related to those of the one-dimensional maps F and G, according to the following results. If $u = \{u_1, \ldots, u_n\}$ is an n-cycle of F for $n \geq 1$ and $v = \{v_1, \ldots, v_m\}$ is an m-cycle of G for $m \geq 1$, then in the phase space (x, y) all the points of the Cartesian product $\Psi = \{u_1, \ldots, u_n\} \times \{v_1, \ldots, v_m\}$ are periodic points of cycles of map T whose stability property is known and follows from that of the cycles u and v;
(7) The *vice-versa* is also true, that is, a point (x_i, y_i) is a periodic point of period n for T if and only if x_i and y_i are periodic points of F and G, respectively, of period n (if n is odd) or a divisor of n (if n is even);
(8) Let Ψ be the set of all periodic points of T. The local stable and unstable sets of a periodic point $(u_i, v_j) \in \Psi$ belong to the eigenvectors of the periodic point (u_i, v_j), which are the horizontal and vertical lines through it;
(9) Properties (1)–(8) also hold assuming that u_i and/or v_j are cyclical chaotic intervals.

Moreover, for any attracting n-cycle of F of period $n > 2$, or for any cyclic chaotic intervals of period $n > 2$, the two-dimensional map T is characterized by multistability, i.e. it has several distinct coexisting attracting sets.

There are also general properties regarding basins of attraction. We know that any n-cycle C of T, of odd or even period n, is necessarily associated with a cycle of F and one of G, say C_x and C_y, respectively. Let us denote by $\mathcal{B}(C)$ the total basin of an attractor C. We have that $\mathcal{B}(C) = \cup_{n=0}^{\infty} T^{-n}(\mathcal{B}_{im}(C))$, where $\mathcal{B}_{im}(C)$ is the immediate basin of C, made up of the connected components of the basin containing C. Analogously, for the one-dimensional map F we have $\mathcal{B}(C_x) = \cup_{n=0}^{\infty} F^{-n}(\mathcal{B}_{im}(C_x))$, where $\mathcal{B}_{im}(C_x)$ is the immediate 1-dimensional basin of C_x, along the x axis.

Regarding the basins of attraction of C, C_x and C_y, we have the following results:

(a) Let C be an attracting cycle of T associated with the cycles C_x and C_z of F, then

$$\begin{aligned} \mathcal{B}(C) &\subseteq [\mathcal{B}(C_x) \cup \mathcal{B}(C_z)] \times R_2([\mathcal{B}(C_x) \cup \mathcal{B}(C_z)]); \\ \mathcal{B}_{im}(C) &\subseteq [\mathcal{B}_{im}(C_x) \cup \mathcal{B}_{im}(C_z)] \times R_2([\mathcal{B}_{im}(C_x) \cup \mathcal{B}_{im}(C_z)]) \end{aligned} \tag{A2}$$

and $\mathcal{B}_{im}(C)$ is made up of rectangles which include the periodic points of C;

(b) For T, the image of a horizontal segment is a vertical segment and *vice-versa*;

(c) For T, the preimages of a horizontal segment, if any, are vertical segments and *vice-versa*;

(d) For any periodic point $P = (x, y)$ of the map T of period $n \geq 1$, the horizontal and vertical lines $y = y_1$ and $x = x_1$, issuing from P, are trapping sets for the map T^n.

Detecting invariant expanding cones for generating word sets to identify chaos in piecewise-linear maps

David J. W. Simpson

ABSTRACT
We show how to formally identify chaotic attractors in continuous, piecewise-linear maps on $\mathbb{R}^N$. For such a map f, this is achieved by constructing three objects. First, $\Omega_{\text{trap}} \subset \mathbb{R}^N$ is trapping region for f. Second, $\mathbf{W}$ is a finite set of words that encodes the forward orbits of all points in Ω_{trap}. Finally, $C \subset T\mathbb{R}^N$ is an invariant expanding cone for derivatives of compositions of f formed by the words in $\mathbf{W}$. The existence of Ω_{trap}, $\mathbf{W}$, and C implies f has a topological attractor with a positive Lyapunov exponent. We develop an algorithm that identifies these objects for two-dimensional homeomorphisms comprised of two affine pieces. The main effort is in the explicit construction of Ω_{trap} and C. Their existence is equated to a set of computable conditions in a general way. This results in a computer-assisted proof of chaos throughout a relatively large region of parameter space. We also observe how the failure of C to be expanding can coincide with a bifurcation of f. Lyapunov exponents are evaluated using one-sided directional derivatives so that forward orbits that intersect a switching manifold (where f is not differentiable) can be included in the analysis.

1. Introduction

Piecewise-linear maps form canonical representations of nonlinear dynamics and provide effective models of diverse physical systems [10]. Much work has been done to identify properties that imply a piecewise-linear map has a chaotic attractor. These include Markov partitions [15], homoclinic connections [26], and situations where the dynamics is effectively one-dimensional [21]. In the context of ergodic theory, an attractor with an absolutely continuous invariant measure exists for piecewise-expanding maps [7,35] and piecewise-smooth maps with certain expansion properties [31,38]. However, when applied to the two-dimensional border-collision normal form (2d BCNF) such properties have only been verified over regions of parameter space that are small in comparison to where numerical simulations suggest chaotic attractors actually exist [12,13]. To address this issue we use invariant expanding cones to bound Lyapunov exponents. We show that this appears to be a highly effective method for identifying chaotic attractors in a formal way.

For a continuous map f on $\mathbb{R}^N$, the Lyapunov exponent $\lambda(x, v)$ characterizes the asymptotic rate of separation of the forward orbits of arbitrarily close points x and $x + \delta v$:

$$\left\| f^n(x + \delta v) - f^n(x) \right\| \sim e^{\lambda n} \|\delta v\|. \tag{1}$$

A positive Lyapunov exponent for bounded orbits is a standard indicator of chaos [24]. Now suppose there exist $N \times N$ matrices A_i such that

$$f^n(x + \delta v) - f^n(x) = \delta A_{n-1} \cdots A_1 A_0 v + o(\delta), \tag{2}$$

where $o(\delta)$ vanishes faster than δ as $\delta \to 0$. These matrices exist if f is C^1: each A_i is the Jacobian matrix $\mathrm{D}f$ evaluated at $f^i(x)$. If $\mathbf{M}$ denotes the set of all A_i and there exists $c > 1$ and a cone $C \subset T\mathbb{R}^N$ with the property

$$Mv \in C \text{ and } \|Mv\| \geq c\|v\|, \text{ for all } v \in C \text{ and all } M \in \mathbf{M}, \tag{3}$$

then immediately we have $\lambda(x, v) > 0$ for any $v \in C$. This idea is attributed to Alekseev [1] (see [5,37]) and is useful for establishing splitting and hyperbolicity in smooth dynamical systems [8,27,34].

Condition (3) is rather strong as it requires C to be invariant and expanding for every matrix in the possibly infinite set $\mathbf{M}$. But if f is piecewise-linear then $\mathbf{M}$ contains only as many matrices as pieces in the map. For this reason invariant expanding cones are perfectly suited, and perhaps under-utilized, for analysing piecewise-linear maps. Invariant cones were central to Misiurewicz's strategy for establishing hyperbolicity and transitivity in the Lozi map [26]. More recently in [17] an invariant expanding cone was constructed for the 2d BCNF to finally prove the widely held conjecture that a chaotic attractor exists throughout a physically-important parameter regime $\mathcal{R}_{\mathrm{BYG}}$ first highlighted in [4].

In this paper we present a simple but powerful generalization of the above approach and use it to show that chaotic attractors persist beyond $\mathcal{R}_{\mathrm{BYG}}$, in fact in some places right up to where there exist stable low-period orbits. The idea is to let $\mathbf{M}$ consist of certain products of the A_i, rather than of the A_i themselves. Each product $M \in \mathbf{M}$ is characterized by a word $\mathcal{W}$ connecting its constitute matrices to the pieces of f. As long as the set of all such words $\mathbf{W}$ generates the symbolic itinerary of the forward orbit of x, then again (3) implies $\lambda(x, v) > 0$. This approach is quite flexible because there is a large amount of freedom in our choice of the set of words $\mathbf{W}$.

To prove f has a chaotic attractor one needs to identify a trapping region $\Omega_{\mathrm{trap}} \subset \mathbb{R}^N$ (which ensures there exists a topological attractor), a set of words $\mathbf{W}$ that generates the symbolic itineraries for all $x \in \Omega_{\mathrm{trap}}$, and an invariant expanding cone C for the matrices corresponding to the words in $\mathbf{W}$. Below we find these objects for the 2d BCNF. More precisely, we propose a way by which Ω_{trap} and C can be constructed for a particular word set $\mathbf{W}$ and prove that all three objects have the required properties if a certain set of computable conditions are met. While, for a given combination of parameter values, these conditions could be checked by hand, it is more appropriate to check them numerically. Below we formulate this as an algorithm (Algorithm 7.1).

The remainder of this paper is organized as follows. We start in Section 2 by showing where Algorithm 7.1 detects a chaotic attractor in a typical two-dimensional slice of the parameter space of the 2d BCNF. Then in Section 3 we clarify technical features mentioned above (cones, words, trapping regions, etc) and express $\lambda(x, v)$ in terms of one-sided directional derivatives in order to accommodate points whose forward orbits intersect a switching manifold (where $\mathrm{D}f$ is not defined). The result $\lambda(x, v) > 0$ is formalized by Theorem 3.9 for N-dimensional, continuous, piecewise-linear maps with two pieces. The theorem is framed in terms of $\mathbf{W}$-recurrent sets (these are sets to which forward orbits return following one or more words in $\mathbf{W}$). Such sets provide a practical way by which the approach can be applied to concrete examples, and in Section 4 we show they imply that $\mathbf{W}$ generates symbolic itineraries as needed. Here we also characterize the matrices A_i in the expression (2). This is quite subtle because if $f^i(x)$ lies on a switching manifold then A_i depends on v as well as x. Section 4 concludes with a proof of Theorem 3.9.

In subsequent sections we work to apply this methodology to the 2d BCNF. In Section 5 we consider sets of 2×2 matrices and devise a set of conditions for the existence of an invariant expanding cone. In Section 6 we connect consecutive points of an orbit to construct a polygon Ω. We then identify conditions implying Ω is forward invariant and can be perturbed into a trapping region $\Omega_{\rm trap}$. In Section 7 we state Algorithm 7.1 and prove its validity. Here we comment further on the application of the algorithm to the 2d BCNF and discuss instances in which failure of the algorithm coincides with a bifurcation of f at which the chaotic attractor is destroyed. Concluding remarks are provided in Section 8.

2. Chaotic attractors in the two-dimensional border-collision normal form

Let f be a continuous map on $\mathbb{R}^2$ that is affine on each side of a line Σ. Assume coordinates $x = (x_1, x_2) \in \mathbb{R}^2$ are chosen so that $\Sigma = \{x \mid x_1 = 0\}$. If f is generic in the sense that $f(\Sigma)$ intersects Σ at exactly one point, and this point is not a fixed point of f, then there exists an affine coordinate change that puts f into the form

$$f(x) = \begin{cases} \begin{bmatrix} \tau_L & 1 \\ -\delta_L & 0 \end{bmatrix} \begin{bmatrix} x_1 \\ x_2 \end{bmatrix} + \begin{bmatrix} 1 \\ 0 \end{bmatrix}, & x_1 \le 0, \\ \begin{bmatrix} \tau_R & 1 \\ -\delta_R & 0 \end{bmatrix} \begin{bmatrix} x_1 \\ x_2 \end{bmatrix} + \begin{bmatrix} 1 \\ 0 \end{bmatrix}, & x_1 \ge 0, \end{cases} \tag{4}$$

where $\tau_L, \tau_R, \delta_L, \delta_R \in \mathbb{R}$. The coordinate change required to arrive at (4) is provided in Appendix A.

The four parameter family (4) is the 2d BCNF of [28] except the value of the border-collision bifurcation parameter (usually denoted μ) is fixed at 1. The family (4) has been studied extensively to understand border-collision bifurcations (where a fixed point collides with a switching manifold) arising in many applications, particularly vibrating mechanical systems with impacts or friction [10,32]. If $\tau_L = -\tau_R$ and $\delta_L = \delta_R$ then (4) reduces to the Lozi map [23].

Figure 1 shows a two-dimensional slice of the parameter space of (4) defined by fixing

$$\delta_L = 0.3, \quad \delta_R = 0.3. \tag{5}$$

In the blue regions (4) has a stable period-n orbit for one value $1 \le n \le 5$ as indicated. To the right of the curve labelled HC, where the fixed point in $x_1 < 0$ attains a homoclinic

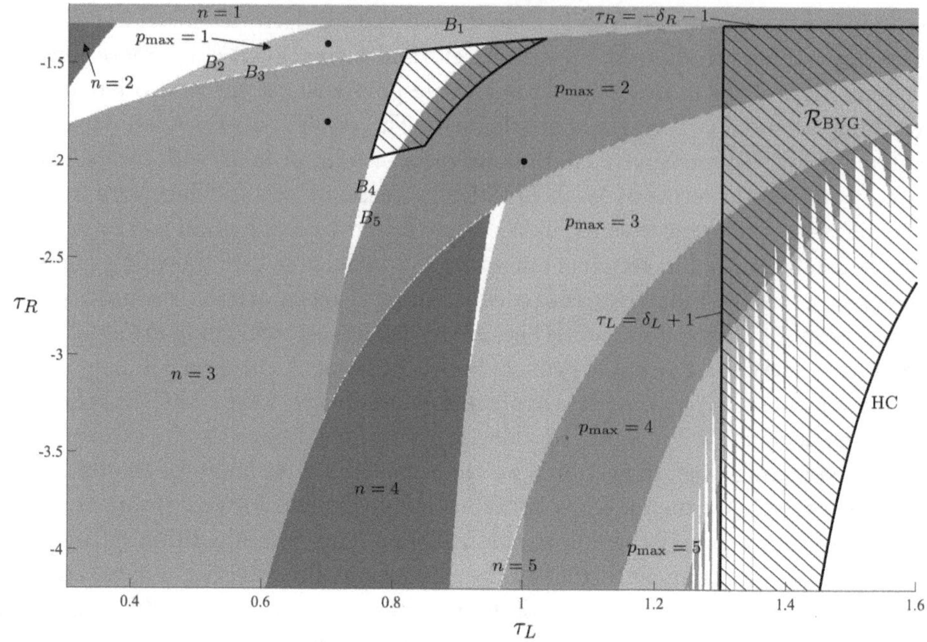

Figure 1. A two-parameter bifurcation diagram of the 2d BCNF (4) with (5). In the regions labelled $n = 1, n = 2, \ldots$ (4) has a stable period-n orbit for the indicated value of n (in places where the $n = 3$ and $n = 4$ regions overlap only the $n = 4$ region is shown). In the regions labelled $p_{\max} = 1, p_{\max} = 2, \ldots$ Algorithm 7.1 establishes the existence of a chaotic attractor by using (6) with the indicated value of $p_{\max}$. The three black dots indicate the parameter combinations examined in Figures 15–17 [see also Figure 2(b)]. The boundaries B_1 to B_5 are discussed in Section 7.

connection, there is usually no attractor because here the stable and unstable manifolds of the fixed point form a homoclinic tangle that provides a path by which orbits can escape. Numerical explorations suggest that in all other areas of Figure 1 (i.e. left of the HC curve and not inside a blue region) (4) has a chaotic attractor [3]. Different values of the determinants $\delta_L, \delta_R \in (0, 1)$ correspond to different rates of contraction for the action of the map, so seem to produce qualitatively similar results in the sense that, as the value of τ_L is increased, the dynamics broadly changes from low-period orbits, to chaos, and then to the absence of an attractor. Note that $\delta_L, \delta_R \in (0, 1)$ implies (4) is orientation-preserving (determinants positive) and dissipative (determinants less than 1) thus can arise as a local approximation to Poincaré map of a ODE model of a dissipative system.

The two striped regions of Figure 1 are of particular interest. In the central striped region numerical simulations reveal that (4) has two chaotic attractors [16]. This shows that the problem of the uniqueness of the chaotic attractor (not addressed in this paper) is not trivial. The region $\mathcal{R}_{\rm BYG}$ (bounded by the HC curve and the lines $\tau_L = \delta_L + 1$ and $\tau_R = -\delta_R - 1$) is the 'robust chaos' parameter regime of [4] [restricted to (5)]. This parameter regime is exactly where (4) has two saddle fixed points, one with positive eigenvalues and one with negative eigenvalues. In $\mathcal{R}_{\rm BYG}$ (4) has a quasi-one-dimensional attractor, Figure 2(a), because the attractor can be characterized as the closure of a one-dimensional unstable manifold [17].

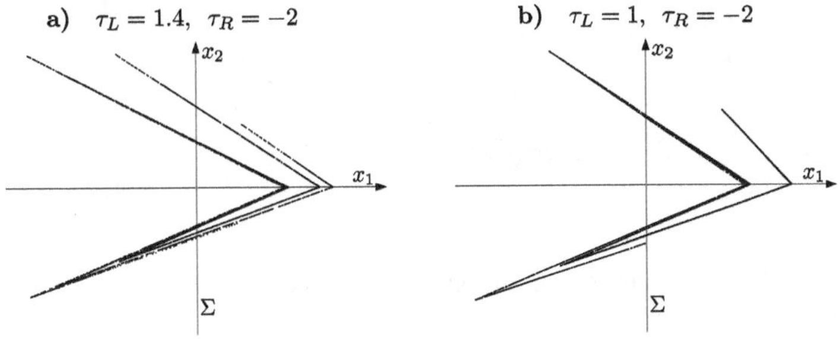

Figure 2. Chaotic attractors of (4) with (5). More precisely, these plots show 5000 iterates of the forward orbit of the origin with some transient points removed.

In [17] it was shown that throughout $\mathcal{R}_{\text{BYG}}$ (4) has an attractor with a positive Lyapunov exponent. This was achieved by constructing a trapping region and an invariant expanding cone for both matrices in (4). That is, the methodology of this paper was used with $\mathbf{W} = \{L, R\}$. In [17] it was not necessary to show that the symbolic itineraries of orbits are generated by $\mathbf{W}$ because in this case $\mathbf{W}$ generates all symbolic itineraries.

The approach of [17] fails to find a chaotic attractor in $\tau_L < \delta_L + 1$ because here both eigenvalues of $A_L = \begin{bmatrix} \tau_L & 1 \\ -\delta_L & 0 \end{bmatrix}$ have modulus less than 1, so there does not exist an invariant expanding cone for A_L. Therefore, if we are to show that (4) has a chaotic attractor with $\tau_L < \delta_L + 1$, we cannot include L in our word set $\mathbf{W}$. We instead use word sets of the form

$$\mathbf{W} = \left\{ RL^p \,\middle|\, 0 \le p \le p_{\max} \right\}. \tag{6}$$

This is clarified in Section 3.5. The red regions of Figure 1 show where Algorithm 7.1 finds an attractor with a positive Lyapunov exponent by using (6) with some $p_{\max} \ge 1$ over a 1024×512 grid of (τ_L, τ_R)-values [Figure 2(b) shows an example]. The algorithm is highly effective in that it establishes chaos over about 90%[1] of parameter space between the blue regions and $\tau_L = \delta_L + 1$ (and also succeeds in part of $\mathcal{R}_{\text{BYG}}$). In fact in some places there is no gap between the blue and red regions meaning that the algorithm finds a chaotic attractor right up to the bifurcation at which the attractor is destroyed (this is discussed further in Section 7).

3. Main definitions and a bound on the Lyapunov exponent

In this section we define the main objects and state the main theoretical result, Theorem 3.9. In order to arrive at Theorem 3.9 quickly some discussion is deferred to Section 4.

3.1. Trapping regions and topological attractors

Here we provide topological definitions for a continuous map f on $\mathbb{R}^N$, see for instance [18,29] for further details.

Definition 3.1: A set $\Omega \subseteq \mathbb{R}^N$ is *forward invariant* if $f(\Omega) \subseteq \Omega$.

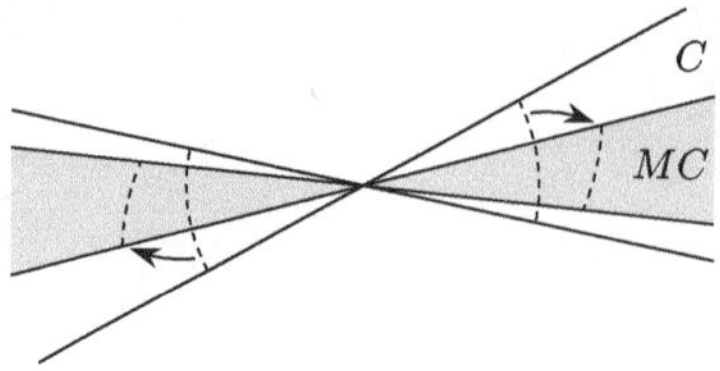

Figure 3. A sketch of an invariant expanding cone C and its image $MC = \{Mv \mid v \in C\}$.

Definition 3.2: A non-empty compact set $\Omega_{\text{trap}} \subset \mathbb{R}^N$ is a *trapping region* if $f(\Omega_{\text{trap}}) \subseteq \text{int}(\Omega_{\text{trap}})$ (where int$(\cdot)$ denotes interior).

Definition 3.3: An *attracting set* is $\Lambda = \cap_{i=0}^{\infty} f^i(\Omega_{\text{trap}})$ for some trapping region $\Omega_{\text{trap}} \subset \mathbb{R}^N$.

Topological attractors are invariant subsets of an attracting set that satisfy some kind of indivisibility condition (e.g. they contain a dense orbit) [24]. In this paper it is not necessary to consider such conditions as we only seek to show that f has an attractor and this is achieved by showing that f has a trapping region.

3.2. Cones

We denote the tangent space to a point $x \in \mathbb{R}^N$ by $T\mathbb{R}^N$. This is the usual tangent space – it is not affected by the piecewise-linear nature of the maps considered below. The tangent space is isomorphic to $\mathbb{R}^N$ and indeed we often treat tangent vectors $v \in T\mathbb{R}^N$ as elements of $\mathbb{R}^N$, as in the expression $x + \delta v$.

Definition 3.4: A non-empty set $C \subseteq T\mathbb{R}^N$ is said be to a *cone* if $tv \in C$ for all $v \in C$ and all $t \in \mathbb{R}$.

Definition 3.5: Let M be an $N \times N$ matrix. A cone $C \subseteq T\mathbb{R}^N$ is said to be *invariant* under M if

$$Mv \in C, \quad \text{for all } v \in C. \tag{7}$$

The cone is said to be *expanding* under M if there exists $c > 1$ such that

$$\|Mv\| \geq c\|v\|, \quad \text{for all } v \in C, \tag{8}$$

see Figure 3. For a finite set of real-valued $N \times N$ matrices $\mathbf{M}$, we say C is *invariant* [resp. *expanding*] if it is invariant [resp. expanding] under every $M \in \mathbf{M}$.

3.3. One-sided directional derivatives and Lyapunov exponents

The *one-sided directional derivative* of a function $\phi : \mathbb{R}^N \to \mathbb{R}^N$ at $x \in \mathbb{R}^N$ in a direction $v \in T\mathbb{R}^N$ is defined as

$$\mathrm{D}_v^+ \phi(x) = \lim_{\delta \to 0^+} \frac{\phi(x + \delta v) - \phi(x)}{\delta}, \tag{9}$$

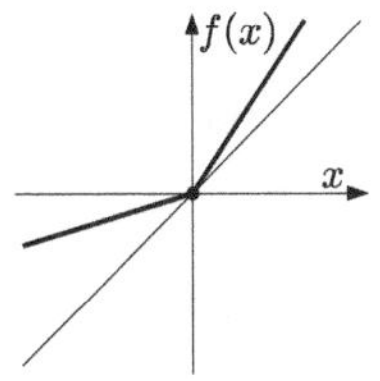

Figure 4. A sketch of the one-dimensional map (11) with $0 < a_L < 1 < a_R$.

if this limit exists [9,30]. If $\mathrm{D}_v^+ f^n(x)$ exists for all $n \geq 1$, then taking $\delta \to 0^+$ in (1) gives

$$\left\| \mathrm{D}_v^+ f^n(x) \right\| \sim \mathrm{e}^{\lambda n} \|v\|.$$

By further taking $n \to \infty$ we arrive at

$$\lambda(x, v) = \limsup_{n \to \infty} \frac{1}{n} \ln \left(\left\| \mathrm{D}_v^+ f^n(x) \right\| \right), \tag{10}$$

where, following the usual convention, the supremum limit is taken because from the point of view of ascertaining stability one wants to record the largest possible fluctuations.

If f is C^1 then the Jacobian matrix $\mathrm{D}f$ exists everywhere and (10) becomes

$$\lambda(x, v) = \limsup_{n \to \infty} \frac{1}{n} \ln \left(\left\| \mathrm{D}f^n(x) v \right\| \right).$$

Oseledet's theorem gives conditions under which $\lambda(x, v)$ takes at most N values for almost all x in an invariant set [6,36]. But this is often not the case for piecewise-linear maps. As a minimal example, consider the one-dimensional map

$$f(x) = \begin{cases} a_L x, & x \leq 0, \\ a_R x, & x \geq 0, \end{cases} \tag{11}$$

shown in Figure 4. If $a_L, a_R > 0$ and $a_L \neq a_R$ then the fixed point $x = 0$ has two different Lyapunov exponents: $\lambda(0, -1) = \ln(a_L)$ and $\lambda(0, 1) = \ln(a_R)$.

3.4. Two-piece, piecewise-linear, continuous maps

For ease of explanation we develop our methodology for piecewise-linear maps comprised of only two pieces. The extension to maps with more pieces is expected to be reasonably straight-forward. Specifically we consider maps of the form

$$f(x) = \begin{cases} A_L x + b, & x_1 \leq 0, \\ A_R x + b, & x_1 \geq 0, \end{cases} \tag{12}$$

where A_L and A_R are $N \times N$ matrices and $b \in \mathbb{R}^N$. The assumption that f is continuous on the switching manifold $\Sigma = \{x \mid x_1 = 0\}$ implies that A_L and A_R differ in only their first columns. That is,

$$A_R - A_L = \zeta e_1^{\mathsf{T}}, \tag{13}$$

for some $\zeta \in \mathbb{R}^N$ and where e_1 is the first standard basis vector of $\mathbb{R}^N$.

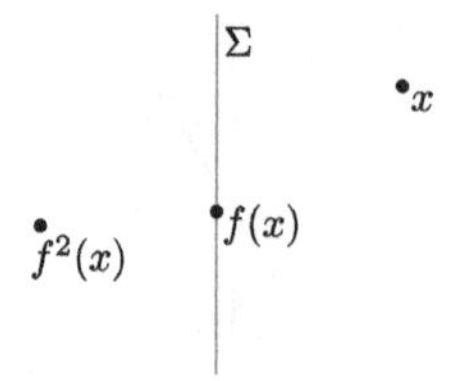

Figure 5. Part of an orbit for which $\Gamma(x;3) = \{RLL, RRL\}$, see (15).

If $\det(A_L) \neq 0$ and $\det(A_R) \neq 0$ then f maps the half-spaces $x_1 \leq 0$ and $x_1 \geq 0$ in a one-to-one fashion to half-spaces with boundary $f(\Sigma)$. If $\det(A_L)\det(A_R) < 0$ then f is not invertible because $x_1 \leq 0$ and $x_1 \geq 0$ are mapped to the same half-space, see [25]. If $\det(A_L)\det(A_R) > 0$ then f is invertible (see [32] for an explicit expression for f^{-1}) and so we have the following result.

Lemma 3.6: *The map* (12) *is invertible if and only if* $\det(A_L)\det(A_R) > 0$.

3.5. Words as symbolic representations of finite parts of orbits

To describe the symbolic itineraries of orbits of (12) relative to Σ we use words (defined here) and symbol sequences (defined in Section 4.1) on the alphabet $\{L, R\}$.

Definition 3.7: A *word* of length $n \geq 1$ is a function $\mathcal{W} : \{0, 1, \ldots, n-1\} \to \{L, R\}$ and we write $\mathcal{W} = \mathcal{W}_0\mathcal{W}_1 \cdots \mathcal{W}_{n-1}$.

Sometimes we abbreviate k consecutive instances of a symbol by putting k as a superscript. For example $\mathcal{W} = RL^3 = RLLL$ is a word of length four with $\mathcal{W}_0 = R$, $\mathcal{W}_1 = L$, $\mathcal{W}_2 = L$, and $\mathcal{W}_3 = L$.

In order to obtain words from orbits we first define the following set-valued function on $\mathbb{R}^N$:

$$\gamma_{\rm set}(x) = \begin{cases} \{L\}, & x_1 < 0, \\ \{L, R\}, & x_1 = 0, \\ \{R\}, & x_1 > 0. \end{cases} \tag{14}$$

Then, given $x \in \mathbb{R}^N$ and $n \geq 1$, we define

$$\Gamma(x;n) = \Big\{\mathcal{W} : \{0, 1, \ldots, n-1\} \to \{L, R\} \,\Big|\, \mathcal{W}_i \in \gamma_{\rm set}\left(f^i(x)\right) \text{ for all } i = 0, 1, \ldots, n-1\Big\}. \tag{15}$$

Notice that if $f^i(x) \in \Sigma$ for m values of $i \in \{0, 1, \ldots, n-1\}$, then $\Gamma(x;n)$ contains 2^m words. For example for Figure 5 we have $\Gamma(x;3) = \{RLL, RRL\}$.

Symbolic representations are often instead defined in a way that produces a unique word for every x and n [19]. In Section 4.2 we will see how the above formulation is particularly convenient for describing $\mathrm{D}_v^+ f^n(x)$.

3.6. Sufficient conditions for a positive Lyapunov exponent

Here we state our main result for obtaining $\lambda(x, v) > 0$. To do this we first provide a few more definitions that are discussed further in Section 4.

Given a finite set of words $\mathbf{W}$, let

$$\mathbf{M} = \left\{ \Phi(\mathcal{W}) \,\middle|\, \mathcal{W} \in \mathbf{W} \right\}, \tag{16}$$

where

$$\Phi(\mathcal{W}) = A_{\mathcal{W}_{n-1}} \cdots A_{\mathcal{W}_1} A_{\mathcal{W}_0}. \tag{17}$$

Roughly speaking, $\Phi(\mathcal{W})$ is equal to Df^n for orbits that map under f following the word $\mathcal{W}$.

Definition 3.8: A set $\Omega_{\rm rec} \subset \mathbb{R}^N$ is said to be $\mathbf{W}$-*recurrent* if for all $x \in \Omega_{\rm rec}$ there exists $n \geq 1$ such that $f^n(x) \in \Omega_{\rm rec}$ and every $\mathcal{W} \in \Gamma(x; n)$ can be written as a concatenation of words in $\mathbf{W}$.

Theorem 3.9: *Suppose $\Omega_{\rm rec}$ is $\mathbf{W}$-recurrent. Suppose there exists an invariant expanding cone for $\mathbf{M}$ given by (16). Then there exists $\lambda_{\rm bound} > 0$ such that for all $x \in \Omega_{\rm rec}$,*

$$\liminf_{n \to \infty} \frac{1}{n} \ln\left(\left\| \mathrm{D}_v^+ f^n(x) \right\| \right) \geq \lambda_{\rm bound}, \quad \textit{for some } v \in T\mathbb{R}^N. \tag{18}$$

Moreover, if f is invertible and $\Omega_{\rm trap} \subset \bigcup_{i=-\infty}^{\infty} f^i(\Omega_{\rm rec})$ is a trapping region for f, then f has an attractor $\Lambda \subset \Omega_{\rm trap}$ and (18) holds for all $x \in \Lambda$.

Observe (18) implies $\lambda(x, v) > 0$ because the supremum limit is greater than or equal to the infimum limit. Theorem 3.9 is proved in the next section.

4. Relationships between words, recurrent sets, and directional derivatives

Here we look deeper at the concepts introduced in the previous section for maps of the form (12). First in Section 4.1 we take the limit $n \to \infty$ in (15) to obtain a set of symbol sequences for the forward orbit of a point $x \in \mathbb{R}^N$. The key observation is that these sequences are generated by $\mathbf{W}$ whenever x belongs to a $\mathbf{W}$-recurrent set. Then in Section 4.2 we show that $D_v^+ f^n(x)$ exists for all x, v, and n, and describe it in terms of the product (17). Finally in Section 4.3 we prove Theorem 3.9.

4.1. Symbol sequences and generating word sets

Definition 4.1: A *symbol sequence* is a function $\mathcal{S} : \mathbb{N} \to \{L, R\}$ and we write $\mathcal{S} = \mathcal{S}_0 \mathcal{S}_1 \mathcal{S}_2 \cdots$.

Definition 4.2: Let $\mathbf{W} = \{\mathcal{W}^{(1)}, \ldots, \mathcal{W}^{(k)}\}$ be a finite set of words. We say that $\mathbf{W}$ *generates* a symbol sequence $\mathcal{S}$ if there exists a sequence $\alpha : \mathbb{N} \to \{1, \ldots, k\}$ such that $\mathcal{S} = \mathcal{W}^{(\alpha_0)} \mathcal{W}^{(\alpha_1)} \mathcal{W}^{(\alpha_2)} \cdots$.

For example the set $\mathbf{W} = \{R, RL, RLL\}$ generates $\mathcal{S}$ if and only if $\mathcal{S}_0 = R$ and $\mathcal{S}$ does not contain the word LLL. In the language of coding LLL is a *forbidden word* and the set of sequences generated by $\mathbf{W}$ is a *run-length limited shift* [22].

Next define

$$\Gamma(x) = \left\{ \mathcal{S} : \mathbb{N} \to \{L, R\} \,\middle|\, \mathcal{S}_i \in \gamma_{\text{set}}\left(f^i(x)\right) \text{ for all } i \geq 0 \right\}, \tag{19}$$

which represents the $n \to \infty$ limit of (15).

Lemma 4.3: *Let Ω_{rec} be $\mathbf{W}$-recurrent. Then $\mathbf{W}$ generates every $\mathcal{S} \in \Gamma(x)$ for all $x \in \Omega_{\text{rec}}$.*

Proof: Choose any $x \in \Omega_{\text{rec}}$ and $\mathcal{S} \in \Gamma(x)$. Let $x_0 = x$ and $\mathcal{S}^{(0)} = \mathcal{S}$. By induction we have that for all $j \geq 0$ there exists $n_j \geq 1$ such that $x_{j+1} = f^{n_j}(x_j) \in \Omega_{\text{rec}}$ and $\mathcal{S}^{(j)} = \mathcal{X}^{(j)}\mathcal{S}^{(j+1)}$, where $\mathcal{S}^{(j+1)} \in \Gamma(x_{j+1})$ and $\mathcal{X}^{(j)}$ is a word of length n_j that can be written as a concatenation of words in $\mathbf{W}$. Then $\mathcal{S} = \mathcal{X}^{(0)}\mathcal{X}^{(1)}\mathcal{X}^{(2)}\cdots$ as required. ■

4.2. A characterization of one-sided directional derivatives

So far we have constructed sets of words $\Gamma(x; n)$ and sets of symbol sequences $\Gamma(x)$ to describe the forward orbit of a point x. These have the advantage that they only depend on x, so they are relatively simple to describe and analyse. However, in order to identify the matrices A_i in the expression (2) (which we use to evaluate $D_v^+ f^n(x)$), we also require knowledge of v (albeit only for forward orbits that intersect Σ). To this end we define

$$\gamma((x, v)) = \begin{cases} L, & x_1 < 0, \text{ or } x_1 = 0 \text{ and } v_1 < 0, \\ R, & x_1 > 0, \text{ or } x_1 = 0 \text{ and } v_1 \geq 0, \end{cases} \tag{20}$$

where (x, v) is an element of the tangent bundle $\mathbb{R}^N \times T\mathbb{R}^N$. The choice of the symbol R when $x_1 = v_1 = 0$ is not important to the results below because in this case $A_L v = A_R v$ by (13). We then have the following result (given also in [33]).

Lemma 4.4: *For any $x \in \mathbb{R}^N$ and $v \in T\mathbb{R}^N$,*

$$D_v^+ f(x) = A_{\gamma((x,v))} v. \tag{21}$$

Proof: If $x_1 > 0$ then $(x + \delta v)_1 > 0$ for sufficiently small values of $\delta > 0$. In this case $f(x + \delta v) - f(x) = f_R(x + \delta v) - f_R(x) = \delta A_R v$, and so $D_v^+ f(x) = A_R v$. The same calculation occurs in the case $x_1 = 0$ and $v_1 \geq 0$ because we may take $f(x) = f_R(x)$ (by the continuity of f on Σ) and $(x + \delta v)_1 \geq 0$ so we may similarly take $f(x + \delta v) = f_R(x + \delta v)$. The same arguments apply to f^L if $x_1 < 0$, or $x_1 = 0$ and $v_1 < 0$, and result in $D_v^+ f(x) = A_L v$. ■

Higher directional derivatives are dictated by the evolution of tangent vectors. For this reason we define the following map on $\mathbb{R}^N \times T\mathbb{R}^N$:

$$h((x,v)) = \left(f(x), D_v^+ f(x)\right). \tag{22}$$

Since D^+ satisfies composition rule

$$D_v^+ f^{i+j}(x) = D^+_{D_v^+ f^i(x)} f^j\left(f^i(x)\right), \quad \text{for any } i,j \geq 1, \tag{23}$$

the nth iterate of h is

$$h^n((x,v)) = \left(f^n(x), D_v^+ f^n(x)\right). \tag{24}$$

Then Lemma 4.4 implies

$$D_v^+ f^n(x) = A_{\gamma(h^{n-1}((x,v)))} \cdots A_{\gamma(h((x,v)))} A_{\gamma((x,v))} v.$$

Finally we can use (17) to write this as

$$D_v^+ f^n(x) = \Phi(\mathcal{W})v, \quad \text{where } \mathcal{W}_i = \gamma\left(h^i((x,v))\right) \text{ for each } i \in \{0,\ldots,n-1\}. \tag{25}$$

We have thus proved the following result.

Proposition 4.5: *The derivative $D_v^+ f^n(x)$ exists for all $x \in \mathbb{R}^N$, $v \in T\mathbb{R}^N$, and $n \geq 1$, and is given by* (25).

4.3. A lower bound on the Lyapunov exponent

Here we work towards a proof of Theorem 3.9. Note that in Lemma 4.7 the bound on the Lyapunov exponent does not require that the cone is expanding, but gives $\lambda_{\text{bound}} > 0$ if $c > 1$.

Lemma 4.6: *The map h is invertible if and only if f is invertible.*

Proof: The first component of h is f. If $\det(A_L)\det(A_R) \leq 0$ then f^{-1} is not well-defined by Lemma 3.6 and thus h^{-1} is also not well-defined.

Now suppose $\det(A_L)\det(A_R) > 0$. By Lemma 3.6 the first component of h is well-defined. By Lemma 4.4 the second component of h is $h_2(x,v) = A_{\gamma((x,v))}v$. If $x_1 < 0$ then $h_2(x,v) = A_L v$ is invertible because $\det(A_L) \neq 0$. Similarly if $x_1 > 0$ then $h_2(x,v) = A_R v$ is invertible because $\det(A_R) \neq 0$. If $x_1 = 0$ then

$$h_2(x,v) = \begin{cases} A_L v, & v_1 < 0, \\ A_R v, & v_1 \geq 0, \end{cases}$$

which is equal to (12) with $b = \mathbf{0}$ (the zero vector) and so is invertible by Lemma 3.6. ■

Lemma 4.7: *Let* **W** *be a finite set of words and* **M** *be given by* (16). *Suppose there exists an invariant cone C satisfying* (8) *with some $c > 0$ for all $M \in$* **M**. *Let $x \in \mathbb{R}^N$ and suppose* **W**

generates every $\mathcal{S} \in \Gamma(x)$. Then (18) is satisfied with $\lambda_{\text{bound}} = \frac{\ln(c)}{L_{\max}}$, where $L_{\max}$ is the length of the longest word(s) in $\mathbf{W}$. *If f is invertible the same bound applies to $f^i(x)$ for any $i \in \mathbb{Z}$.*

Proof: Let $\mathcal{W}^{(1)}, \ldots, \mathcal{W}^{(k)}$ be the words in $\mathbf{W}$ and let $L_1, \ldots, L_k$ be their lengths. Let $v \in C$ with $v \neq \mathbf{0}$. Define the sequence

$$\mathcal{S} = \gamma((x, v))\gamma\,(h((x, v)))\,\gamma\left(h^2((x, v))\right)\cdots .$$

Then $\mathcal{S} \in \Gamma(x)$ thus there exists a sequence α_j such that

$$\mathcal{S} = \mathcal{W}^{(\alpha_0)}\mathcal{W}^{(\alpha_1)}\mathcal{W}^{(\alpha_2)}\cdots .$$

For each $j \geq 1$, let $n_j = L_{\alpha_0} + L_{\alpha_1} + \cdots + L_{\alpha_{j-1}}$ and let

$$v_j = \mathrm{D}_v^+ f^{n_j}(x). \tag{26}$$

By (25) we have $v_j = \Phi(\mathcal{W}^{(\alpha_0)}\mathcal{W}^{(\alpha_1)}\cdots\mathcal{W}^{(\alpha_{j-1})})v$. By (17) we have

$$\Phi\left(\mathcal{W}^{(\alpha_0)}\mathcal{W}^{(\alpha_1)}\cdots\mathcal{W}^{(\alpha_{j-1})}\right) = \Phi\left(\mathcal{W}^{(\alpha_{j-1})}\right)\Phi\left(\mathcal{W}^{(\alpha_0)}\mathcal{W}^{(\alpha_1)}\cdots\mathcal{W}^{(\alpha_{j-2})}\right)$$

and thus, for all $j \geq 1$,

$$v_j = \Phi\left(\mathcal{W}^{(\alpha_{j-1})}\right)v_{j-1}, \tag{27}$$

where $v_0 = v$. Since $v_0 \in C$ and C is invariant under each $\Phi(\mathcal{W}^{(\alpha_i)})$, we have $v_j \in C$ for all $j \geq 1$. Moreover $\|v_j\| \geq c\|v_{j-1}\|$ and so $\|v_j\| \geq c^j\|v_0\|$. Then by (26)

$$\begin{aligned}
\liminf_{n\to\infty}\frac{1}{n}\ln\left(\left\|\mathrm{D}_v^+ f^n(x)\right\|\right) &= \liminf_{j\to\infty}\frac{1}{n_j}\ln\left(\|v_j\|\right)\\
&\geq \liminf_{j\to\infty}\frac{1}{jL_{\max}}\ln\left(c^j\|v\|\right)\\
&= \lambda_{\text{bound}}.
\end{aligned}$$

Finally suppose f is invertible and choose any $i \in \mathbb{Z}$. Write $(\tilde{x}, \tilde{v}) = h^i(x, v)$. In the case $i < 0$ this is well-defined by Lemma 4.6. By the composition rule (23) we have

$$\mathrm{D}_{\tilde{v}}^+ f^n(\tilde{x}) = \mathrm{D}_v^+ f^{n+i}(x),$$

for any $n \geq 0$ for which $n + i \geq 0$. Thus by taking $n \to \infty$ we obtain at the same bound for $\tilde{x} = f^i(x)$. ■

Proof of Theorem 3.9: Choose any $x \in \Omega_{\text{rec}}$. By Lemma 4.3, $\mathbf{W}$ generates every $\mathcal{S} \in \Gamma(x)$. Thus by Lemma 4.7, inequality (18) holds for some $\lambda_{\text{bound}} > 0$. For the second part of the theorem, an attractor $\Lambda \subset \Omega_{\text{trap}}$ exists because Ω_{trap} is a trapping region. Choose any $y \in \Lambda$. We can write $y = f^i(x)$ for some $i \in \mathbb{R}$, thus (18) holds for y by the second part of Lemma 4.7. ■

5. Cones for sets of 2 × 2 matrices

For the remainder of the paper we apply the above methodology to maps on $\mathbb{R}^2$ with the Euclidean norm $\|\cdot\|$.

Let M be a real-valued 2×2 matrix. We are interested in the behaviour of $v \mapsto Mv$, where $v \in \mathbb{R}^2$, in regards to cones. We start in Section 5.1 by deriving properties of this map for vectors of the form $v = (1, m)$, where $m \in \mathbb{R}$ is the slope of v. The behaviour of scalar multiples of $(1, m)$ follows trivially by linearity. The behaviour of $e_2 = (0, 1)$ can be inferred by considering the limit $m \to \infty$. However, this vector will not be of interest to us because if $\mathbf{W}$ is given by (6), then it contains R. Notice $\|A_R e_2\| = \|e_2\|$ so e_2 cannot belong to an invariant expanding cone for $\mathbf{M}$ given by (16).

In Section 5.2 we consider several matrices $M^{(i)}$. We use fixed points of $v \mapsto M^{(i)} v$ to construct a cone and derive conditions sufficient for the cone to be invariant and expanding.

5.1. Results for a single matrix M

Write $v = (1, m)$ and

$$M = \begin{bmatrix} a & b \\ c & d \end{bmatrix}. \tag{28}$$

We first decompose $v \mapsto Mv$ into two real-valued functions G and H. Let

$$H(m) = \|Mv\|^2 - \|v\|^2. \tag{29}$$

This function is particularly amenable to analysis because it is quadratic in m:

$$H(m) = \left(b^2 + d^2 - 1\right) m^2 + 2(ab + cd)m + a^2 + c^2 - 1. \tag{30}$$

The factor $\frac{\|Mv\|}{\|v\|}$ by which the norm of v changes under multiplication by M is less than 1 if $H(m) < 0$, and greater than 1 if $H(m) > 0$.

The slope of Mv is

$$G(m) = \frac{c + dm}{a + bm}, \tag{31}$$

assuming $a + bm \neq 0$. That is, Mv is a scalar multiple of $(1, G(m))$. From (31),

$$\frac{dG}{dm} = \frac{\det(M)}{(a + bm)^2}, \tag{32}$$

and so if $\det(M) > 0$ (which is the case below where M is a product instances of A_L and A_R with $\det(A_L) > 0$ and $\det(A_R) > 0$) then G is increasing on any interval for which $a + bm \neq 0$. Fixed points of G satisfy

$$bm^2 + (a - d)m - c = 0. \tag{33}$$

Note that $m \in \mathbb{R}$ is a fixed point of G if and only if $v = (1, m)$ is an eigenvector of M.

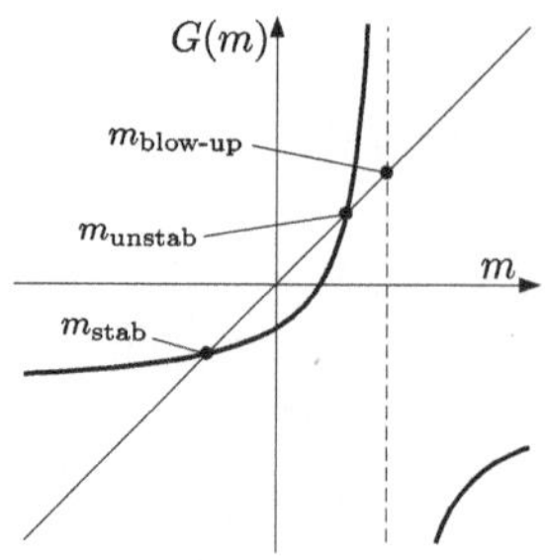

Figure 6. A sketch of the slope map (31) when the stable fixed point m_{stab} has a smaller value than the unstable fixed point m_{unstab}.

Lemma 5.1: *Suppose*

$$0 < \det(M) < \tfrac{1}{4}\,\mathrm{trace}(M)^2 \text{ and } b \neq 0. \tag{34}$$

Then G has exactly two fixed points. At one fixed point, call it m_{stab}, we have $\frac{dG}{dm} = \eta$, for some $0 < \eta < 1$, and at the other fixed point, call it m_{unstab}, we have $\frac{dG}{dm} = \frac{1}{\eta}$. Moreover, m_{unstab} lies between m_{stab} and $m_{blow-up} = -\frac{a}{b}$ (see for example Figure 6).

Proof: By (34) M has distinct eigenvalues $\lambda_1, \lambda_2 \in \mathbb{R}$ with $\lambda_1\lambda_2 = \det(M) > 0$. Without loss of generality assume $|\lambda_1| > |\lambda_2|$. Since λ_1 and λ_2 satisfy the characteristic equation for M, we find that $m_{\mathrm{stab}} = \frac{\lambda_1 - a}{b}$ and $m_{\mathrm{unstab}} = \frac{\lambda_2 - a}{b}$ satisfy the fixed point equation (33). These are only fixed points of $G(m)$ because (33) is quadratic. By evaluating (32) at $m = m_{\mathrm{stab}}$ we obtain $\frac{dG}{dm} = \frac{\det(M)}{\lambda_1^2} = \frac{\lambda_2}{\lambda_1}$. So $\eta = \frac{\lambda_2}{\lambda_1}$ and indeed $\eta \in (0, 1)$. Similarly at $m = m_2$ we have $\frac{dG}{dm} = \frac{\lambda_1}{\lambda_2} = \frac{1}{\eta}$.

Now suppose $m_{\mathrm{stab}} < m_{\mathrm{blow-up}}$. By the intermediate value theorem, $G(m)$ has a fixed point between m_{stab} and $m_{\mathrm{blow-up}}$ because $\frac{dG}{dm} < 1$ at m_{stab} whereas $G(m) \to \infty$ as m converges to $m_{\mathrm{blow-up}}$ from the right. This fixed point must be m_{unstab}, thus we have $m_{\mathrm{stab}} < m_{\mathrm{unstab}} < m_{\mathrm{blow-up}}$. If instead $m_{\mathrm{stab}} > m_{\mathrm{blow-up}}$, an analogous argument produces $m_{\mathrm{blow-up}} < m_{\mathrm{unstab}} < m_{\mathrm{stab}}$. ■

5.2. Results for a set of matrices M

Let $\mathbf{M} = \{M^{(1)}, \ldots, M^{(k)}\}$ be a set of real-valued 2×2 matrices. Write

$$M^{(j)} = \begin{bmatrix} a_j & b_j \\ c_j & d_j \end{bmatrix}, \tag{35}$$

for each $j \in \{1, \ldots, k\}$, and let

$$G_j(m) = \frac{c_j + d_j m}{a_j + b_j m}, \tag{36}$$

$$H_j(m) = \left(b_j^2 + d_j^2 - 1\right) m^2 + 2(a_j b_j + c_j d_j)m + a_j^2 + c_j^2 - 1, \tag{37}$$

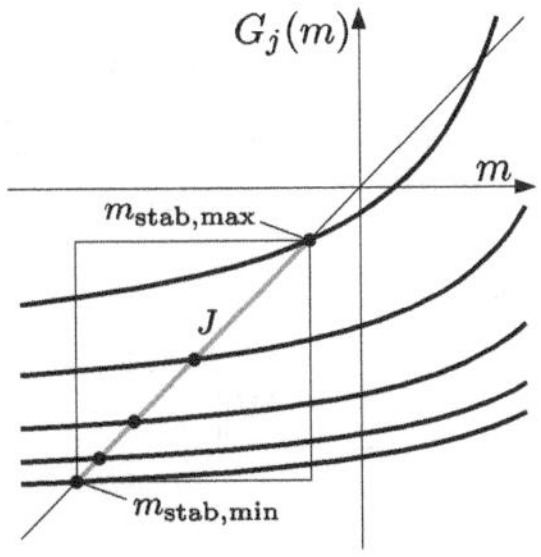

Figure 7. A sketch of five slope maps (36). Here condition (40) is satisfied, thus $G_j(J) \subset J$, for each j, and so the cone C_J (39) is invariant under **M**, by Proposition 5.2.

denote (30) and (31) applied to (35). Assume

$$0 < \det\left(M^{(j)}\right) < \tfrac{1}{4}\,\mathrm{trace}\left(M^{(j)}\right)^2 \text{ and } b_j \neq 0, \tag{38}$$

for each j so that each $M^{(j)}$ satisfies the assumptions of Lemma 5.1. Let $m^{(j)}_{\mathrm{stab}}$ and $m^{(j)}_{\mathrm{unstab}}$ denote the stable and unstable fixed points of (36) and let $m_{\mathrm{stab,min}}$ and $m_{\mathrm{stab,max}}$ denote the minimum and maximum values of $\{m^{(1)}_{\mathrm{stab}}, \ldots, m^{(k)}_{\mathrm{stab}}\}$. Define the interval $J = [m_{\mathrm{stab,min}}, m_{\mathrm{stab,max}}]$, see Figure 7, and the cone

$$C_J = \left\{t\,(1,m) \,\middle|\, t \in \mathbb{R}, m \in J\right\}. \tag{39}$$

Proposition 5.2: *If*

$$m^{(j)}_{\mathrm{unstab}} \notin J, \quad \textit{for each } j \in \{1,\ldots,k\}, \tag{40}$$

then C_J is invariant under **M**. *If also*

$$H_j(m) > 0, \quad \textit{for each } j \in \{1,\ldots,k\} \textit{ and all } m \in J, \tag{41}$$

then C_J is expanding under **M**.

Proof: Choose any $j \in \{1,\ldots,k\}$. Let $m^{(j)}_{\mathrm{blow-up}}$ be the value of m at which (36) is undefined. By assumption $m^{(j)}_{\mathrm{unstab}} \notin J$, so by Lemma 5.1 if $m^{(j)}_{\mathrm{unstab}} < m_{\mathrm{stab,min}}$, then $m^{(j)}_{\mathrm{blow-up}} < m^{(j)}_{\mathrm{unstab}}$, while if $m^{(j)}_{\mathrm{unstab}} > m_{\mathrm{stab,max}}$, then $m^{(j)}_{\mathrm{blow-up}} > m^{(j)}_{\mathrm{unstab}}$. In either case $m^{(j)}_{\mathrm{blow-up}} \notin J$, thus $G_j|_J$ (the restriction of G_j to J) is continuous.

We now show that

$$\min_{m \in J} G_j(m) \geq m_{\mathrm{stab,min}}. \tag{42}$$

By (32) $G_j|_J$ is increasing because $\det(M^{(j)}) > 0$. Thus $G_j|_J$ achieves its minimum at $m = m_{\mathrm{stab,min}}$. If $m^{(j)}_{\mathrm{stab}} \neq m_{\mathrm{stab,min}}$ (otherwise (42) is trivial) then, since $\frac{dG_j}{dm}\big|_{m=m^{(j)}_{\mathrm{stab}}} < 1$ by Lemma 5.1, we have $G_j(m) > m$ for some values of $m < m^{(j)}_{\mathrm{stab}}$ close to $m^{(j)}_{\mathrm{stab}}$. Thus $G_j(m_{\mathrm{stab,min}}) > m_{\mathrm{stab,min}}$, for otherwise G_j would have another fixed point in J by the immediate value theorem and this is not possible because $m^{(j)}_{\text{stab}}$ is unique fixed point of $G_j|_J$. This verifies (42). We similarly have $\max_{m\in J} G_j(m) \le m_{\text{stab,max}}$. Thus $G_j(m) \in J$ for all $m \in J$. Thus $M^{(j)}u \in C_J$ for all $u \in C_J$. Since j is arbitrary, C_J is forward invariant under $\mathbf{M}$.

Finally, with $v = (1, m)$ we have

$$\min_{u\in C_J\setminus\{\mathbf{0}\}} \frac{\|M^{(j)}u\|}{\|u\|} = \min_{m\in J} \frac{\|M^{(j)}v\|}{\|v\|} = \min_{m\in J} \sqrt{\frac{H_j(m)}{\|v\|^2} + 1} = c_j,$$

where the minimum value c_j is achieved because J is compact. If (41) is satisfied then $c_j > 1$. Thus with $c = \min\limits_j c_j$, C_J is expanding under $\mathbf{M}$. ■

We complete this section by showing how the expansion condition (41) can be checked with a finite set of calculations based on the fact that each H_j is quadratic in m. If H_j has two distinct real roots, then H_j is increasing at one root, call it $m^{(j)}_{\text{incr}}$, and decreasing at the other root, call it $m^{(j)}_{\text{decr}}$.

Proposition 5.3: *Suppose H_j has two distinct real roots. Then*

$$H_j(m) > 0, \quad \textit{for all } m \in J, \tag{43}$$

if and only if two of the following inequalities are satisfied

$$m_{\text{stab,max}} < m^{(j)}_{\text{decr}}, \tag{44}$$

$$m^{(j)}_{\text{decr}} < m^{(j)}_{\text{incr}}, \tag{45}$$

$$m^{(j)}_{\text{incr}} < m_{\text{stab,min}}. \tag{46}$$

Notice (44)–(46) cannot all be satisfied because $m_{\text{stab,min}} < m_{\text{stab,max}}$.

Proof: The proof is achieved by brute-force; there are six cases to consider.

If (44) and (45) are true, then $H_j(m) > 0$ for all $m < m^{(j)}_{\text{decr}}$, which includes all $m \in J$, thus (43) is true. If (44) and (46) are true, then $H_j(m) > 0$ for all $m \in (m^{(j)}_{\text{incr}}, m^{(j)}_{\text{decr}})$ so (43) is true. If (45) and (46) are true, then $H_j(m) > 0$ for all $m > m^{(j)}_{\text{incr}}$ so (43) is true.

If (44) and (45) are false, then $H_j(m) < 0$ at $m = m_{\text{stab,max}}$ so (43) is false. If (44) and (46) are false then $H_j(m) \le 0$ at $m = \min[m^{(j)}_{\text{incr}}, m_{\text{stab,max}}]$, which belongs to J, thus (43) is false. If (45) and (46) are false, then $H_j(m) < 0$ at $m = m_{\text{stab,min}}$ so (43) is false. ■

Remark 5.1: In the special case that $b_j^2 + d_j^2 - 1 = 0$ and $a_j b_j + c_j d_j \neq 0$, the function H_j has exactly one root:

$$m^{(j)}_{\text{root}} = -\frac{a_j^2 + c_j^2 - 1}{2(a_j b_j + c_j d_j)}. \tag{47}$$

If $a_j b_j + c_j d_j < 0$ [resp. $a_j b_j + c_j d_j > 0$] then $H_j(m) > 0$ for all $m \in J$ if and only if $m_{\text{root}} > m_{\text{stab,max}}$ [resp. $m_{\text{root}} < m_{\text{stab,min}}$]. This case arises in the implementation below because with $M^{(1)} = A_R$ we have $b_1 = 1$ and $d_1 = 0$.

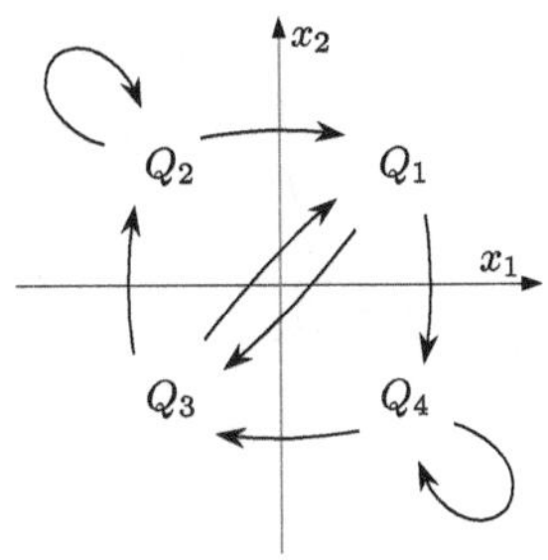

Figure 8. The action of the map (4) between the quadrants of $\mathbb{R}^2$ in the orientation-preserving setting, $\delta_L > 0$ and $\delta_R > 0$. Here Q_i denotes the closure of the ith quadrant of $\mathbb{R}^2$. For example if $x \in Q_1$ then $f(x) \in Q_3 \cup Q_4$.

6. The construction of a trapping region

In this section we study the 2d BCNF (4) in the orientation-preserving case: $\delta_L > 0$ and $\delta_R > 0$. In this case the sign of $f(x)_2$ is opposite to the sign of x_1. This implies points map between the quadrants of $\mathbb{R}^2$ as shown in Figure 8. For example if x belongs to the first quadrant, then $f(x)$ belongs to either the third quadrant or the fourth quadrant.

Let

$$\Pi_L = \left\{x \in \mathbb{R}^2 \,\middle|\, x_1 \le 0\right\},$$
$$\Pi_R = \left\{x \in \mathbb{R}^2 \,\middle|\, x_1 \ge 0\right\},$$

denote the closed left and right half-planes. Also write

$$f_L(x) = A_L x + b,$$
$$f_R(x) = A_R x + b,$$

for the left and right half-maps of (4).

6.1. Preimages of the switching manifold under f_L

With $\delta_L \neq 0$ the set $f_L^{-p}(\Sigma)$ is a line for all $p \ge 1$. Below we use these lines to partition Π_L into regions D_p whose points escape Π_L after exactly p iterations of f, see already Figure 11.

If $f_L^{-p}(\Sigma)$ is not vertical, let m_p denote its slope and let c_p denote its x_2-intercept with Σ. That is,

$$f_L^{-p}(\Sigma) = \left\{x \in \mathbb{R}^2 \,\middle|\, x_2 = m_p x_1 + c_p\right\}. \tag{48}$$

By iterating backwards under f_L we obtain $m_1 = -\tau_L$, $c_1 = -1$, and for all $p \ge 1$,

$$m_{p+1} = -\frac{\delta_L}{m_p} - \tau_L, \tag{49}$$

$$c_{p+1} = -\frac{c_p}{m_p} - 1, \tag{50}$$

whenever $m_p \neq 0$. As an example Figure 9 shows (49) with $p^* = 3$.

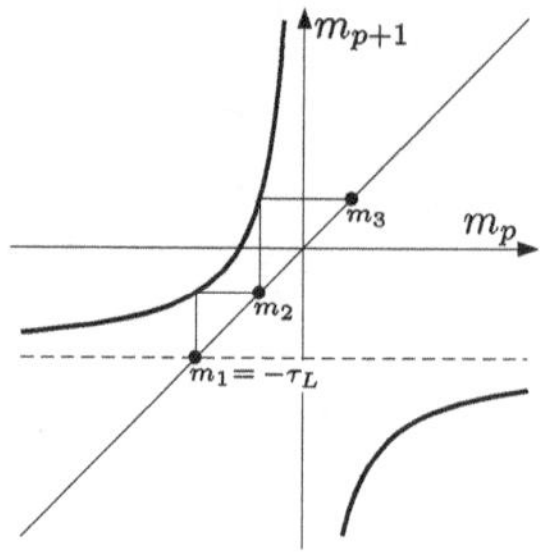

Figure 9. A typical plot of (49). Here $p^* = 3$ (see Definition 6.1).

Definition 6.1: Let p^* be the smallest $p \geq 1$ for which $m_p \geq 0$, with $p^* = \infty$ if $m_p < 0$ for all $p \geq 1$.

Next we establish monotonicity of $f_L^{-p}(\Sigma)$ in Π_L and provide an explicit expression for p^*.

Lemma 6.2: *The sequence* $\{m_p\}_{p=1}^{p^*}$ *is increasing; the sequence* $\{c_p\}_{p=1}^{p^*}$ *is decreasing.*

With p^* finite this can be more suitably written as $m_1 < m_2 < \cdots < m_{p^*}$ and $c_1 > c_2 > \cdots c_{p^*}$.

Proof: Since $m_1 = -\tau_L$, if $\tau_L \leq 0$ then $p^* = 1$ and the result is trivial. Suppose $\tau_L > 0$ (for which $p^* \geq 2$). Let $g(m) = -\frac{\delta_L}{m} - \tau_L$ denote the map (49). Since $g(m_1) - m_1 = \frac{\delta_L}{\tau_L} > 0$ and $\frac{dg}{dm} = \frac{\delta_L}{m^2} > 0$ for all $m < 0$, the forward orbit of m_1 under g is increasing while $m < 0$. That is, $\{m_p\}_{p=1}^{p^*}$ is increasing.

We now prove $\{c_p\}_{p=1}^{p^*}$ is decreasing by induction. Observe $c_1 > c_2$ because $c_1 = -1$ and $c_2 = -\frac{1}{\tau_L} - 1$ (where $\tau_L > 0$). Thus it remains to consider $p^* \geq 3$. Suppose $c_p > c_{p+1}$ for some $p \in \{1, \ldots, p^* - 2\}$ (this is our induction hypothesis). It remains to show $c_{p+1} > c_{p+2}$. Rearranging (50) produces

$$\frac{c_{p+1} + 1}{-c_p} = \frac{1}{m_p}.$$

But $c_{p+1} < c_p < 0$ and $m_p < m_{p+1} < 0$ thus

$$\frac{c_{p+1} + 1}{-c_{p+1}} > \frac{1}{m_{p+1}},$$

which is equivalent to $c_{p+1} > -\frac{c_{p+1}}{m_{p+1}} - 1 = c_{p+2}$. ■

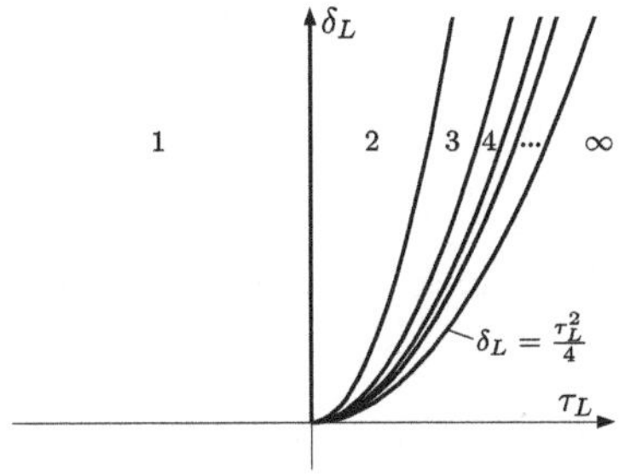

Figure 10. The value p^* of Definition 6.1 as given by Lemma 6.2.

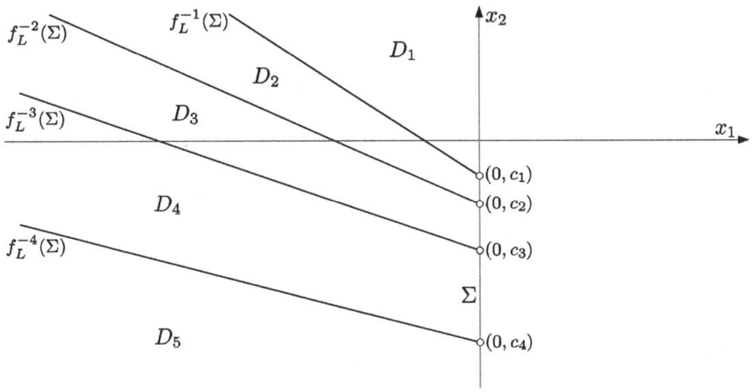

Figure 11. An example of the first few preimages of the switching manifold Σ under the left half-map f_L. These lines are described by (48) and bound the regions D_p of Definition 6.4.

Proposition 6.3: *The value p^* of Definition 6.1 is given by*

$$p^* = \begin{cases} 1, & \tau_L \le 0, \\ \left\lceil \dfrac{\pi}{\phi} - 1 \right\rceil, & 0 < \tau_L < 2\sqrt{\delta_L}, \\ \infty, & \tau_L \ge 2\sqrt{\delta_L}, \end{cases} \tag{51}$$

where $\phi = \cos^{-1}\left(\frac{\tau_L}{2\sqrt{\delta_L}}\right) \in \left(0, \frac{\pi}{2}\right)$, see Figure 10.

Proof: Since $m_1 = -\tau_L$, if $\tau_L \le 0$ then $p^* = 1$. If $\tau_L \ge 2\sqrt{\delta_L}$ then $m_\infty = \frac{1}{2}(-\tau_L - \sqrt{\tau_L^2 - 4\delta_L})$ is a fixed point of (49), call this map $g(m)$. Moreover $m_1 < m_\infty < 0$ and $g(m) > m$ for all $m_1 \le m < m_\infty$ so $m_p \to m_\infty$ as $p \to \infty$. Thus $p^* = \infty$ in this case.

If $0 < \tau_L < 2\sqrt{\delta_L}$ then the eigenvalues of A_L are $\lambda_1 = \sqrt{\delta_L}\,e^{i\phi}$ and $\lambda_2 = \sqrt{\delta_L}\,e^{-i\phi}$. Then $m_1 = -(\lambda_1 + \lambda_2)$ and (49) can be written as $m_{p+1} = -\frac{\lambda_1\lambda_2}{m_p} - (\lambda_1 + \lambda_2)$. From these we obtain, by induction on p, that while $\lambda_1^p \ne \lambda_2^p$ we have

$$m_p = \frac{\lambda_1^{p+1} - \lambda_2^{p+1}}{\lambda_2^p - \lambda_1^p},$$

which we can rewrite as

$$m_p = -\frac{\sin((p+1)\phi)}{\sin(p\phi)}\sqrt{\delta_L}. \tag{52}$$

By (52), if $\frac{\pi}{p+1} \le \phi < \frac{\pi}{p}$, then $m_p \ge 0$ and $m_j < 0$ for all $j \in \{1, \dots, p-1\}$, so $p^* = p$. Since $\frac{\pi}{p+1} \le \phi < \frac{\pi}{p}$ is equivalent to $p = \lceil \frac{\pi}{\phi} - 1 \rceil$, the proof is completed. ■

6.2. Partitioning the left half-plane by the number of iterations required to escape

By Lemma 6.2, in Π_L each $f_L^{-p}(\Sigma)$ is located below $f_L^{-(p-1)}(\Sigma)$, for $p = 2, \dots, p^*$, see Figure 11. This implies that the regions $D_p \subset \Pi_L$, defined below, are disjoint.

Definition 6.4: Let

$$D_1 = \left\{x \in \Pi_L \,\middle|\, x_2 > m_1 x_1 + c_1\right\}. \tag{53}$$

For all finite $p \in \{2, \dots, p^*\}$ let

$$D_p = \left\{x \in \Pi_L \,\middle|\, m_p x_1 + c_p < x_2 \le m_{p-1} x_1 + c_{p-1}\right\}. \tag{54}$$

If $p^* < \infty$ also let

$$D_{p^*+1} = \left\{x \in \Pi_L \,\middle|\, x_2 \le m_{p^*} x_1 + c_{p^*}\right\}. \tag{55}$$

Notice that if $p^* < \infty$ then $D_1, \dots, D_{p^*+1}$ partition Π_L. We now look the number of iterations of f_L required for $x \in \Pi_L$ to escape Π_L.

Definition 6.5: Given $x \in \Pi_L$ let $\chi_L(x)$ be the smallest $p \ge 1$ for which $f_L^p(x) \notin \Pi_L$, with $\chi_L(x) = \infty$ if there exists no such p.

Proposition 6.6: *Let $x \in \Pi_L$ and $p \in \{1, \dots, p^*+1\}$ be finite. Then $x \in D_p$ if and only if $\chi_L(x) = p$.*

Proof: We first show that $x \in D_p$ implies $\chi_L(x) = p$ for all $p \in \{1, \dots, p^*+1\}$ by induction on p. If $x \in D_1$, then $x_2 > -\tau_L x_1 - 1$ (recalling $m_1 = -\tau_L$ and $c_1 = -1$) and so $f_L(x)_1 = \tau_L x_1 + x_2 + 1 > 0$, hence $\chi_L(x) = 1$. Suppose $x \in D_p$ implies $\chi_L(x) = p$ for some $p \in \{1, \dots, p^*\}$ (this is our induction hypothesis). Choose any $x \in D_{p+1}$. Then $m_{p+1}x_1 + c_{p+1} < x_2 \le m_p x_1 + c_p$. By using (4), (49), and (50) we obtain $m_p f_L(x)_1 + c_p < f_L(x)_2 \le m_{p-1} f_L(x)_1 + c_{p-1}$, except if $p = 1$ the latter inequality is absent. Thus $f_L(x) \in D_p$ and so $\chi_L(f_L(x)) = p$ by the induction hypothesis. Also $f_L(x)_1 \le 0$ (because $x \notin D_1$), therefore $\chi_L(x) = p + 1$.

If $p^* < \infty$ the converse is true because $D_1, \dots, D_{p^*+1}$ partition Π_L. It remains show in the case $p^* = \infty$ that if $x \in E = \Pi_L \setminus \bigcup_{p=1}^{\infty} D_p$, then $\chi_L(x) = \infty$. We have

$$E = \left\{x \in \Pi_L \,\middle|\, x_2 \le m_\infty x_1 + c_\infty\right\},$$

where $m_\infty = \lim_{p\to\infty} m_p$ is given in the proof of Proposition 6.3 and $c_\infty = \lim_{p\to\infty} c_p = \frac{-m_\infty}{m_\infty+1}$. If $x \in E$ then $f_L(x)_2 \le m_\infty f_L(x)_1 + c_\infty$ (obtained by repeating the calculations in the above induction step) and $f_L(x)_1 = \tau_L x_1 + x_2 + 1 \le c_\infty + 1$ (obtained by using also $x_1 \le 0$). Thus $f_L(x)_1 < 0$ (because $c_\infty < -1$ by Lemma 6.2). Therefore $f_L(x) \in E$ which shows that E is forward invariant under f_L. Therefore $\chi_L(x) = \infty$ for any $x \in E$. ■

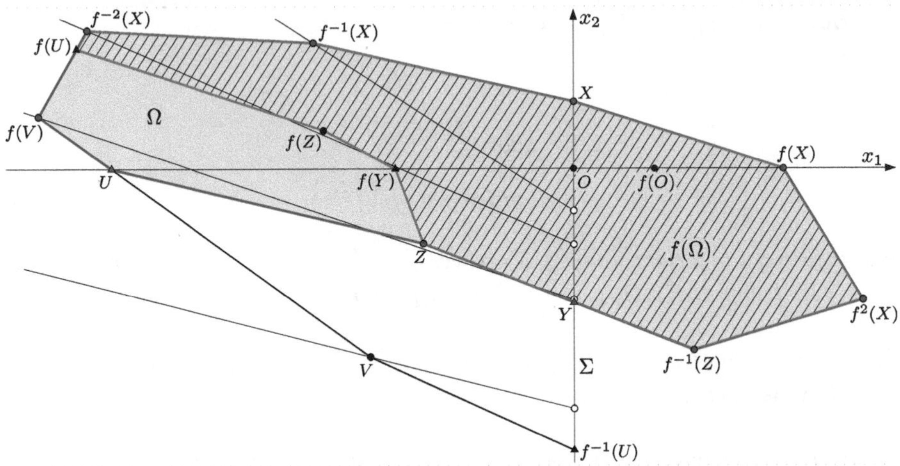

Figure 12. A plot of Ω (shaded) and $f(\Omega)$ (striped). The values in Definition 6.7 are $r = 4$ (so $Z = f^4(X)$) and $\ell = 5$ (so $V = f^{-4}(X)$). Here (58)–(60) are satisfied so $f(\Omega) \subseteq \Omega$ by Proposition 6.10.

6.3. *A forward invariant region and a trapping region*

Let $\beta > 0$ and $X = (0, \beta)$. Here we use the first few images and preimages of X to form a polygon Ω. To do this we require the following assumption on X:

$$\text{There exist } i, j \geq 1 \text{ such that } f^i(X) \in \Pi_L \text{ and } f^{-j}(X) \in \Pi_R. \tag{56}$$

Definition 6.7: Let r and ℓ be the smallest values of i and j satisfying (56), respectively.

That is, $f^i(X)_1 > 0$ for all $i = 1, \ldots, r-1$ and $f^r(X)_1 \leq 0$. Also $f^{-j}(X)_1 < 0$ for all $j = 1, \ldots, \ell - 1$ and $f^{-\ell}(X)_1 \geq 0$. Notice $r \geq 2$ because $f(X)_1 = \beta + 1 > 0$. Also $\ell \geq 2$ because $f^{-1}(X)_1 = -\frac{\beta}{\delta_L} < 0$.

In what follows $\overleftrightarrow{PQ}$ denotes the line through distinct points $P, Q \in \mathbb{R}^2$. Proofs of the next three results are deferred to the end of this section.

Lemma 6.8: *Suppose* (56) *is satisfied. Let* $Z = f^r(X)$ *and let* Y *denote the intersection of* $\overleftrightarrow{Zf^{-1}(Z)}$ *with* Σ. *Let* $V = f^{-(\ell-1)}(X)$ *and let* U *denote the intersection of* $\overleftrightarrow{Vf(V)}$ *with* $f(\Sigma)$, *see Figure* 12. *The closed polygonal chain formed by connecting the points*

$$U, f^{-(\ell-2)}(X), \ldots, f^{-1}(X), X, f(X), \ldots, f^{r-1}(X), Z, \tag{57}$$

in order, and then from Z *back to* U, *has no self-intersections (it is a Jordan curve).*

Definition 6.9: Let Ω be the polygonal chain of Lemma 6.8 and its interior.

Now we show that if three conditions are satisfied then Ω is forward invariant and we can shrink it by an arbitrarily small amount to obtain a trapping region, Ω_{trap}.

Proposition 6.10: *Suppose* (56) *is satisfied and*

$$Y \text{ lies above } f^{-1}(U), \tag{58}$$

$$Z \text{ lies above } \overleftrightarrow{f^{-1}(U)V}, \tag{59}$$

$$Z \text{ lies to the right of } \overleftrightarrow{Vf(V)}. \tag{60}$$

Then Ω is forward invariant under f. Moreover, for all $\varepsilon > 0$ there exists a trapping region $\Omega_{\text{trap}} \subset \Omega$ for f with $d_H(\Omega, \Omega_{\text{trap}}) \leq \varepsilon$, where d_H denotes Hausdorff distance.[2]

The next result allows us to apply Theorem 3.9. The actual application to Theorem 3.9 is detailed in the next section.

Proposition 6.11: *Suppose* (56), (58), (59), *and* (60) *are satisfied. Let*

$$\Omega_{\text{rec}} = \left\{ x \in \Omega \,\middle|\, x_1 > 0 \right\}, \tag{61}$$

and **W** *be given by* (6) *with*

$$p_{\max} = \max\left[\chi_L(Y), \chi_L(Z)\right]. \tag{62}$$

Then Ω_{rec} is **W**-*recurrent and $\Omega \subset \bigcup_{i=-\ell}^{0} f^i(\Omega_{\text{rec}})$.*

Proof of Lemma 6.8: Here we use the notation $[A, B)$ to denote the line segment $\{(1-s)A + sB \,|\, 0 \leq s < 1\}$ for points $A, B \in \mathbb{R}^2$.

The points X and $f(X)$ belong to Π_R, thus the line segment $[X, f(X))$, call it ζ, is contained in Π_R. The points $f^{-(\ell-1)}(X), \ldots, f^{-2}(X), f^{-1}(X)$ all belong to $\text{int}(\Pi_L)$, thus for each $i \in \{1, \ldots, \ell-1\}$ the line segment $L_i = [f^{-i}(X), f^{-(i-1)}(X))$ is contained in $\text{int}(\Pi_L)$. These line segments are mutually disjoint because if L_i and L_j, with $i < j$, intersect at some point P, then $P \in L_i$ implies $f^i(P) \in \zeta$ and so $f^i(P)_1 \geq 0$, while $P \in L_j$ implies $f^i(P) \in L_{j-i}$ and so $f^i(P)_1 < 0$, which is a contradiction. By a similar argument the line segments $[f^i(X), f^{i+1}(X))$, for $i \in \{1, \ldots, r-1\}$, are mutually disjoint and contained in $x_2 \leq 0$ (whereas ζ is contained in $x_2 > 0$). This shows that the chain formed by connecting the points (57) has no self-intersections. The addition of $[Z, U)$ introduces no intersections because $[Z, U)$ and part of $[f_R^{r-1}(X), Z)$ are the only components of the chain that belong to the third quadrant. ∎

Proof of Proposition 6.10: **Step 1** — Interior angles of Ω.

Define $\theta : \mathbb{R}^2 \to [0, 2\pi)$ as follows: $\theta(x)$ is the angle at x from $[f_L^{-1}(x), x]$ anticlockwise to $[x, f_L(x)]$, see Figure 13. Also define $g(x) = f_L(x) - x$ and

$$S(x) = g(x) \wedge g\left(f_L^{-1}(x)\right), \tag{63}$$

where we have defined the wedge product $A \wedge B = A_1 B_2 - A_2 B_1$. Through a little algebraic manipulation we can obtain[3]

$$\sin(\theta(x)) = \frac{S(x)}{\|g(x)\| \left\|g\left(f_L^{-1}(x)\right)\right\|}. \tag{64}$$

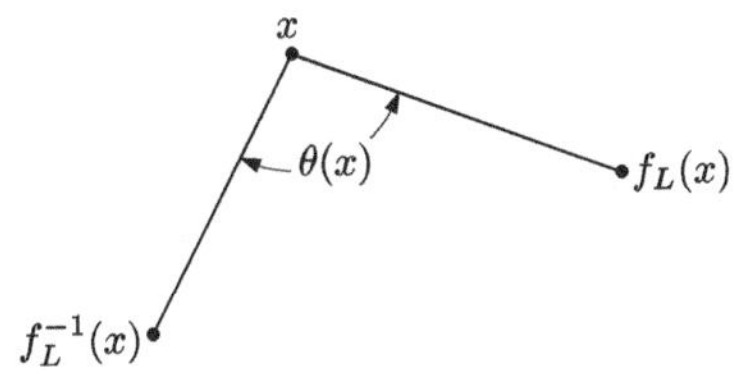

Figure 13. The angle $\theta(x)$ formed at a point x by its preimage and image under f_L.

From $f_L(x) = A_L x + b$ we obtain the formula $g(f_L(x)) = A_L g(x)$. From this and (63) we obtain

$$S(f_L(x)) = \det(A_L) S(x). \tag{65}$$

Since $\det(A_L) > 0$ we can conclude that sign of S (and thus also the sign of $\sin(\theta)$) is constant along orbits of f_L. In particular, $\sin(\theta(f_L^{-i}(X)))$ has the same sign for each $i \in \{0, 1, \ldots, \ell - 2\}$. Thus the angles $\theta(f_L^{-i}(X))$ must be all less than π, all equal to π, or all greater than π. But the path connecting $X, f^{-1}(X), \ldots, f^{-\ell}(X)$ (where $f = f_L$) includes X on the positive x_2-axis, U on the negative x_1-axis, and $f_L^{-1}(U)$ on the negative x_2-axis, and has no self-intersections (Lemma 6.8). Therefore the angles are all less than π. By applying a similar argument to f_R we conclude that all interior angles of Ω are less than π, except possibly at the points U and Z.

Step 2 — Convex subsets of Ω.

We now define two convex subsets of Ω (one in $x_2 \leq 0$ and one in $x_2 \geq 0$). Let Ω_{upper} be the polygon formed by connecting the points

$$U, f^{-(\ell-2)}(X), \ldots, f^{-1}(X), X, f(X), \tag{66}$$

in order, and from $f(X)$ back to U. Let Ω_{lower} be the polygon formed by connecting the points

$$f(X), f^2(X), \ldots, f^{r-1}(X), Z, f(Y), \tag{67}$$

in order, and from $f(Y)$ back to $f(X)$. Since U and $f(X)$ lie on $x_2 = 0$ while all other vertices of Ω_{upper} lie in $x_2 \geq 0$, the interior angles of Ω_{upper} at U and $f_R(x)$ are less than π. Thus by the previous result Ω_{upper} is convex. For similar reasons, Ω_{lower} is also convex.

Step 3 — Consequences of assumptions (58)–(60).

All points between $U = (U_1, 0)$, where $U_1 < 0$, and $f_R(X) = (\beta + 1, 0)$, where $\beta > 0$, belong to $\text{int}(\Omega)$. This includes $O = (0, 0)$ and $f(O) = (1, 0)$. Also $f(Y) \in \text{int}(\Omega)$ because (58) implies that $f(Y)$ lies between U and $f(O)$. We now show $f(Z) \in \text{int}(\Omega)$. Assumptions (59) and (60) imply that either $Z = Y$ (in which case $f(Z) \in \text{int}(\Omega)$ is immediate) or Z belongs to the interior of the quadrilateral $\mathcal{Q} = f^{-1}(U)VUO$. Each vertex of $\mathcal{Q}$ belongs to Π_L where $f = f_L$ is affine, thus $f(\mathcal{Q})$ is the quadrilateral $Uf(V)f(U)f(O)$. Each vertex of $f(\mathcal{Q})$ belongs to Ω_{upper}, which is convex, thus $f(\mathcal{Q}) \subset \Omega_{\text{upper}}$. Since $Z \in \text{int}(\mathcal{Q})$ we have that $f(Z) \in \text{int}(\Omega_{\text{upper}}) \subset \text{int}(\Omega)$.

Step 4 — Forward invariance of Ω.

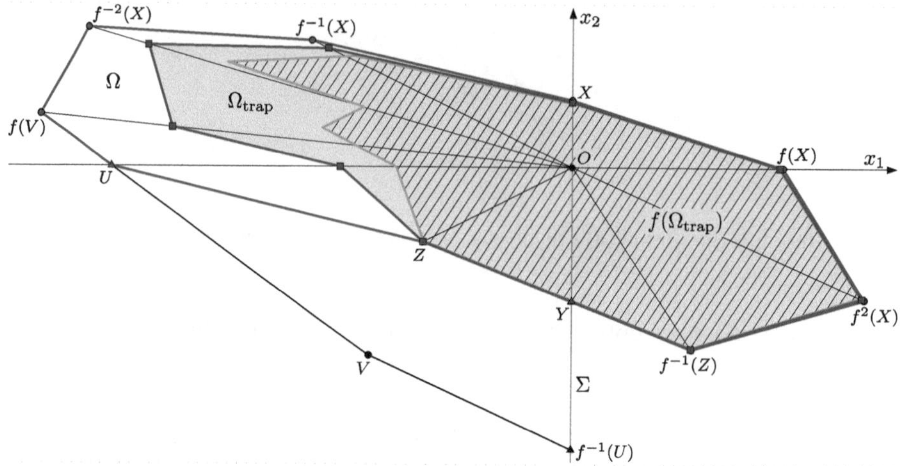

Figure 14. A plot of Ω (unshaded), Ω_{trap} (shaded), and $f(\Omega_{trap})$ (striped). Here $r = 4$ and $\ell = 5$ as in Figure 12. The set Ω_{trap} is plotted by using $\varepsilon = 0.5$ in (68).

Write $\Omega = \Omega_L \cup \Omega_R$ where

$$\Omega_L = \Omega \cap \Pi_L,$$
$$\Omega_R = \Omega \cap \Pi_R.$$

Observe $f(\Omega) = f_L(\Omega_L) \cup f_R(\Omega_R)$. Since f_L is affine, $f_L(\Omega_L)$ is a polygon. Evidently its vertices all belong to Ω_{upper}. Since Ω_{upper} is convex, $f_L(\Omega_L) \subset \Omega_{upper} \subset \Omega$. By a similar argument, $f_R(\Omega_R) \subset \Omega_{lower} \subset \Omega$, thus Ω is forward invariant.

Step 5 — Define Ω_{trap}.

Let $P^{(1)}, \ldots, P^{(\ell+r)}$ denote the points (57) in order. That is, $P^{(1)} = U$ around to $P^{(\ell+r)} = Z$. For each $j \in \{1, \ldots, \ell + r\}$ define

$$P_\varepsilon^{(j)} = \left(1 - \frac{\varepsilon^j}{\|P^{(j)}\|}\right) P^{(j)}, \tag{68}$$

and assume ε is small enough that $1 - \frac{\varepsilon^j}{\|P^{(j)}\|} > 0$ for each j. Each $P_\varepsilon^{(j)}$ is the result of moving from $P^{(j)}$ a distance ε^j towards O, see Figure 14. Let Ω_{trap} be the polygon with vertices $P_\varepsilon^{(j)}$ (connected in the same order as for Ω). Immediately we have $d_H(\Omega, \Omega_{trap}) \leq \varepsilon$. Also $\Omega_{trap} \subset \Omega$ because $[P^{(j)}, O] \subset \Omega$ for each j.

Step 6 — Convex subsets of Ω_{trap}.

Analogous to Ω_{upper} and Ω_{lower}, let $\Omega_{trap,upper} \subset \Omega_{trap}$ be the polygon in $x_2 \geq 0$ formed by connecting the points

$$P_\varepsilon^{(1)}, P_\varepsilon^{(2)}, \ldots, P_\varepsilon^{(\ell+1)}, \tag{69}$$

in order, and from $P_\varepsilon^{(\ell+1)}$ back to $P_\varepsilon^{(1)}$. Notice $P_\varepsilon^{(1)}$ and $P_\varepsilon^{(\ell+1)}$ lie on $x_2 = 0$. Also let $Y_\varepsilon \in \Sigma$ denote the intersection of $[P_\varepsilon^{(\ell+r)}, P_\varepsilon^{(\ell+r-1)}]$ with Σ and let $\Omega_{trap,lower} \subset \Omega_{trap}$ be the

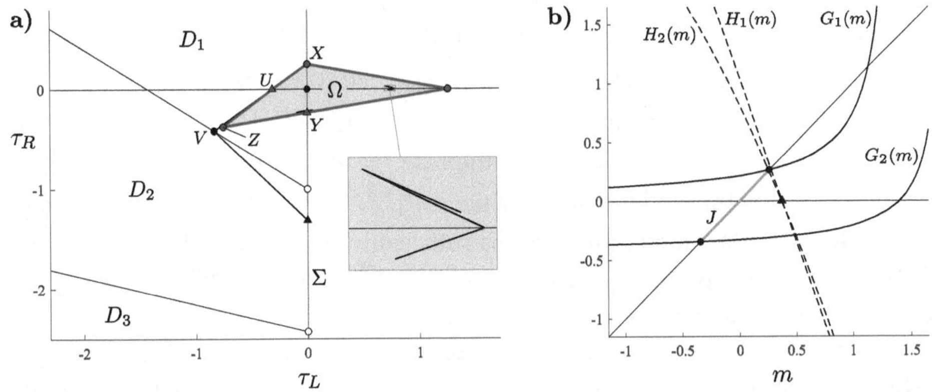

Figure 15. Constructive elements produced by Algorithm 7.1 for (4) with (5) and $(\tau_L, \tau_R) = (0.7, -1.4)$. Here the algorithm obtains $\beta = 0.25$ and returns $\chi_{\texttt{chaos}} = \texttt{true}$. Panel (a) shows the forward invariant region Ω (see Figure 12), the regions D_p (see Figure 11), and a numerically computed attractor. Panel (b) shows the slope maps G_j (36) and the functions H_j (37) for $j = 1, 2$.

polygon in $x_2 \leq 0$ formed by connecting the points

$$P_\varepsilon^{(\ell+1)}, P_\varepsilon^{(\ell+2)}, \ldots, P_\varepsilon^{(\ell+r)}, f(Y_\varepsilon), \tag{70}$$

in order, and from $f(Y_\varepsilon)$ back to $P_\varepsilon^{(\ell+1)}$. Notice $f(Y_\varepsilon)$ lies on $x_2 = 0$ close to $f(Y)$. Each interior angle of $\Omega_{\text{trap,upper}}$ and $\Omega_{\text{trap,lower}}$ is at most an order-ε perturbation of the corresponding interior angle of Ω_{upper} or Ω_{lower}. All interior angles of Ω_{upper} and Ω_{lower} are less than π, so the same is true for $\Omega_{\text{trap,upper}}$ and $\Omega_{\text{trap,lower}}$ assuming ε is sufficiently small. That is, $\Omega_{\text{trap,upper}}$ and $\Omega_{\text{trap,lower}}$ are convex.

Step 7 — The set Ω_{trap} is a trapping region.

Write $\Omega_{\text{trap}} = \Omega_{\text{trap},L} \cup \Omega_{\text{trap},R}$ where

$$\Omega_{\text{trap},L} = \Omega_{\text{trap}} \cap \Pi_L,$$
$$\Omega_{\text{trap},R} = \Omega_{\text{trap}} \cap \Pi_R,$$

and observe $f(\Omega_{\text{trap}}) = f_L(\Omega_{\text{trap},L}) \cup f_R(\Omega_{\text{trap},R})$.

The vertices of $\Omega_{\text{trap},L}$ are Y_ε, $P_\varepsilon^{(\ell+r)}$, and $P_\varepsilon^{(j)}$ for $j = 1, \ldots, \ell$. We now show that, if ε is sufficiently small, then these vertices all map under $f = f_L$ to either $\text{int}(\Omega_{\text{trap,upper}})$ or to a point on $x_2 = 0$ in the open line segment $I = (P_\varepsilon^{(1)}, P_\varepsilon^{(\ell+1)})$. Since $\Omega_{\text{trap,upper}}$ is convex this implies $f_L(\Omega_{\text{trap},L}) \subset \text{int}(\Omega_{\text{trap,upper}}) \cup I$. Consequently $f_L(\Omega_{\text{trap},L}) \subset \text{int}(\Omega_{\text{trap}})$ because $I \subset \text{int}(\Omega_{\text{trap}})$.

Certainly $f(Y_\varepsilon), f(P_\varepsilon^{(\ell)}) \in I$, assuming ε is sufficiently small. If $Z \neq Y$ then $f(P_\varepsilon^{(\ell+r)}) \in \text{int}(\Omega_{\text{trap,upper}})$, assuming ε is sufficiently small, because $f(Z) \in \text{int}(\Omega)$. If $Z = Y$ then $f(P_\varepsilon^{(\ell+r)}) \in I$. Now choose any $j = 1, \ldots, \ell - 1$. By definition, $P_\varepsilon^{(j)} = (1-s)P^{(j)} + sO$, with $s = \frac{\varepsilon^j}{\|P^{(j)}\|}$. In $\Pi_L, f = f_L$ is affine, so we have

$$f\left(P_\varepsilon^{(j)}\right) = \begin{cases} (1-s)f(U) + sf(O), & j = 1, \\ (1-s)P^{(j+1)} + sf(O), & j = 2, \ldots, \ell - 1, \end{cases} \tag{71}$$

Therefore, for $j \neq 1$, $f(P_\varepsilon^{(j)})$ is the result of moving from $P^{(j+1)}$ a distance $\frac{\varepsilon^j \|P^{(j+1)} - f(O)\|}{\|P^{(j)}\|}$ towards $f(O)$. Thus $f(P_\varepsilon^{(j)})$ belongs to the triangle $P^{(j+1)}P^{(j+2)}O$, assuming ε is sufficiently small, because $P^{(j+1)}$ and $P^{(j+2)}$ lie in $x_2 \geq 0$ with $P^{(j+2)}$ located clockwise (with respect to O) from $P^{(j+1)}$. This is true in the case $j = 1$ also. The distance from $f(P_\varepsilon^{(j)})$ to $[P^{(j+1)}, P^{(j+2)}]$ is proportional to ε^j, while $d_H([P_\varepsilon^{(j+1)}, P_\varepsilon^{(j+2)}], [P^{(j+1)}, P^{(j+2)}])$ is proportional to ε^{j+1}. Therefore, assuming ε is sufficiently small, $f(P_\varepsilon^{(j)})$ lies in the triangle $P_\varepsilon^{(j+1)}P_\varepsilon^{(j+2)}O$ and not on the line segment $[P_\varepsilon^{(j+1)}, P_\varepsilon^{(j+2)}]$. Thus $f(P_\varepsilon^{(j)}) \in \mathrm{int}(\Omega_{\mathrm{trap,upper}})$ and this completes our demonstration that $f_L(\Omega_{\mathrm{trap},L}) \subset \mathrm{int}(\Omega_{\mathrm{trap}})$.

From similar arguments it follows that $f_R(\Omega_{\mathrm{trap},R}) \subset \mathrm{int}(\Omega_{\mathrm{trap,lower}}) \cup I$, assuming ε is sufficiently small, and so $f_R(\Omega_{\mathrm{trap},R}) \subset \mathrm{int}(\Omega_{\mathrm{trap}})$. Hence Ω_{trap} is a trapping region for f. ■

Proof of Proposition 6.11: We first show $\Omega \subset \bigcup_{i=-\ell}^{0} f^i(\Omega_{\mathrm{rec}})$. Choose any $x \in \Omega$. If $x_1 > 0$ then $x \in \Omega_{\mathrm{rec}}$. If $x_1 \leq 0$ then $x \in D_i$, for some $i \in \{1, \ldots, \ell\}$. The upper bound $i = \ell$ is a consequence of (58)–(60) because V lies above $f_L^{-\ell}(\Sigma)$ and $f^{-1}(U)$ lies on or above $f_L^{-\ell}(\Sigma)$. Thus by Proposition 6.6, $f^i(x)_1 > 0$, and so $f^i(x) \in \Omega_{\mathrm{rec}}$ because Ω is forward invariant.

Now choose any $y \in \Omega_{\mathrm{rec}}$. If $f(y) \in \Omega_{\mathrm{rec}}$, let $n = 1$ and observe that the first symbol of any $\mathcal{S} \in \Gamma(y)$ is R, which belongs to $\mathbf{W}$. So now suppose $f(y) \notin \Omega_{\mathrm{rec}}$. Also suppose $f(Y)_1 < 0$, so then $f(y)$ belongs to the quadrilateral $YZf(Y)O$ (if instead $f(Y)_1 \geq 0$ then the following arguments can be applied to the part of $YZf(Y)O$ that belongs to Π_L). We have $\chi_L(f(y)) \leq \max[\chi_L(Y), \chi_L(Z), \chi_L(f(Y)), \chi_L(O)]$ (this follows from the linear ordering of the regions D_p). But $\chi_L(f(Y)) = \chi_L(Y) - 1$ and $\chi_L(O) = 1$, thus $\chi_L(f(y)) \leq p_{\max}$.

Let $n = \chi_L(f(y)) + 1$. Then $f^n(y)_1 > 0$ and so $f^n(y) \in \Omega_{\mathrm{rec}}$ because Ω is forward invariant. Also $f^j(y)_1 \leq 0$ for all $j = 1, \ldots, n-1$ with $f^j(y)_1 = 0$ only possible for $j = 1$ and $j = n-1$. In summary, $y_1 > 0, f(y)_1 \leq 0, f^j(y)_1 < 0$ for all $j = 2, \ldots, n-2, f^{n-1}(y)_1 \leq 0$, and $f^n(y) \in \Omega_{\mathrm{rec}}$. Thus there are four possibilities for the first n symbols of $\mathcal{S} \in \Gamma(y)$: RL^{n-1}, RRL^{n-2}, $RL^{n-2}R$, and $RRL^{n-3}R$ (the last possibility can only arise if $f(y)_1 = 0$ and $f^{n-1}(y)_1 = 0$). All four words can be expressed as a concatenation of words in $\mathbf{W}$ (because $n - 1 \leq p_{\max}$). Thus Ω_{rec} is $\mathbf{W}$-recurrent. ■

7. An algorithm for detecting a chaotic attractor

In the previous two sections we obtained sufficient conditions for the assumptions of Theorem 3.9 to hold with a trapping region for the 2d BCNF (4). In Section 7.1 we summarize these conditions and state Algorithm 7.1 (in pseudo-code) for testing their validity. In Section 7.2 we further discuss the application of the algorithm to the slice of parameter space shown in Figure 1.

7.1. Statement and proof of the algorithm

The polygon Ω constructed in Section 6.3 typically satisfies (58)–(60) for some interval of β-values (where $X = (0, \beta)$). Within this interval, smaller values of β tend to correspond to smaller values of $\chi_L(Y)$ and $\chi_L(Z)$ and so produce a smaller value for $p_{\max}$, (62). Smaller values of $p_{\max}$ are more favourable for the cone C_J (39) to be well-defined, invariant, and expanding. This is because with a smaller value of p_{max} there are less matrices in $\mathbf{M}$ and therefore fewer inequalities that need to be satisfied.

For these reasons we search for a suitable value of β by iteratively increasing its value in steps of size β_{step} from β_{min} up to (at most) β_{max}. To produce Figure 1 we used

$$\beta_{step} = 0.01, \quad \beta_{min} = 0.01, \quad \beta_{max} = 5. \tag{72}$$

For a given value of β there are five groups of conditions that need to be checked. These are labelled (C1)–(C5) in Algorithm 7.1. First we require Ω to be well-defined. This is established by showing that r and ℓ of Definition 6.7 exist. To produce Figure 1 this was implemented by iterating X backwards and forwards up to maximum allowed values

$$r_{max} = 15, \quad \ell_{max} = 15. \tag{73}$$

Second we check conditions (58)–(60). If these are satisfied then Ω is forward invariant and in Algorithm 7.1 this fixes the value of β. We then evaluate p_{max} by iterating Y and Z under f, (62). The remaining three conditions are that the cone C_J is well-defined, that C_J is invariant, and that C_J is expanding. The computations involved in the last two steps are elementary because $G_j(m)$ and $H_j(m)$ are polynomials of degree two or less. Algorithm 7.1 registers its success or failure by the termination value of the Boolean variable $\chi_{\texttt{chaos}}$.

The theorem below assumes calculations are done exactly. For Figure 1 calculations were performed with rounding at 16 digits.

Theorem 7.1: *Let f be a map of the form* (4) *with $\delta_L, \delta_R > 0$. If Algorithm* 7.1 *outputs $\chi_{chaos} =$ `true` then f has an attractor with a positive Lyapunov exponent.*

Proof: Suppose Algorithm 7.1 outputs $\chi_{\texttt{chaos}} =$ `true`. Then (C1)–(C5) all hold for some fixed $\beta > 0$. Since (C1) holds, Ω is well-defined by Lemma 6.8. Since (C2) holds, f has a trapping region $\Omega_{trap} \subset \Omega$ by Proposition 6.10. Thus f has an attractor $\Lambda \subset \Omega_{trap}$. Let $\mathbf{W}$ be given by (6) and Ω_{rec} be given by (61). Then, by Proposition 6.11, $\Lambda \subset \bigcup_{i=-\infty}^{\infty} f^i(\Omega_{rec})$ and $\mathbf{W}$ generates $\Gamma(y)$ for all $y \in \Omega_{rec}$.

Since (C3) holds, the cone C_J is well-defined. Since (C4) holds, C_J is forward invariant under $\mathbf{M} = \{\Phi(\mathcal{W}) \mid \mathcal{W} \in \mathbf{W}\}$ by Proposition 5.2. Since (C5) holds, C_J is also expanding under $\mathbf{M}$ by Propositions 5.2 and 5.3. Then by Theorem 3.9 for all $x \in \Lambda$ there exists $v \in T\mathbb{R}^2$ such that $\lambda(x, v) > 0$. ∎

7.2. Comments on the results of Algorithm 7.1

As mentioned in Section 2, for (4) with $\delta_L = \delta_R = 0.3$, Algorithm 7.1 outputs $\chi_{\texttt{chaos}} =$ `true` throughout the red regions of Figure 1. Here we examine three sample parameter combinations in detail. For each of the three black dots in Figure 1, the polygon Ω is well-defined and forward invariant. Figures 15(a)–17(a) show Ω using the value of $\beta > 0$ generated by Algorithm 7.1.

In Figure 15(a) we have $Y, Z \in D_1$, so $p_{max} = 1$ and $\mathbf{W} = \{R, RL\}$. Figure 15(b) shows how (C5) is satisfied. With $j = 1$ we have $\mathcal{W} = R$, so $H_1(m)$ is linear, see Remark 5.1, and $m_{root} > m_{stab,max}$. With $j = 2$ we have $\mathcal{W} = RL$ with which $H_2(m)$ is quadratic and (44) and (46) are satisfied. Numerical simulations suggest that at these parameter values f has

set $\chi_{\text{chaos}} = \text{false}$
set $\beta = \beta_{\min} > 0$
While $\chi_{\text{chaos}} = \text{false}$ and $\beta \leq \beta_{\max}$
(C1) If r or ℓ do not exist
        set $\beta = \beta + \beta_{\text{step}}$
    else
(C2)   If any of (58)–(60) are false
            set $\beta = \beta + \beta_{\text{step}}$
        else
            set $\chi_{\text{chaos}} = \text{true}$
        end
    end
end
If $\chi_{\text{chaos}} = \text{true}$
    Evaluate $p_{\max}$ (62).
(C3) If (38) is false for some $M^{(j)} = A_L^{j-1} A_R$ with $j \in \{1, \ldots, p_{\max} + 1\}$
        set $\chi_{\text{chaos}} = \text{false}$
    else
        Evaluate $m_{\text{stab}}^{(j)}$ and $m_{\text{unstab}}^{(j)}$ for each j.
        Evaluate $m_{\text{stab,min}}$ and $m_{\text{stab,max}}$.
(C4)   If $m_{\text{stab,min}} \leq m_{\text{unstab}}^{(j)} \leq m_{\text{stab,max}}$ for some $j \in \{1, \ldots, p_{\max} + 1\}$
            set $\chi_{\text{chaos}} = \text{false}$
        else
(C5)   If, for some $j \in \{1, \ldots, p_{\max} + 1\}$, H_j does not have two
                distinct real roots or two of (44)- -(46) are
                false (or, if $b_j^2 + d_j^2 = 1$, the condition in
                Remark 5.1 is false)
                set $\chi_{\text{chaos}} = \text{false}$
            end
        end
    end
end

a unique two-piece chaotic attractor with one piece intersecting $f(\Sigma)$ [as shown in the magnification of Figure 15(a)], and its image intersecting Σ.

In Figure 16(a) we have $Y \in D_1$ and $Z \in D_2$, so $p_{\max} = 2$ and $\mathbf{W} = \{R, RL, RL^2\}$. Here Algorithm 7.1 returns $\chi_{\texttt{chaos}} = \texttt{false}$ because (C5) is not satisfied. This is because $H_3(m)$ has no real roots, and this is evident in Figure 16(b). Indeed at these parameter values f has a stable period-3 orbit corresponding to the word RL^2. Nevertheless, f does appear to have a chaotic attractor contained in $D_1 \cup \Pi_R$. It may be possible to prove this attractor has a positive Lyapunov exponent by constructing a trapping region in $D_1 \cup \Pi_R$. For the parameter values of Figure 17 we again have $p_{\max} = 2$ but now $H_3(m)$ has real roots satisfying (44) and (46) and Algorithm 7.1 terminates with $\chi_{\texttt{chaos}} = \texttt{true}$.

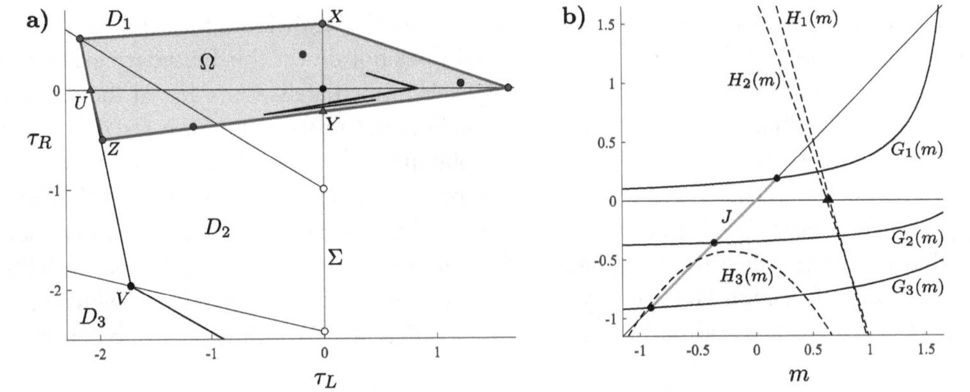

Figure 16. Constructive elements produced by Algorithm 7.1 for (4) with (5) and $(\tau_L, \tau_R) = (0.7, -1.8)$. Here the algorithm obtains $\beta = 0.65$ and returns $\chi_{\texttt{chaos}} = \texttt{false}$. Panel (a) shows the forward invariant region Ω, the regions D_p, a numerically computed attractor, and a stable period-3 orbit (indicated with circles). Panel (b) shows G_j and H_j for $j = 1, 2, 3$.

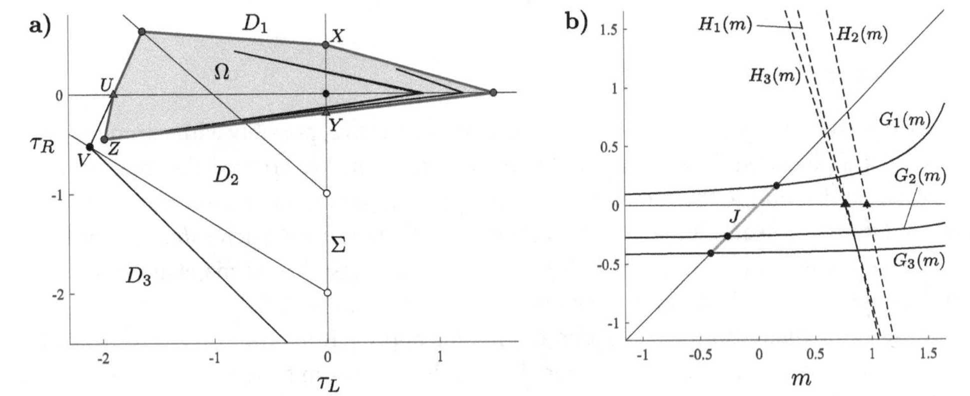

Figure 17. Constructive elements produced by Algorithm 7.1 for (4) with (5) and $(\tau_L, \tau_R) = (1, -2)$. Here the algorithm obtains $\beta = 0.49$ and returns $\chi_{\texttt{chaos}} = \texttt{true}$. Panel (a) shows the forward invariant region Ω, the regions D_p, and a numerically computed attractor [shown also in Figure 2(b)]. Panel (b) shows G_j and H_j for $j = 1, 2, 3$.

We now discuss the region boundaries in Figure 1 labelled B_1 to B_5. Boundary B_1 is the horizontal line $\tau_R = -\delta_R - 1$. Below B_1, and for $\tau_L > 0.7$ approximately, Algorithm 7.1 terminates with $\chi_{\texttt{chaos}} = \texttt{true}$. On B_1 (C5) is not satisfied because A_R has an eigenvalue of -1 so $H_1(m) = 0$ at $m = m_{\rm stab}^{(1)}$. Indeed above B_1 the map f has an asymptotically stable fixed point in $x_1 > 0$. In this way Algorithm 7.1 detects a true bifurcation boundary between chaotic and non-chaotic dynamics.

On B_2 (C5) is not satisfied because $H_2(m) = 0$ at $m = m_{\rm stab}^{(1)}$. Thus to the left of B_2, Algorithm 7.1 terminates with $\chi_{\texttt{chaos}} = \texttt{false}$ because some $v \in C_J$ do not expand when multiplied by $M = A_L A_R$. Nevertheless numerical results suggest f has a chaotic attractor here. It may be possible to prove this by using a different word set $\mathbf{W}$.

Boundary B_3 is the upper boundary of the blue region in which there exists a stable period-3 orbit of period $n = 3$ [corresponding to the word RL^2, see Figure 16(a)]. On this boundary the periodic orbit is destroyed in a border-collision bifurcation by having one of its points collide with Σ. Algorithm 7.1 does not detect this boundary exactly as evident in Figure 1 by the presence of white pixels immediately above B_3. At these pixels Algorithm 7.1 obtains $p_{\max} = 2$ with which (C5) is not satisfied because $H_3(m)$ has no real roots. In nearby red pixels Algorithm 7.1 obtains $p_{\max} = 1$ with which the behaviour of $H_3(m)$ is irrelevant. The number of white pixels appears to tend to zero in the limit $\beta_{\text{step}} \to 0$ because the size of the interval of β-values for which Ω is forward invariant with $p_{\max} = 1$ vanishes as we approach B_3 from above. In a similar way as we approach the homoclinic bifurcation HC from above the size of the interval of β-values for which (C1) and (C2) are satisfied approaches zero (in [17] a different approach was used to construct a trapping region).

On B_4 the period-three orbit loses stability by attaining an eigenvalue of -1. For $\tau_R < -2.6$, approximately, Algorithm 7.1 detects this boundary exactly. On B_4 (C5) is not satisfied because $H_3(m) = 0$ at $m = m_{\text{stab}}^{(3)}$. That is, $\|Mv\| = \|v\|$ for the eigenvector $v = (1, m_{\text{stab}}^{(3)})$ of $M = A_L^2 A_R$ corresponding to the eigenvalue -1. Finally, boundary B_5 is analogous to boundary B_2. On B_5 we have $H_3(m) = 0$ at $m = m_{\text{stab}}^{(1)}$.

8. Discussion

We have presented a general method by which one can prove, possibly with computer assistance, that a piecewise-linear map has a chaotic attractor. We applied the method to the 2d BCNF and found a chaotic attractor throughout a parameter regime that, unlike the logistic family for example, does not contain periodic windows. Such *robust chaos* is typical for piecewise-linear maps and for this reason piecewise-linear maps are desirable in applications that use chaos such as chaos-based cryptography [20].

In our implementation we considered only one approach for the construction of Ω_{trap} and only word sets of the form (6). There is considerable room to generalize these, such as by defining Ω_{trap} be to the union of a polygon and its images under f [33].

A major next step would be the application of this method to families of higher-dimensional maps, such the N-dimensional border-collision normal form. Results of this nature have already been achieved in [11,14]. To construct a trapping region and a cone it may be helpful to work with convex polytopes [2]. It would also be useful to obtain a converse to Theorem 3.9: if f has a topological attractor with a positive Lyapunov exponent, must some Ω_{trap}, $\mathbf{W}$, and C (satisfying the required properties) exist?

Notes

1. To the left of $\tau_L = \delta_L + 1$, Figure 1 contains 173779 red pixels and 21462 white pixels, giving 89.01%.
2. The *Hausdorff distance* between sets Ω_1 and Ω_2 is defined as

$$d_H(\Omega_1, \Omega_2) = \max\left[\sup_{x\in\Omega_1}\inf_{y\in\Omega_2}\|x-y\|, \sup_{y\in\Omega_2}\inf_{x\in\Omega_1}\|x-y\|\right].$$

3. Equation (64) is a version of the well-known formula $\sin(\theta) = \frac{\|u\times v\|}{\|u\|\|v\|}$ for the sine of the angle between $u, v \in \mathbb{R}^3$.

Acknowledgements

The author thanks Paul Glendinning whose May 2017 visit to Massey University stimulated many of the ideas presented here.

Disclosure statement

No potential conflict of interest was reported by the author(s).

Funding

This work was supported by Marsden Fund contract MAU1809, managed by Royal Society Te Apārangi.

ORCID

D. J. W. Simpson http://orcid.org/0000-0002-0284-6283

References

[1] V.M. Alekseev, *Quasirandom dynamical systems. I. Quasirandom diffeomorphisms*, Math. USSR Sb. 5 (1968), pp. 73–128.
[2] N. Athanasopoulos and M. Lazar, *Alternative stability conditions for switched discrete time linear systems*, IFAC Proceedings Volumes 47 (2014), pp. 6007–6012.
[3] S. Banerjee and C. Grebogi, *Border collision bifurcations in two-dimensional piecewise smooth maps*, Phys. Rev. E 59(4) (1999), pp. 4052–4061.
[4] S. Banerjee, J.A. Yorke, and C. Grebogi, *Robust chaos*, Phys. Rev. Lett. 80(14) (1998), pp. 3049–3052.
[5] L. Barreira, D. Dragičević, and C. Valls, *Positive top Lyapunov exponents via invariant cones: Single trajectories*, J. Math. Anal. Appl. 423 (2015), pp. 480–496.
[6] L. Barreira and Y. Pesin, *Introduction to Smooth Ergodic Theory*, Vol. 148, American Mathematical Society, Providence, RI, 2013.
[7] J. Buzzi, *Absolutely continuous invariant measures for generic multi-dimensional piecewise affine expanding maps*, Int. J. Bifurcation Chaos 9(9) (2011), pp. 1743–1750.
[8] S. Das and J.A. Yorke, *Multichaos from quasiperiodicity*, SIAM J. Appl. Dyn. Syst. 16(4) (2017), pp. 2196–2212.
[9] A. Dhara and J. Dutta, *Optimality Conditions in Convex Optimization. A Finite-Dimensional View*, CRC Press, Boca Raton, FL, 2012.
[10] M. di Bernardo, C.J. Budd, A.R. Champneys, and P. Kowalczyk, *Piecewise-Smooth Dynamical Systems. Theory and Applications*, Springer-Verlag, New York, 2008.
[11] P. Glendinning, *Bifurcation from stable fixed point to N-dimensional attractor in the border collision normal form*, Nonlinearity 28 (2015), pp. 3457–3464.
[12] P. Glendinning, *Bifurcation from stable fixed point to 2D attractor in the border collision normal form*, IMA J. Appl. Math. 81(4) (2016), pp. 699–710.
[13] P. Glendinning, *Robust chaos revisited*, Eur. Phys. J. Spec. Top. 226(9) (2017), pp. 1721–1738.
[14] P. Glendinning and M.R. Jeffrey, *Grazing-sliding bifurcations, border collision maps and the curse of dimensionality for piecewise smooth bifurcation theory*, Nonlinearity 28 (2015), pp. 263–283.
[15] P. Glendinning and C.H. Wong, *Two dimensional attractors in the border collision normal form*, Nonlinearity 24 (2011), pp. 995–1010.
[16] P.A. Glendinning and D.J.W. Simpson, *Robust chaos and the continuity of attractors*, Trans. Math. Appl. 4(1) (2020), pp. tnaa002.
[17] P.A. Glendinning and D.J.W. Simpson, *A constructive approach to robust chaos using invariant manifolds and expanding cones*, Discrete Contin. Dyn. Syst. 41(7) (2021), pp. 3367–3387.

[18] J. Guckenheimer and P.J. Holmes, *Nonlinear Oscillations, Dynamical Systems, and Bifurcations of Vector Fields*, Springer-Verlag, New York, 1986.
[19] B. Hao and W. Zheng, *Applied Symbolic Dynamics and Chaos*, World Scientific, Singapore, 1998.
[20] L. Kocarev and S. Lian (eds.), *Chaos-based Cryptography. Theory, Algorithms and Applications*, Springer, New York, 2011.
[21] P. Kowalczyk, *Robust chaos and border-collision bifurcations in non-invertible piecewise-linear maps*, Nonlinearity 18 (2005), pp. 485–504.
[22] D. Lind and B. Marcus, *An Introduction to Symbolic Dynamics and Coding*, Cambridge University Press, New York, 1995.
[23] R. Lozi, *Un attracteur étrange(?) du type attracteur de Hénon*, J. Phys. (Paris) 39(C5) (1978), pp. C5-9–C5-10. In French.
[24] J.D. Meiss, *Differential Dynamical Systems*, SIAM, Philadelphia, 2007.
[25] C. Mira, L. Gardini, A. Barugola, and J. Cathala, *Chaotic Dynamics in Two-Dimensional Noninvertible Maps*, Nonlinear Science Vol. 20, World Scientific, Singapore, 1996.
[26] M. Misiurewicz, *Strange attractors for the Lozi mappings*, in *Nonlinear Dynamics, Annals of the New York Academy of Sciences*, R.G. Helleman, ed., Wiley, New York, 1980, pp. 348–358.
[27] S. Newhouse, *Cone-fields, domination, and hyperbolicity*, in *University in Modern Dynamical Systems and Applications*, M. Brin, B. Hasselblatt, and Y. Pesin, eds., Cambridge University Press, New York, 2004.
[28] H.E. Nusse and J.A. Yorke, *Border-collision bifurcations including 'period two to period three' for piecewise smooth systems*, Physica D 57 (1992), pp. 39–57.
[29] R.C. Robinson, *An Introduction to Dynamical Systems. Continuous and Discrete*, Prentice Hall, Upper Saddle River, NJ, 2004.
[30] R.T. Rockafellar, *Convex Analysis*, Princeton University Press, Princeton, NJ, 1970.
[31] M. Rychlik, *Invariant Measures and the Variational Principle for Lozi Mappings*, Springer, New York, 2004, pp. 190–221.
[32] D.J.W. Simpson, *Border-collision bifurcations in* $\mathbb{R}^n$, SIAM Rev. 58(2) (2016), pp. 177–226.
[33] D.J.W. Simpson, *Chaotic attractors from border-collision bifurcations: Stable border fixed points and determinant-based Lyapunov exponent bounds*, NZJM 50 (2020), pp. 71–91.
[34] Ł. Struski and J. Tabor, *Expansivity and cone-fields in metric spaces*, J. Dyn. Differ. Equ. 26 (2014), pp. 517–527.
[35] M. Tsujii, *Absolutely continuous invariant measures for expanding piecewise linear maps*, Invent. Math. 143 (2001), pp. 349–373.
[36] M. Viana, *Lectures on Lyapunov Exponents*, Cambridge Studies in Advanced Mathematics Vol. 145, Cambridge University Press, Cambridge, 2014.
[37] M. Wojtkowski, *Invariant families of cones and Lyapunov exponents*, Ergod. Theory Dyn. Syst. 5 (1985), pp. 145–161.
[38] L.-S. Young, *Bowen-Ruelle measures for certain piecewise hyperbolic maps*, Trans. Am. Math. Soc. 287(1) (1985), pp. 41–48.

Appendix. The significance of the 2d BCNF

Let f be a continuous map on $\mathbb{R}^2$ that is affine on each side of $\Sigma = \{x \mid x_1 = 0\}$. Then f has the form

$$f(x) = \begin{cases} \begin{bmatrix} a_L & b \\ c_L & d \end{bmatrix} \begin{bmatrix} x_1 \\ x_2 \end{bmatrix} + \begin{bmatrix} p \\ q \end{bmatrix}, & x_1 \le 0, \\ \begin{bmatrix} a_R & b \\ c_R & d \end{bmatrix} \begin{bmatrix} x_1 \\ x_2 \end{bmatrix} + \begin{bmatrix} p \\ q \end{bmatrix}, & x_1 \ge 0, \end{cases} \tag{A1}$$

for some $a_L, a_R, b, c_L, c_R, d, p, q \in \mathbb{R}$. By computing the image of Σ under f we find that $f(\Sigma)$ intersects Σ at a unique point if and only if $b \neq 0$. Moreover, if $b \neq 0$ then this point is not a fixed point of (A1) if and only if $\xi = (1 - d)p + bq \neq 0$.

Now suppose $b \neq 0$ and $\xi \neq 0$. Then the coordinate change

$$\tilde{x} = \frac{1}{\xi}\left(\begin{bmatrix} 1 & 0 \\ -d & b \end{bmatrix} x + \begin{bmatrix} 0 \\ dp - bq \end{bmatrix}\right), \tag{A2}$$

is well-defined and invertible. Also notice it leaves Σ unchanged. By directly applying (A2) to (A1) we find that if $\xi > 0$ then f is transformed to (4) with $\tilde{x}$ in place of x and $\tau_L = a_L + d$, $\delta_L = a_L d - bc_L$, $\tau_R = a_R + d$, and $\delta_R = a_R d - bc_R$. If instead $\xi < 0$ then $\tau_L = a_R + d$, $\delta_L = a_R d - bc_R$, $\tau_R = a_L + d$, and $\delta_R = a_L d - bc_L$.

Index

Note: Figures are indicated by *italics*. Tables are indicated by **bold**. Endnotes are indicated by the page number followed by 'n' and the endnote number e.g., 20n1 refers to endnote 1 on page 20.

FSC
www.fsc.org
MIX
Papier | Fördert
gute Waldnutzung
FSC® C083411

Kolibri 360 GmbH
Ferdinand-Jühlke-Straße 7
99095 Erfurt, Deutschland
produktsicherheit@kolibri360.de